Basic Blueprint Reading and Sketching

Basic Blueprint Reading and Sketching

Seventh Edition

Dr. C. Thomas Olivo
Thomas P. Olivo

Albany • Bonn • Boston • Cincinnati • Detroit • London • Madrid
Melbourne • Mexico City • New York • Pacific Grove • Paris • San Francisco
Singapore • Tokyo • Toronto • Washington

NOTICE TO THE READER

Publisher does not warrant or guarantee any of the products described herein or perform any independent analysis in connection with any of the product information contained herein. Publisher does not assume, and expressly disclaims, any obligation to obtain and include information other than that provided to it by the manufacturer.

The reader is expressly warned to consider and adopt all safety precautions that might be indicated by the activities herein and to avoid all potential hazards. By following the instructions contained herein, the reader willingly assumes all risks in connection with such instructions.

The publisher makes no representation or warranties of any kind, including but not limited to, the warranties of fitness for particular purpose or merchantability, nor are any such representations implied with respect to the material set forth herein, and the publisher takes no responsibility with respect to such material. The publisher shall not be liable for any special, consequential, or exemplary damages resulting, in whole or part, from the readers' use of, or reliance upon, this material.

Cover Design by Brucie Rosch

Delmar Staff
Publisher: Alar Elken
Acquisitions Editor: Tom Schin
Production Coordinator: Andrew Crouth
Art & Design Coordinator: Nicole Reamer
Editorial Assistant: Fionnuala McAvey

I(T)P®

an International Thomson Publishing company

The ITP logo is a trademark under license
Printed in the United States of America

A service of I(T)P®

For more information contact:

Delmar Publishers
3 Columbia Circle, Box 15015
Albany, New York 12212-5015

International Thomson Publishing Europe
Berkshire House
168-173 High Holborn
London, WC1V 7AA
United Kingdom

Nelson ITP, Australia
102 Dodds Street
South Melbourne,
Victoria, 3205 Australia

Nelson Canada
1120 Birchmont Road
Scarborough, Ontario
M1K 5G4, Canada

International Thomson Publishing France
Tour Maine-Montparnasse
33 Avenue du Maine
75755 Paris Cedex 15, France

International Thomson Editores
Seneca 53
Colonia Polanco
11560 Mexico D. F. Mexico

International Thomson Publishing GmbH
Königswinterer Strasse 418
53227 Bonn
Germany

International Thomson Publishing Asia
60 Albert Street
#15-10 Albert Complex
Singapore 189969

International Thomson Publishing Japan
Hirakawa-cho Kyowa Building, 3F
2-2-1 Hirakawa-cho, Chiyoda-ku,
Tokyo 102, Japan

ITE Spain Paraninfo
Calle Magallanes, 25
28015-Madrid, Espana

1 2 3 4 5 6 7 8 9 10 XXX 03 02 01 00 99 98

Library of Congress Cataloging-in-Publication Data

Olivo, C. Thomas.
Basic blueprint reading and sketching / C. Thomas Olivo, Thomas P. Olivo.— 7th ed.
p. cm.
Includes bibliographical references and index.
ISBN 0-7668-0841-6
1. Blueprints. 2. Freehand technical sketching. I. Olivo, Thomas P. II. Title.
T379.04 1998
604.2'5—dc21 98-44492
CIP

PREFACE

The strongest point of Basic Blueprint Reading & Sketching, a world sales leader since its introduction nearly 50 years ago, is that the book was originally developed through an extensive occupational and trade analysis conducted by Dr. C. T. Olivo when he was New York State Director of Industrial Education. Occupational study and textbook revisions reflecting contemporary manufacturing, construction, and transportation needs related to print reading and sketching are on-going. As identified by current occupational task analysis, each unit of instruction in this text builds sequentially, usually in order of difficulty.

The current edition carries on the tradition of providing a blueprint reading book whose content covers all the necessary print reading skills for individuals whose companies currently operate nationally and/or internationally. This book is geared specifically toward those who have newly entered or are preparing to enter into the world of manufacturing and construction or transportation technology.

The seventh revision of *Basic Blueprint Reading and Sketching* reflects T. P. Olivo's continuing experience as a high school and university teacher of blueprint reading, architectural drawing, engineering drawing, and CAD/drafting. Mr. Olivo continually updates the content of the book by using it in his own teaching and making appropriate changes in its content. In addition, this revision includes standards recommended by organizations such as ANSI and ISO.

This 7th edition provides print reading for several skill levels, from machine operator to professionals. In addition, greater emphasis has been placed on the impact of computers from CAD sketching of ideas of products through CADD/CAM improvements that permit direct transfer from design to manufacturing. It also reflects much greater emphasis on decimal and metric dimensioning with a small retention of fractional dimensions as they are currently used.

The three most significant changes in **Part 1** are the almost exclusive use of decimal angles, the greater usage of decimal and metric dimensions, and, following the recommendation of industrial users of blueprints, broader applications of the term *blueprint reading.* This has been brought about by greater reliance on computers from CAD to CAM in an interdependent world economy. Tying together single part and detail drawings into assembly and working drawings has resulted in a new Unit 25, with two blueprint assignments.

In **Part 2,** the technical sketching assignments have undergone a major overhaul. All assignments have new numbers. Ten of the retained sketching assignments are changed; each is now to be drawn in decimal-inch or metric grids rather than using fractional inches.

The two lettering units, 36 and 37, have been reworked to include capital letters and printing in sentences as well as material on lettering for photocopying.

An exciting, new Unit 43 has been created to develop new basic skills in an ever-increasing reliance upon CAD/CADD drawings and Computer Assisted Manufacturing.

Unit 44, a new sketching unit, reinforces and expands content and the print reading skills learned in Unit 25. Students now have to sketch an assembled object in correct proportions.

A special commendation is given to Dr. C. Thomas Olivo to whom this seventh edition is dedicated. Without his vision in founding Delmar Publishers, VICA, NOCTI; as president of many vocational-education associations; as author and editor of 60 books and 500 publications; and as a consultant to many foreign governments, vocational/occupational education may never have taken hold.

C. Thomas Olivo (deceased)
Thomas P. Olivo

GUIDELINES FOR STUDY

BASIC BLUEPRINT READING AND SKETCHING is designed as a basic text-workbook for individuals who must develop skills in reading and accurately interpreting industrial drawings and in preparing simple technical sketches. When instruction in blueprint reading and sketching is integrated with manufacturing or construction drawing courses, the ability of the student to prepare orthographic and other drawings and working sketches is enhanced and accelerated.

The use of a unit pattern of organization permits the contents to be readily adaptable to functional courses within educational, business and industry, government and other special training programs. This text-workbook is widely used in class, group, and individualized/self-paced instruction. This is not a stand-alone text; it requires a knowledgeable instructor or mentor.

ORGANIZATION OF THE TEXTBOOK

BASIC BLUEPRINT READING AND SKETCHING contains two major parts. *Part 1* covers basic principles of blueprint reading, with companion drawings and assignments relating to applications of each new principle. *Part 2* deals with principles and techniques of preparing technical sketches, without the use of instruments.

The instructional units in Part 1 and Part 2 are arranged in a natural sequence of teaching/learning difficulty. Each unit includes *basic principles, a blueprint* and an *assignment* as listed in the Table of Contents.

UNIT BASIC PRINCIPLES

Blueprint reading and sketching principles, concepts, ANSI and SI Metric drafting symbols and standards, terminology, manufacturing process notes, and other related technical information contained on a mechanical or CAD-produced engineering drawing, are all described in detail in the *basic principles*. The teaching of *basic principles* is enhanced through the use of the revised *two-colored transparencies,* following the sequencing in the text.

UNIT BLUEPRINT(S)

Each new principle is applied on one or more industrial drawings. Many of the drawings in the beginning units are simplified by removing the Title Block. The representative drawings appear in the *blueprint(s)*. While the drawings are based on ANSI and SI Metric measurement systems and representation practices, there are adaptations of the standards as found in industry. CAD pictorials of finished parts are now provided on many of the *blueprint(s)*. The pictorials provide a learning tool by identifying the shape, features, dimensions, and other characteristics of each part.

The *multicolored, hinged transparencies* provide a teaching tool for increasing instructional efficiency and accelerating the student's mastery in reading and interpreting each *blueprint.*

UNIT ASSIGNMENT(S)

Problems and other test items are provided in the unit *assignment,* covering all basic principles that are applied in the *blueprints*. The assignments are stated in the terminology of the workplace and deal with principles, practices, and the kinds of experiences needed to read drawings for dimensions, shape descriptions, machining operations and other essential data as required of technicians, craftpersons, related skilled workers, and professionals.

Use is made of encircled letters, numbers, and leaders (callouts) on each drawing in the unit blueprint(s) and assignment(s). The *callouts* simplify the statement of problems. A ruled space is included with the unit assignment for ease in recording each answer.

The sketching assignments deal with principles and applications of sketching, lettering, and techniques used in making orthographic, oblique, isometric, and perspective sketches.

INSTRUCTOR'S RESOURCE GUIDE

Instructional Supplement: Transparencies

Instructional learning enrichment are made possible through the use of a series of transparencies for overhead projection. All illustrations (figures) within the contents of each unit which relate to basic principles may be visually displayed on a screen by using the two-color *basic principle transparencies.* All *blueprints* for Parts 1 and 2 may also be projected using the multicolor transparency series. These transparencies are designed so that each layer is projected in a different color to show each part drawing at three or four successive stages. Layer 1 provides a pictorial; layer 2 views the external shape and features of each part; layer 3 adds internal details, dimensioning and notes; and layer 4 contains the assignment callouts. As each layer is added, cumulative information is provided to assist the student in developing print reading and sketching skills.

INDUSTRIAL DRAWINGS SUPPLEMENT

An additional series of twelve current industrial prints complement the basic principles and blueprint applications and assignments in the text-workbook. The complementary *blueprints* and *assignments* in the INDUSTRIAL DRAWINGS SUPPLEMENT provide additional, more advanced drawings and assignments. The experiences gained are designed to develop a higher level of competency to interpret drawings and prepare sketches at *an intermediate level.*

INSTRUCTOR'S RESOURCE GUIDE

The INSTRUCTOR'S RESOURCE GUIDE serves the following major functions

- an *Answer Book* for all problems in the *assignments* of the textbook and the INDUSTRIAL DRAWINGS SUPPLEMENT
- a *teaching resource* when using the transparencies. Complete *Teaching Plans* are provided for the *basic principles and blueprint* transparencies for each instructional unit. *Student Objectives* and step-by-step *Teaching Sequences* are provided in each *Teaching Plan.*
- sample *Progress Charts* are included for recording group progress and individual achievement.

THE INSTRUCTIONAL-LEARNING "PACKAGE"

The combination of text-workbook for students, complementary resource teaching/learning aids provided by the transparency series, the industrial drawing supplements and the resource guide for the instructor, are crafted to maximize teaching/learning effectiveness in preparing students/trainees with blueprint reading and sketching skills that meet business-industry needs. When integrated within CAD/Technical Drawing courses, the blueprint reading and sketching skills reinforce the development of competency skills in drafting and design.

ACKNOWLEDGEMENTS

For the 7th edition of BASIC BLUEPRINT READING AND SKETCHING, special credit is given to:

- Eric L. Baker, CAD designer, Dresser/Rand Corporation, Wellsville, NY
- Industrial Advisory Committee:
  Eric Achman, HVAC systems designer, T. P. Woodside, Buffalo, NY
  Steven Dunn, engineer, Knolls Atomic Research Laboratory, Schenectady, NY
  Jennifer Ebanks, owner, Keifito's Plantation Resort, Roatan, Honduras
  Alicia Lowe, television art department coordinator, Wilmington, NC
  Noel Minbiole, student, Hudson Valley Community College, Troy, NY

Appreciation is expressed to the following industry representatives and their companies for "going that extra mile" to provide technical resource materials and industrial drawings. • Joseph D. Harnak, Consulting Engineer, Sperry Rail Service, Inc., for unusually fine technical illustrations, pictorial, and other section drawings. • Alan W. Wysocki, The Barden Corp. for the excellent art copy that has appeared on the cover and for other technical data. • Pat Dalton, David Smith, and Gerald Walden, Caterpillar Corp. for Metric CAD engineering drawings with geometric dimensioning and tolerancing. • Joe Guarino and Robert Dovidio, The Charles Stark Draper Laboratories for CAD bearing drawings with GD & T applications and high precision measurements. • Kevin Mangovan, Clausing Industrial, Inc. for assistance with exploded assembly drawings. • Carl Houck, Hamilton Standard Division of United Technologies Corp. for exceptional technical assistance and CAD drawings with GD & T applications. • Donald Viscio, Heli-Coil, a Black and Decker Company, for representative industrial working drawings and assembly and other detail drawings. • Thomas J. Arconti, Hughes Danbury Optical Systems, Inc. for mechanical-engineering drawings with GD & T applications. • Dan Brookmen, Newport News Shipbuilding Company for prints on welded structures and section drawings.

Special recognition to Stephanie B. Olivo for editorial/writing services and indexing this text. Chad Thomas and the rest of the staff at Carlisle Communications, Dubuque, Iowa, also deserve much credit for their efforts in the production of this edition.

CONTENTS

PART 1 BASIC BLUEPRINT READING xviii

PART 1

BASIC BLUEPRINT READING

Unit 1

BASES FOR BLUEPRINT READING AND SKETCHING

Technical information about the shape and construction of a simple part may be conveyed from one person to another by the spoken or written word. As the addition of details makes the part or mechanism more complex, the designer, draftsperson, engineer, technician, and mechanic must use precise methods to describe the object adequately.

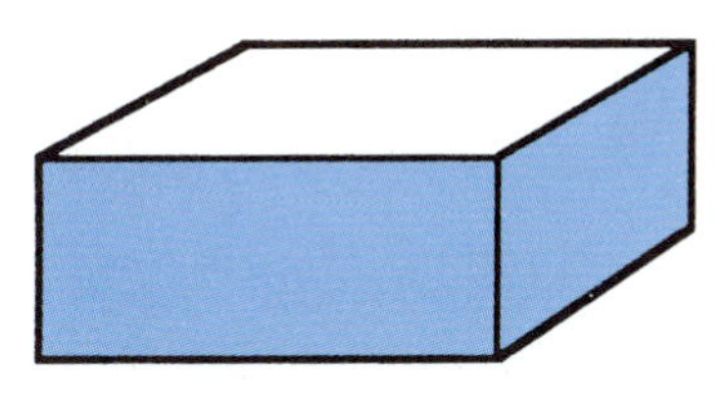

FIGURE 1-1 A picture of a rectangular block.

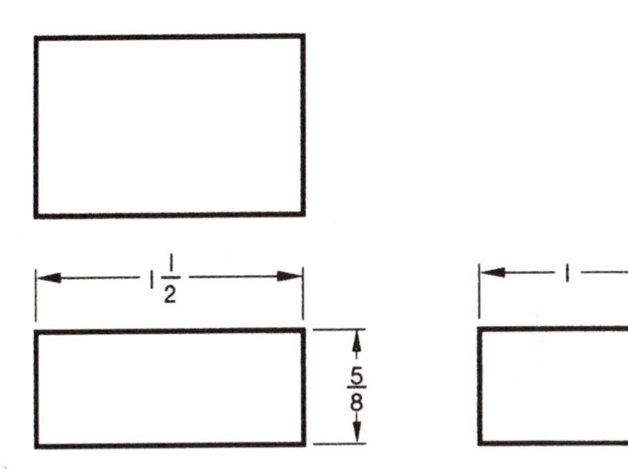

FIGURE 1-2 Mechanical drawing of the rectangular block.

Although a picture drawing (figure 1-1) or a photograph may be used to describe a part, neither method of representing an object shows the exact sizes, interior details, or machining operations that are required. On the other hand, a drawing made accurately with instruments (figure 1-2), or produced by computer or a shop sketch, meets the requirements for an accurate description of shape, construction, size, and other details.

BLUEPRINT READING AS A UNIVERSAL LANGUAGE

Blueprints provide a universal language by which all information about a part is furnished to the craftsperson, technician, designer, and others. Blueprint reading refers to the process of interpreting a drawing. An accurate mental picture of how the object will look when completed can be formed from the information presented.

Training in blueprint reading includes developing the ability to visualize various manufacturing and fabricating processes required to make a part and to apply the basic principles of drafting that underlie the use of different lines, surfaces, and views. The training also includes how to apply dimensions, take measurements, and visualize how the inside of a part or mechanism looks. An understanding of universal measurements and other standards, symbols, signs, and techniques the draftsperson/designer uses to describe a part, unit, or mechanism completely must be developed. Technicians also develop fundamental skills in making sketches so that data relating to dimensions, notes, and other details needed to construct or assemble a part can be recorded on a sketch.

STANDARDS GOVERNING THE PREPARATION OF DRAWINGS

Since all industrialized nations prepare technical drawings according to universally adopted standards, symbols, technical data, and principles of graphic representation, blueprints may be uniformly interpreted throughout the world. This fact translates to mean that parts, structures, machines, and all other products (that are designed according to the same system of measurement) may be accurately manufactured and are interchangeable.

Standards-Setting Organization

The American National Standards Institute (ANSI) and the International Organization for Standardization (metric; ISO) have developed drafting standards that are voluntarily accepted and widely used throughout the world. These systems incorporate and complement other engineering standards that are generated and accepted by such professional organizations as the American Society of Mechanical Engineers (ASME), the American Welding Society (AWS), the American Institute for Architects (AIA), U.S. Military (MIL), and others. These standards-setting bodies deal with specific branches of engineering, science, and technology. Some large companies have adopted their own standards to suit their individual needs. All blueprints in this text closely follow the ANSI and ISO standards and the current industrial practices.

Standards for Sheets

Standards have been established by ANSI for the sheets onto which blueprints are made (figures 1-3; 1-4)

Nearest International Size* (millimeter)	Standard USA Size* (inch)
A4 210 × 297	A 8.5 × 11.0
A3 297 × 420	B 11.0 × 17.0
A2 420 × 594	C 17.0 × 22.0
A1 594 × 841	D 22.0 × 34.
A0 841 × 1189	E 34.0 × 44.0

*ANSI Y14.1–1980

FIGURE 1-3

COMMON ELEMENTS IN DRAFTING

Close examination of a technical drawing reveals that a part or an assembly of many parts must be clearly and accurately represented by applying principles, techniques, and other common elements of drafting, as follows:

- **Lines** that, when combined with each other, provide graphic information about the external shape and details of an object.
- **Dimensions** that give the size and location of parts, shapes, and assemblies.
- **Sections** that are used to expose internal construction details when an object is cut apart by imaginary cutting planes.
- **Processes** that are represented by symbols and notes for purposes of identifying how each part is to be produced.
- **Geometric positioning** that deals with placing and accuracy of geometric objects in relationship to one another.
- **Notes & other non-graphical information** clearly lettered (printed), not written.
- **Tolerances** that deal with accuracy related to form and size and position of finished surfaces, where high precision is required.

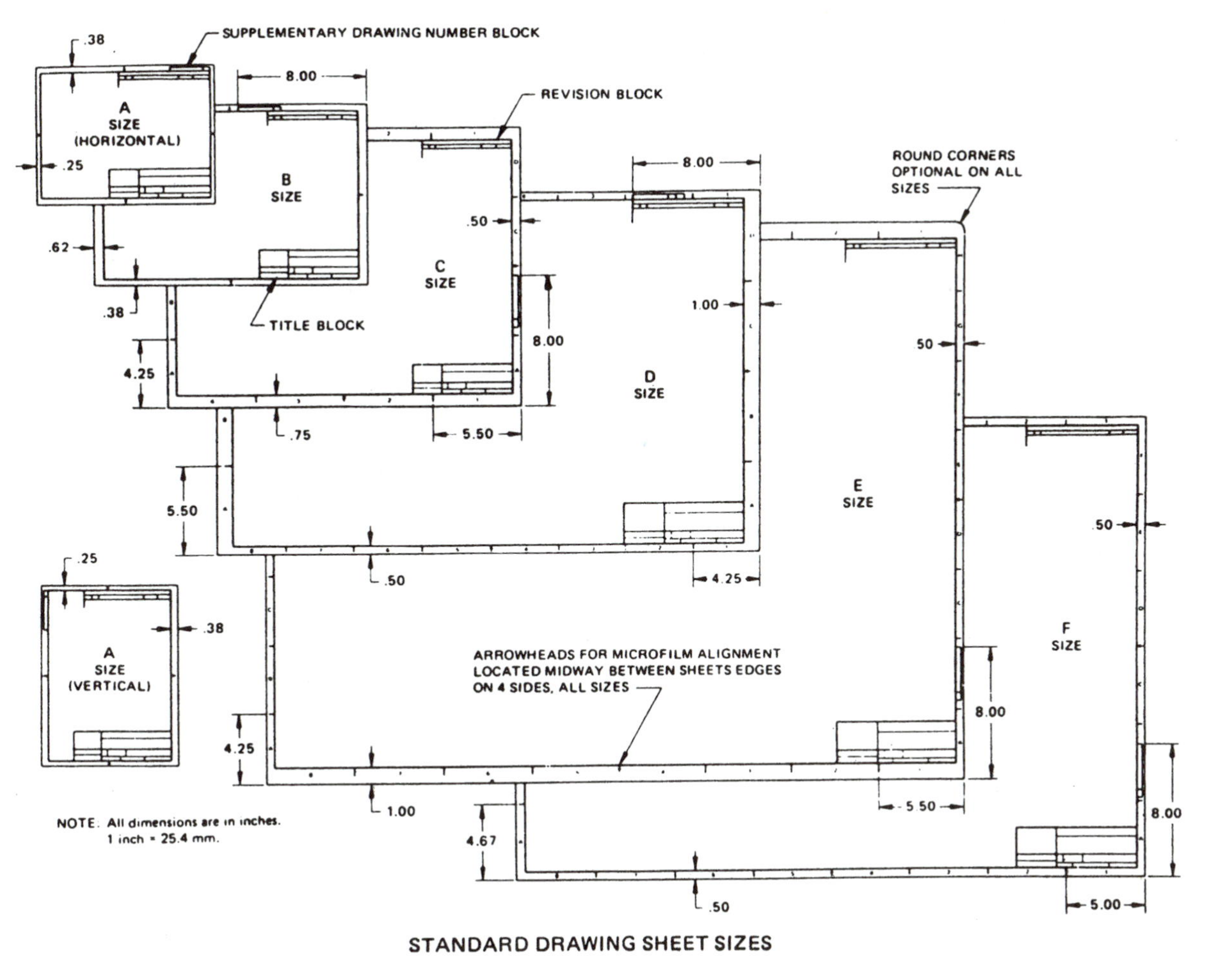

STANDARD DRAWING SHEET SIZES

FIGURE 1-4

- **Materials** that are used for making parts including information about weight, strength, and hardness.
- **Computer-aided drafting** applications, where appropriate.
- **Technical sketching** techniques for preparing line and form engineering and technical sketches to complement precise engineering drawings.

INDUSTRIAL PRACTICES RELATING TO DRAWINGS

Drawing Reproduction Processes

Traditionally, a blueprint of an original drawing, which may be made on high-quality paper, vellum, and acetate or polyester film, is produced by exposing the drawing to a strong light source. Some of the light is held back by the drawn lines as it passes through a transparent tracing onto a sensitized blueprint paper. After exposure, the blueprint paper is passed through a developer fluid, washed by water, moved through a fixative (to set the color), rewashed, and dried. This blueprint (except for slight shrinkage during the drying stage) is an exact duplicate but in negative form (figure 1-5).

The term *blueprint* originally referred to a blue and white reproduction of a drawing. Today it is used loosely in modern industrial language to include many different types of reproductions (figure 1-6).

The highest quality reproductions are made by plotters of various types: the **flatbed plotter** which holds the paper (usually A-size or B-size) on a flat surface while a pen moves in two directions; a **drum plotter** where the paper is attached to a rotating drum producing

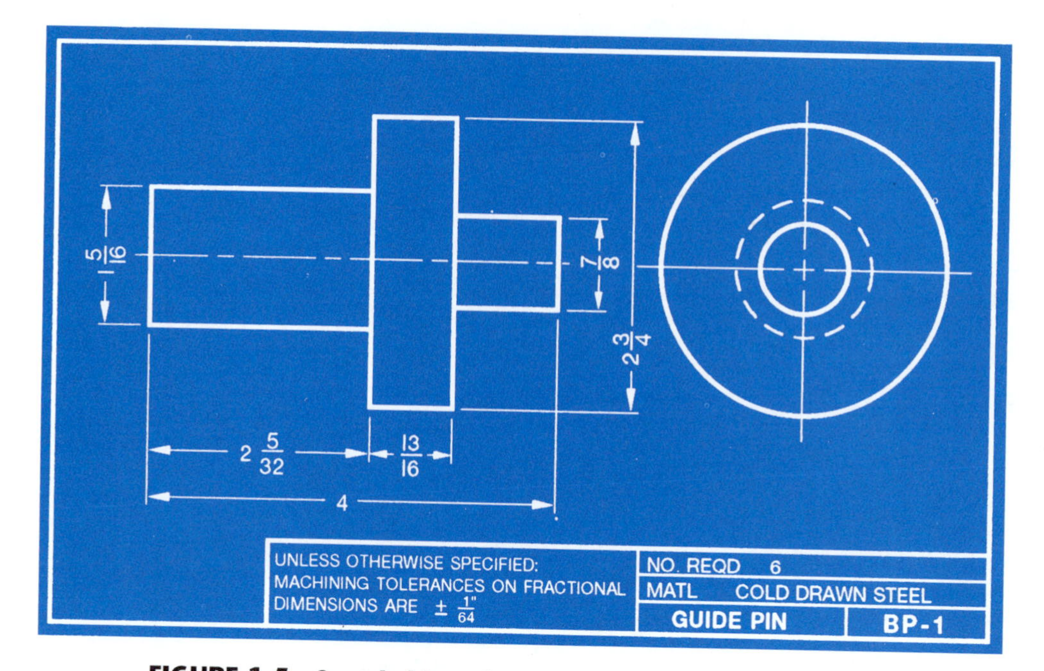

FIGURE 1-5 Sample blueprint (negative of original drawing).

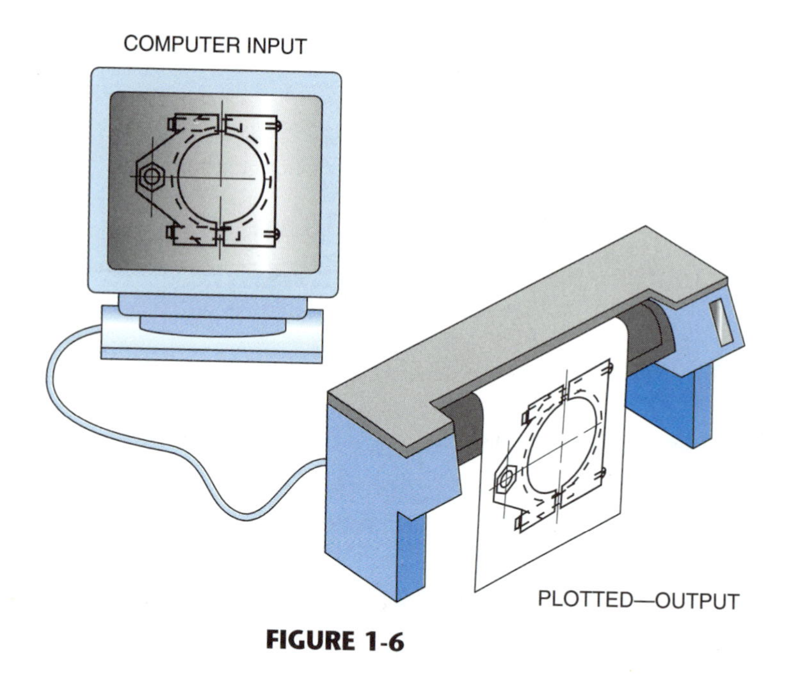

FIGURE 1-6

horizontal ("X" axis) lines as a pen moves across the drum to reproduce vertical ("Y" axis) lines; or a **multigrip plotter** which uses small rubber rollers to grip the paper at the edges as opposed to having the paper attached to the circumference of the drum. Simultaneous moving of the drum and the pen on both of these produces curves and angle lines. The size of the paper used in the drum plotter is limited by the width and the circumference of the drum, which can be designed to use up through E-size or AO drawings.

Today, most plotters operate by commands from a computer. They can hold any number of pens up to 20 and produce very accurate continuous line drawings with varying lineweight and color.

Printers are also used to reproduce drawings. They are generally less expensive than plotters but are not as accurate because they can only simulate lines by printing a closely spaced series of dots.

Another method of reproduction known as **diazo** continues to be popular. This process projects light through the original drawing which is in direct contact with the recipient paper or transparency. That paper is then developed with an ammonia process to show the image, usually in black, brown, or blue. A variety of paper weights and finishes as well as transparencies can be used in diazo.

Copying machines, similar to ones used for typical office copying, have also been designed to handle blueprints and engineering prints of varying sizes. There are several advantages to these machines—they are relatively inexpensive, they do not require coated copy materials or chemicals, and the size of the drawing can be enlarged or reduced as desired.

Drawing Storage Procedures

Original drawings (in modern industrial practice) are seldom sent to the shop, job site, or laboratory. Instead, exact reproductions (called *blueprints* or *whiteprints*) are made of the original. Duplicate copies are usually distributed among all individuals who are involved in planning, manufacturing, assembling, or other work with a part or unit. The original drawings are filed for record purposes and for protection.

The options available for storage of drawings and blueprints have increased as technology has advanced. *Microfilm* has been one of the most widely used procedures for many years and is still used in many places today. In this method, a greatly reduced photographic reproduction is made on film and then mounted into cardboard aperture cards. These cards are easy to file and keep track of. Many feel that they are also more durable.

Blueprints are also saved in special storage devices which include flat-drawer files, files designed for vertical hanging or in tubes. Drawings stored in these ways must also be available for rapid and convenient retrieval; therefore, prints are often folded and stored in standard office file cabinets. If properly organized, this method does provide rapid retrieval and security.

For a number of people, the preferred storage method is the floppy disk whether the drawing is produced by hand or by CAD (computer-aided drawing). With CAD the drawing can be saved directly to the disk or it can be produced as *hardcopy* (a sheet of material with the drawing upon it) by one of several machines. Scanners of various types, including digitizing tablets, convert hand-made drawings into computer language for disk storage.

Another developing technology which is being applied to information storage is the optical disk, which can store a trillion bytes of information that can be accessed in a few seconds.

Some consider this disk storage to be superior to former methods because it reduces deterioration and the possibility of loss. Also, when drawings are saved as hardcopy (prints), there is inevitable wear and tear on them, which compromises the quality of any reproduction. Drawings saved on disks can be consistently reproduced with the same quality. It is also possible to revise and correct drawings which have been saved by this procedure. However, disks and computer systems can develop problems and the longevity of floppy disks or optical disks has not been proven; therefore, there continues to be a need for a choice.

Hardcopy and Softcopy Drawings

The term *blueprint* or *print* identifies drawings that may be produced by conventional drafting room techniques or are computer generated. Blueprint reading traditionally involves the physical handling of a sheet on which a drawing is reproduced, but it also refers to drawings as they appear on computer screens. The latter is called *softcopy*. The drawing sheet is sometimes referred to as *hardcopy* and is a permanent copy of a drawing. Hardcopy may be produced on conventional blueprinting equipment or it may be computer generated. Drawings in hardcopy form may be reproduced in one color or an assortment of colors.

CAD/CAM

Some industries are converting to a CAD/CAM link-up (CAM stands for computer-aided manufacturing). In this process, the drafting department designs, on the computer, a product. The drawing is then converted to coordinate data which is transferred to a numerical control computer tape. Through a software program known as a *post processor,* this information is translated into usable information for the machines in the manufacturing department. No hardcopy is ever produced unless someone needs to study the drawing away from a computer terminal.

SOME TECHNICAL INFORMATION SUPPLIED ON DRAWINGS

The positive form blueprint of the GUIDE PIN (figure 1-7) illustrates other information that generally appears on a drawing. A *title block* is included that contains the following. (See page 115 for a more detailed title block.)

- Name of the part (Guide Pin)
- Quantity needed (6)
- Drawing number (BP-1)
- Dimensional tolerance (± 1/64")
- Material (cold drawn steel)

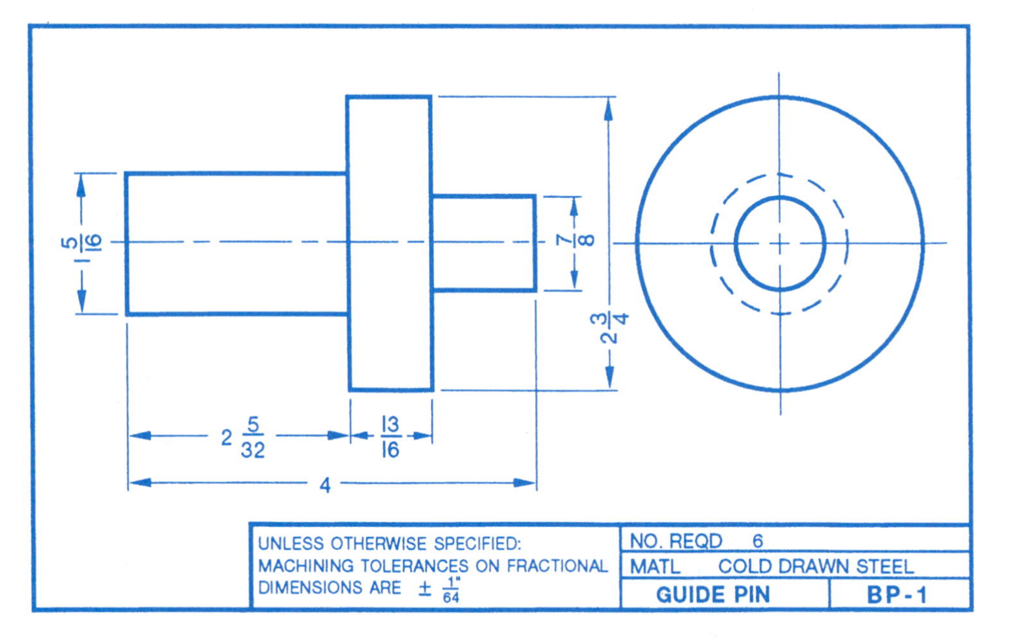

FIGURE 1-7 (Positive) blueprint of original drawing.

UNIVERSAL SYSTEMS OF MEASUREMENT

Drawings contain dimensions that identify measurements of straight and curved lines, surfaces, areas, angles, and solids. *Linear* or *straight line measurements* between two points, lines, or surfaces are most widely used. Every linear measurement begins at a particular *reference point* and ends at a *measured point.* These *references* identify particular features as illustrated in figure 1-8. Characteristics of linear measurements are summarized in the second drawing at (B).

The straight line distance between these two points is the *line of measurement.* It represents the actual required dimension and is often referred to as the *nominal* or *base size.* A nominal (base) size is generally represented on a drawing by a rectangular symbol ▭. The

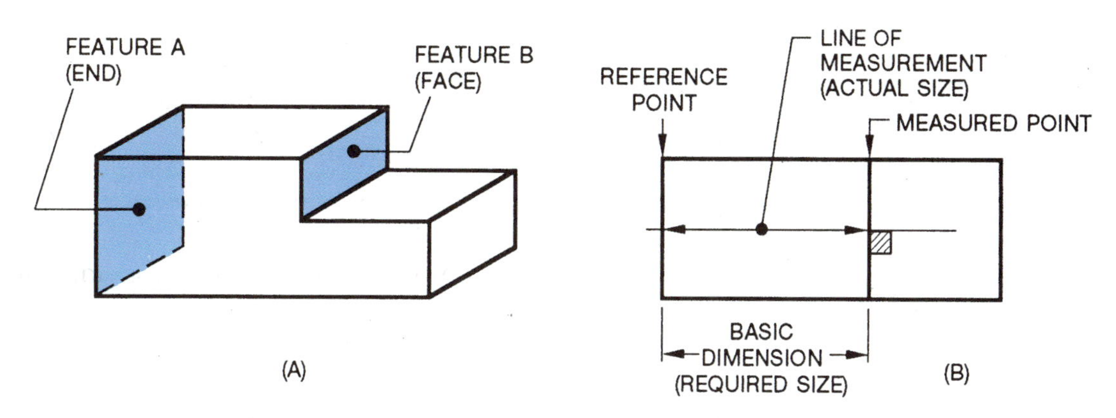

FIGURE 1-8 Features and characteristics of linear measurements.

nominal (base) size is enclosed within the symbol as, for example, 1.625. The degree of accuracy to which a part is to be machined, fitted, or formed, relates to its nominal size; often referred to as a *basic dimension.*

Basic Units of Linear Measurement

Dimensions appear on drawings in one or a combination of two different units of measurement, as in the case where inch and metric units are used together. Customary inch (British-United States standard) dimensions are given as fractional parts of an inch or as decimal inch values. The inch is subdivided into equal parts, called *fractional parts* or *decimal* (*mils*) *parts*). Figure 1-9A illustrates commonly used fractional parts of an inch (1/64", 1/32", 1/16", 1/8", 1/4", and 1/2"). The graduations on the steel rule in figure 1-9B show decimal parts of an inch (.010" (ten mils), .050" (fifty mils), .100" (100 mils), and .500" (500 mills).

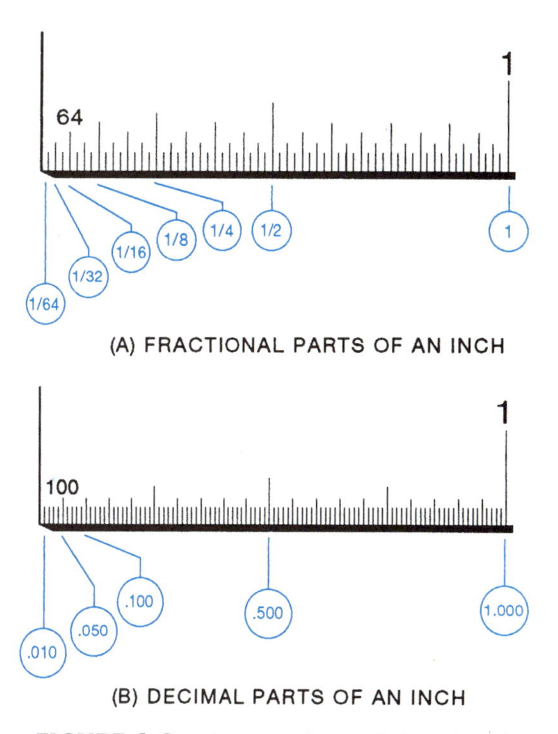

FIGURE 1-9 Commonly used fractional graduations (enlarged) in Customary Inch measurements.

SI Metric Linear Units of Measurement

Standard units of linear measurement in the SI metric system are all related to the *meter.* For most practical purposes, the meter is equal to 39.37 inches. Metric dimensions are generally given on a drawing in millimeters (and sometimes in centimeters) as a multiple of the meter. Common steel rules are graduated in whole and half millimeters as shown in figure 1-10. The millimeter (mm) is the most widely used SI metric linear unit of measurement for engineering drawings.

Conversion tables are readily available to change a dimension in one system to a mathematical equivalent in the second system.

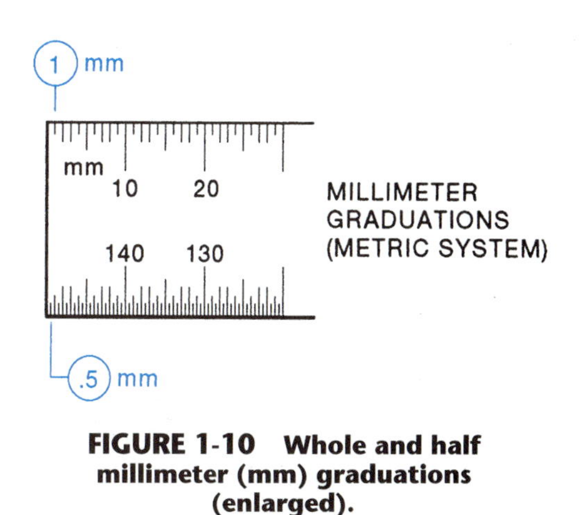

FIGURE 1-10 Whole and half millimeter (mm) graduations (enlarged).

BASES FOR BLUEPRINT READING AND SKETCHING

1. Name one difference between picture drawings or photographs and a technical drawing of an object.
2. Tell briefly why blueprints provide a universal language.
3. Name two organizations who develop drafting standards.
4. List three common elements of drafting that are used to produce technical drawings.
5. Explain how curve and angle lines are produced with a drum plotter.
6. State one reason why printers are not as accurate as plotters for reproducing technical drawings.
7. Name two reproduction methods other than plotters and printers.
8. Provide two advantages of disk storage over storage of original drawings on paper.
9. Explain the difference between hardcopy and softcopy.
10. Name the end points to every linear measurement.
11. Describe what the term "line of measurement" means.
12. Tell what function a nominal dimension serves.
13. Show how the nominal dimension **2.375"** may be represented on a drawing.
14. Name the two major linear measurement systems.
15. Give two examples of fractional parts of the basic unit of measurement in each system.

ASSIGNMENT—UNIT 1

Student's Name Cornelius Pope

1. Tech drawing will show exact size and details
2. There are universally adopted standards
3. (a) American National standars Institute
   (b) International organization for standardization (metric)
4. (a) ______
   (b) ______
   (c) ______
5. ______
6. ______
7. (a) ______
   (b) ______
8. (a) ______
   (b) ______
9. ______
10. ______
11. ______
12. ______
13. ______
14. (a) ______
    (b) ______
15. (a) ______
    (b) ______

Unit 2

THE ALPHABET OF LINES AND OBJECT LINES

The line is the basis of all industrial drawings. By combining lines of different thicknesses, types, and lengths, it is possible to describe graphically any object in sufficient detail so that persons with a basic understanding of blueprint reading can accurately visualize the shape of the part.

THE ALPHABET OF LINES

The American National Standards Institute (ANSI) has adopted and recommended certain drafting techniques and standards for lines. The types of lines commonly found on drawings are known as the *alphabet of lines.* The six types of lines which are most widely used from this alphabet include: (1) object lines, (2) hidden lines, (3) center lines, (4) extension lines, (5) dimension lines, and (6) projection lines. A brief description and examples of each of the six types of lines are given in this section. These lines are used in combination with each other on all the prints in the Blueprint Series. Problem material on the identification of lines is included in the Assignment Series. Other lines, such as those used for showing internal part details or specifying materials, are covered in later units.

The thickness (weights) of lines are relative because they depend largely on the size of the drawing and the complexity of each member of an object. For this reason, comparative thicknesses (weights) of lines are used.

OBJECT LINES

The shape of an object is described on a drawing by thick (dark) lines known as *visible edge* or *object lines.* An object line, figure 2-1, is always drawn thick (dark) and solid so that the outline or shape of the object is clearly emphasized on the drawing, figure 2-2.

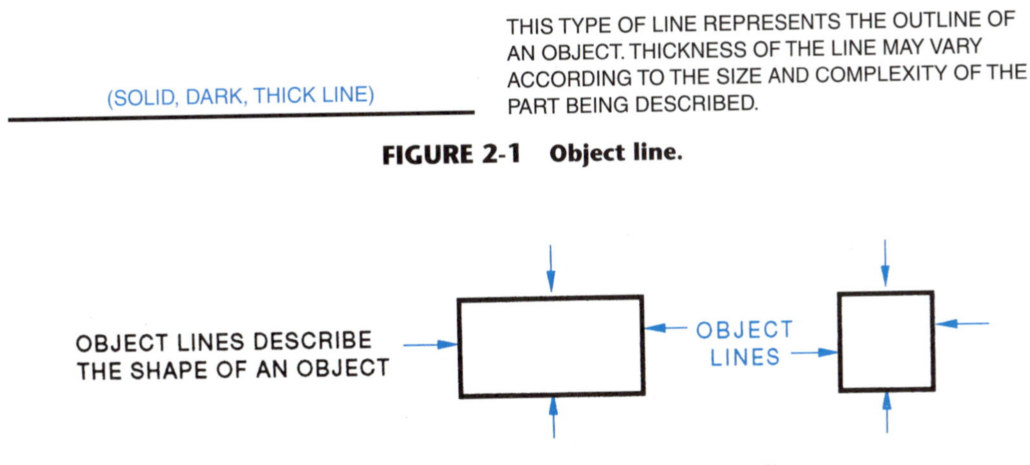

FIGURE 2-1 Object line.

FIGURE 2-2 Application of the object line.

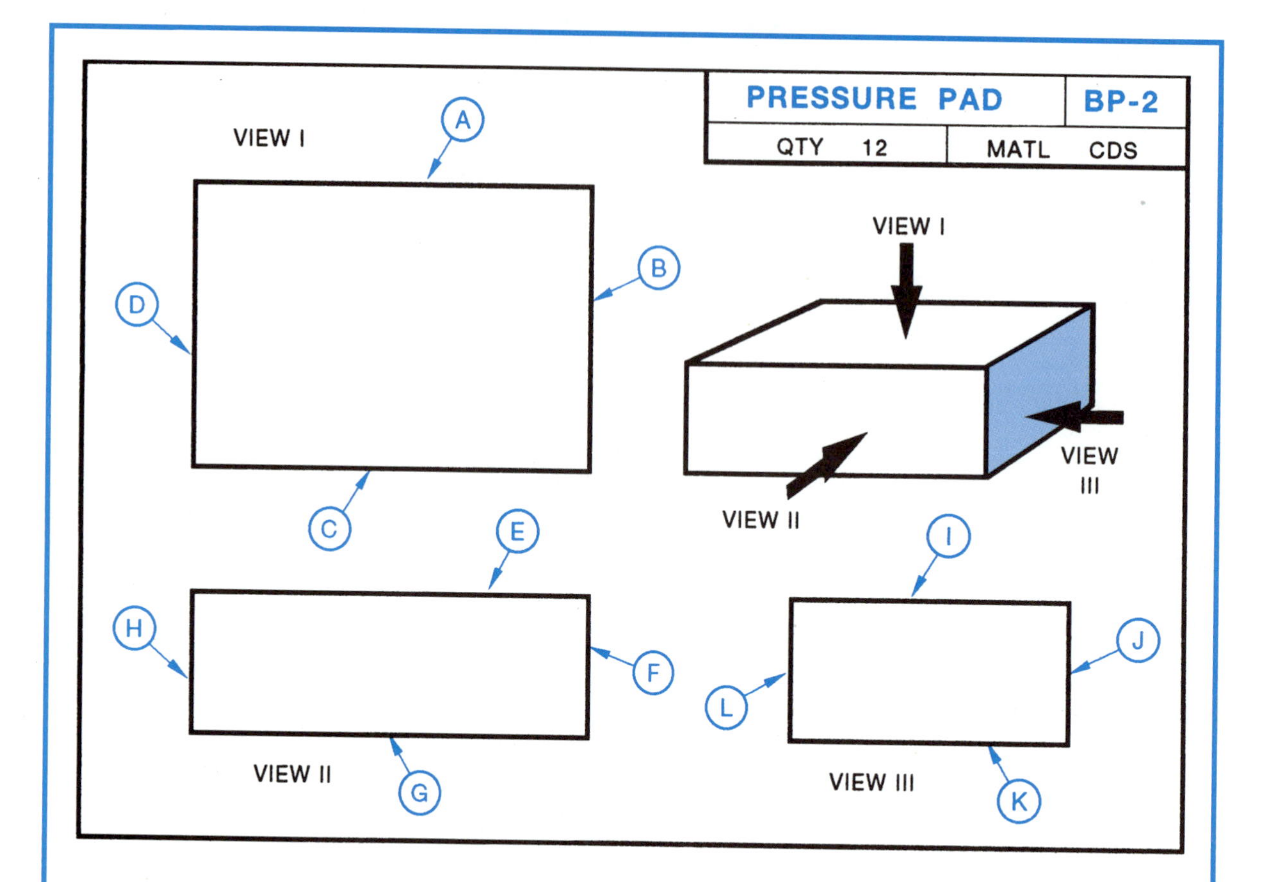

PRESSURE PAD (BP-2)

1. Give the name of the part.
2. What is the number of the blueprint?
3. How many Pressure Pads are needed?
4. What name is given to the thick (dark) line which shows the shape of the part?
5. What lettered lines show the shape of the part in:
   a. View I
   b. View II
   c. View III
6. Identify two standard measurement systems that may be used to dimension the Pressure Pad drawing.
7. Name two basic processes for producing multiple copies of a finished drawing.

ASSIGNMENT—UNIT 2

Student's Name ________________

1. ________________
2. ________________
3. ________________
4. ________________
5. View I ________________
   II ________________
   III ________________
6. a. ________________
   b. ________________
7. a. ________________
   b. ________________

Unit 3

HIDDEN LINES AND CENTER LINES

HIDDEN LINES

To be complete, a drawing must include lines which represent all the edges and intersections of surfaces in the object. Many of these lines are invisible to the observer because they are covered by other portions of the object. To show that a line is hidden, the draftsperson usually uses a thin, broken line of short dashes, figure 3-1. Figure 3-2 illustrates the use of hidden lines.

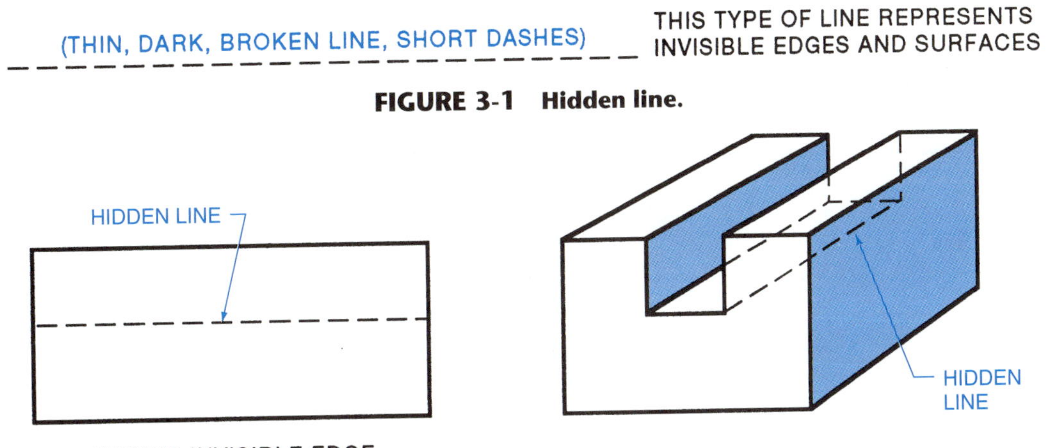

FIGURE 3-1 Hidden line.

FIGURE 3-2 Application of hidden lines.

CENTER LINES

A center line, figure 3-3, is drawn as a thin (light), broken line of long and short dashes, spaced alternately. Center lines are used to indicate the center of a whole circle or a part of a circle, and also to show that an object is symmetrical about a line, figure 3-4. The symbol ℄ is often used with a center line.

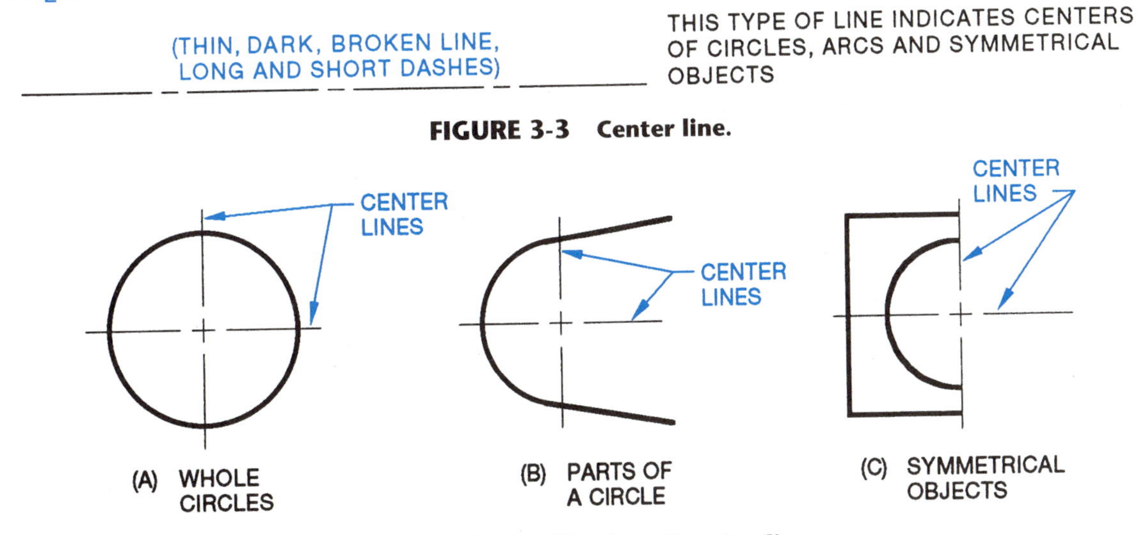

FIGURE 3-3 Center line.

FIGURE 3-4 Application of center lines.

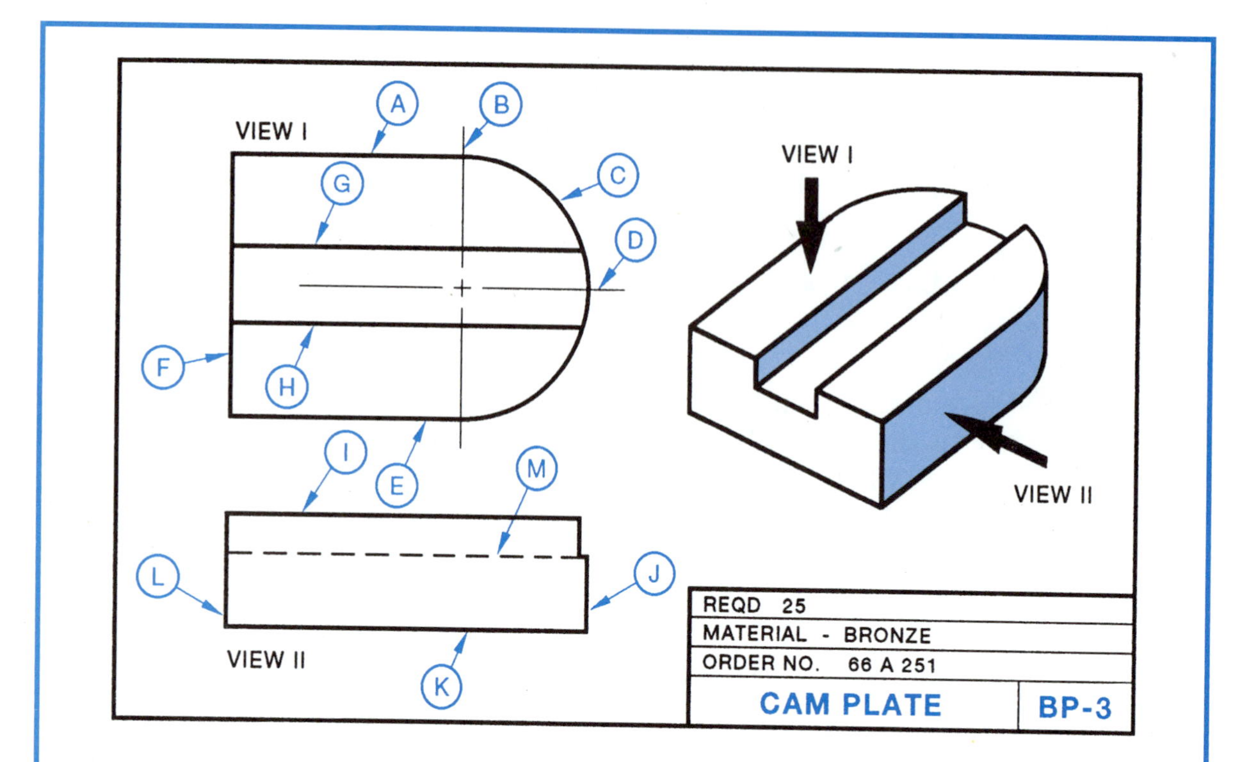

ASSIGNMENT—UNIT 3

CAM PLATE (BP-3)

1. How many Cam Plates are required?
2. Name the material for the parts.
3. What type of line is used to describe the shape of the part?
4. Give the letters of all the lines which show the outside shape of the part.
5. Name the kind of line which represents an invisible edge.
6. What lettered lines in View I show the slot in the Cam Plate?
7. What line in View II shows an invisible surface?
8. What kind of line indicates the center of a circle, or part of a circle?
9. What lettered line in View I shows the circular end?
10. What lettered lines in View I are used to locate the center of the circular end?

Student's Name ____________________

1. ____________________
2. ____________________
3. ____________________
4. ____________________
5. ____________________
6. ____________________
7. ____________________
8. ____________________
9. ____________________
10. ____________________

EXTENSION LINES AND DIMENSION LINES

EXTENSION LINES

Extension lines are used in dimensioning to show the size of an object. Extension lines, figure 4-1, are thin, dark, solid lines which extend away from an object at the exact places between which dimensions are to be placed.

FIGURE 4-1 Extension lines used for dimensions.

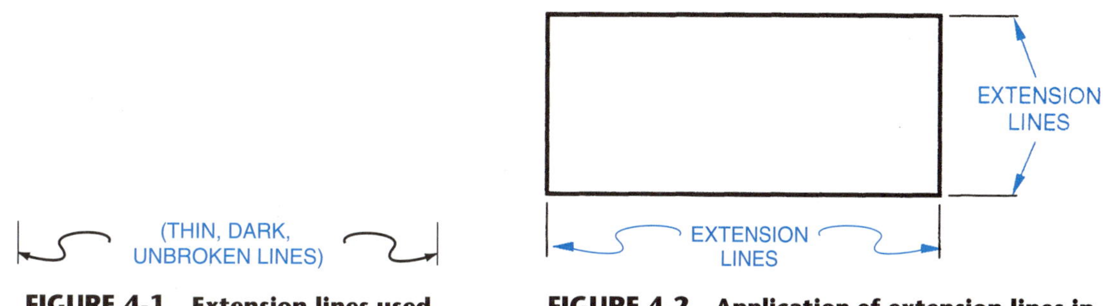

FIGURE 4-2 Application of extension lines in dimensioning.

A space of one-sixteenth inch is usually allowed between the object and the beginning of the extension line, figure 4-2.

DIMENSION LINES

Once the external shape and internal features of a part are represented by a combination of lines, further information is provided by dimensions. Fractional, decimal, and metric dimensions are used on drawings to give size descriptions. Each of these three systems of dimensioning is used throughout this text.

Dimension lines, figure 4-3, are thin lines broken at the dimension and ending with arrowheads. The tips or points of these arrowheads indicate the exact distance referred to by a dimension placed at a break in the line.

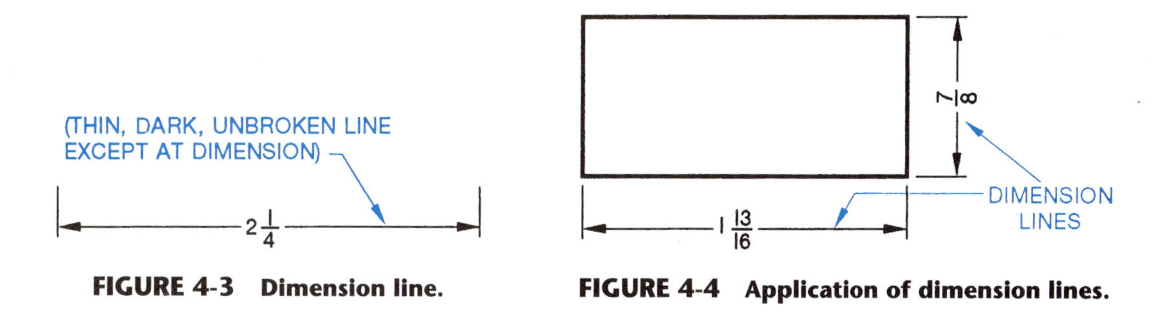

FIGURE 4-3 Dimension line.

FIGURE 4-4 Application of dimension lines.

The point or tip of the arrowhead touches the extension line. The size of the arrow is determined by the thickness of the dimension line and the size of the drawing. Closed (←) and open (→) arrowheads are the two shapes generally used. The closed arrowhead is preferred. The extension line usually projects 1/16 inch beyond a dimension line, figure 4-4. Any additional length to the extension line is of no value in dimensioning.

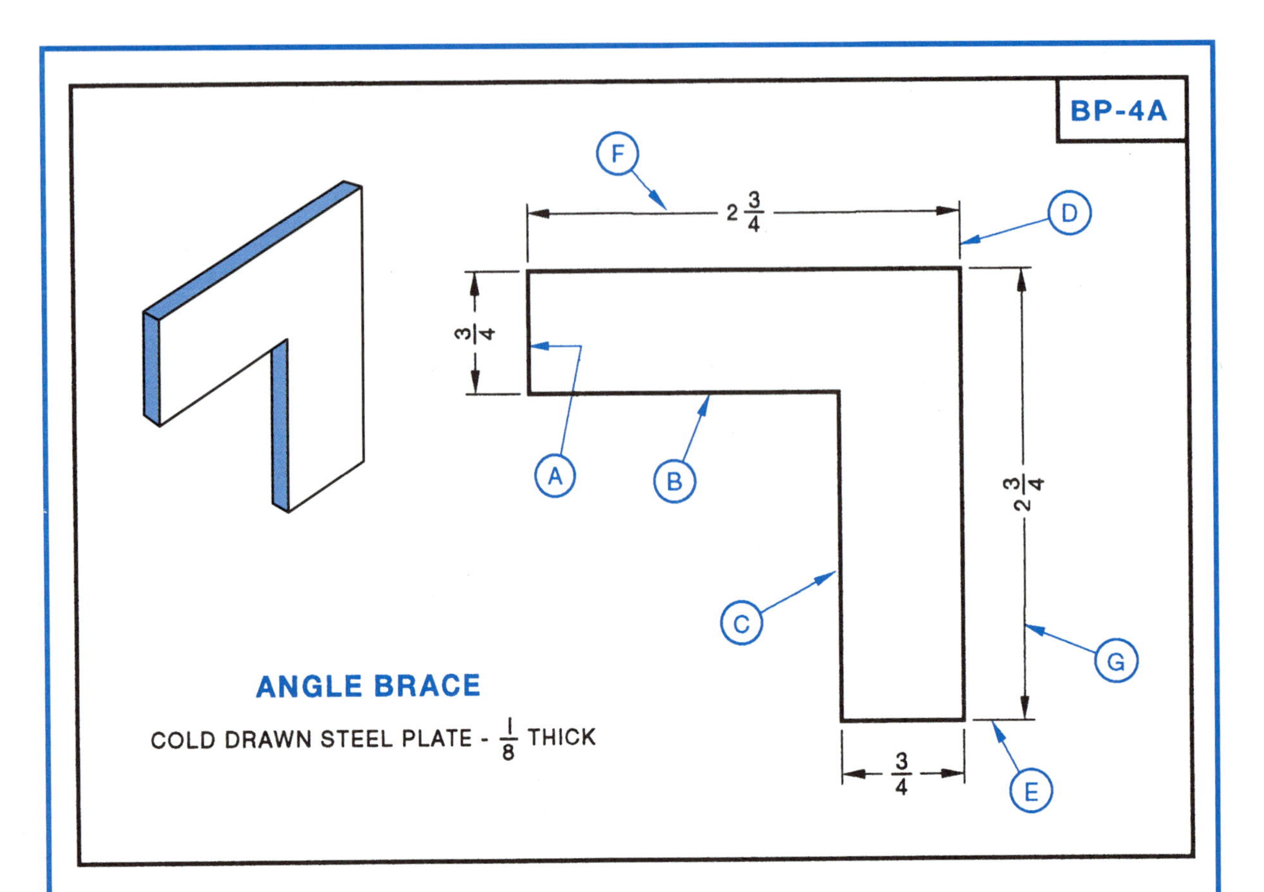

ASSIGNMENT A—UNIT 4

ANGLE BRACE (BP-4A)

1. Name the material specified for the Angle Brace.
2. What is the overall length of the brace from left to right?
3. What is the overall height from the top of the brace to the bottom?
4. What is the width of each leg of the Brace?
5. What is the thickness of the metal in the Brace?
6. What is the name given to the kind of line marked Ⓐ, Ⓑ and Ⓒ?
7. What kind of lines are Ⓓ and Ⓔ?
8. What kind of lines are Ⓕ and Ⓖ?
9. Why are object lines made thicker than extension and dimension lines?

Student's Name ________________

1. ________________
2. ________________
3. ________________
4. ________________
5. ________________
6. ________________
7. ________________
8. ________________
9. ________________

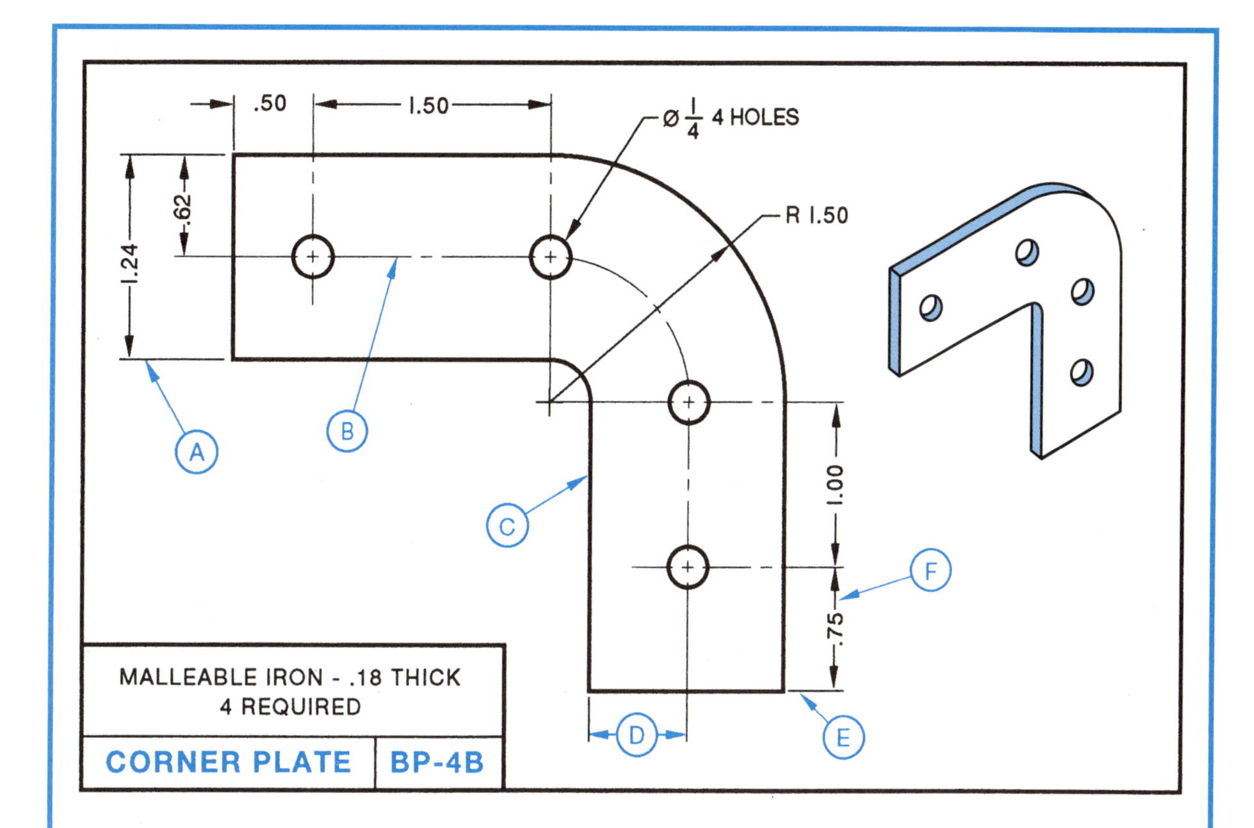

CORNER PLATE (BP-4B)

1. What kind of line is (A)?
2. What kind of line is (B)?
3. What kind of line is (C)?
4. Determine the overall length of the Plate from left to right.
5. Determine the overall height of the Plate from top to bottom.
6. Give the center distance between the two upper holes.
7. Determine distance (D).
8. What kind of line is (E)?
9. What kind of line is (F)?
10. What radius forms the rounded corner of the Plate?
11. Name the material specified for the Plate.
12. How many Corner Plates are required?

ASSIGNMENT B—UNIT 4

Student's Name ______________________

1. ______________________
2. ______________________
3. ______________________
4. ______________________
5. ______________________
6. ______________________
7. ______________________
8. ______________________
9. ______________________
10. ______________________
11. ______________________
12. ______________________

Unit 5

PROJECTION LINES, OTHER LINES, AND LINE COMBINATIONS

PROJECTION LINES

Projection lines are used by draftspersons and designers to establish the relationship of lines and surfaces in one view with corresponding points in other views. *Projection lines,* figure 5-1, are thin, unbroken lines projected from a point in one view to locate the same point in another view. Projection lines do not appear on finished drawings except where a part is complicated and it becomes necessary to show how certain details on a drawing are obtained.

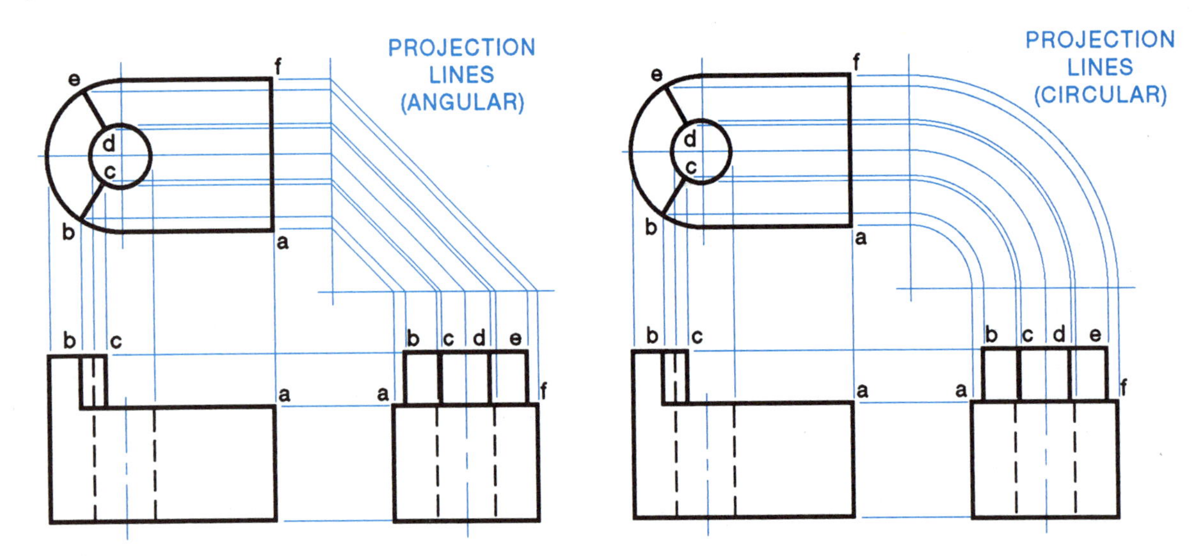

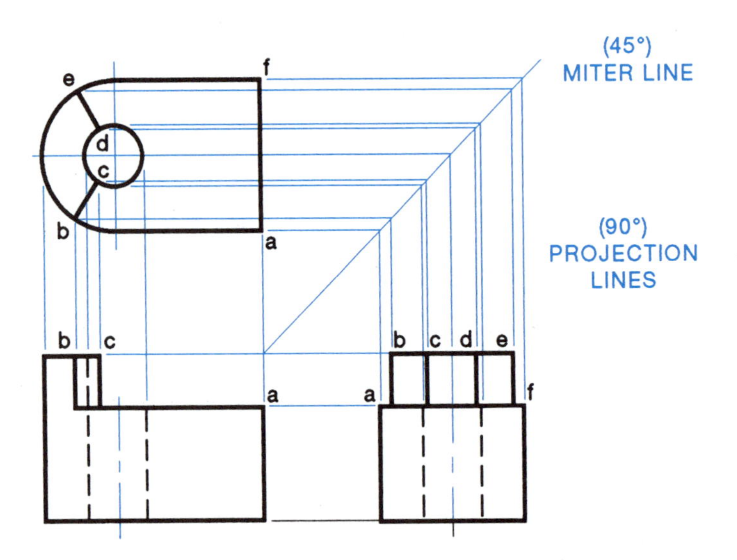

FIGURE 5-1 Application of projection lines.

OTHER LINES

In addition to the six common types of lines, the alphabet of lines also includes other types such as the cutting plane and section lines, break lines, phantom lines to indicate adjacent parts and alternate positions, and lines for repeated detail. These less frequently used lines, figure 5-2, are found in more advanced drawings, and will be described in greater detail as they are used in later drawings in this text.

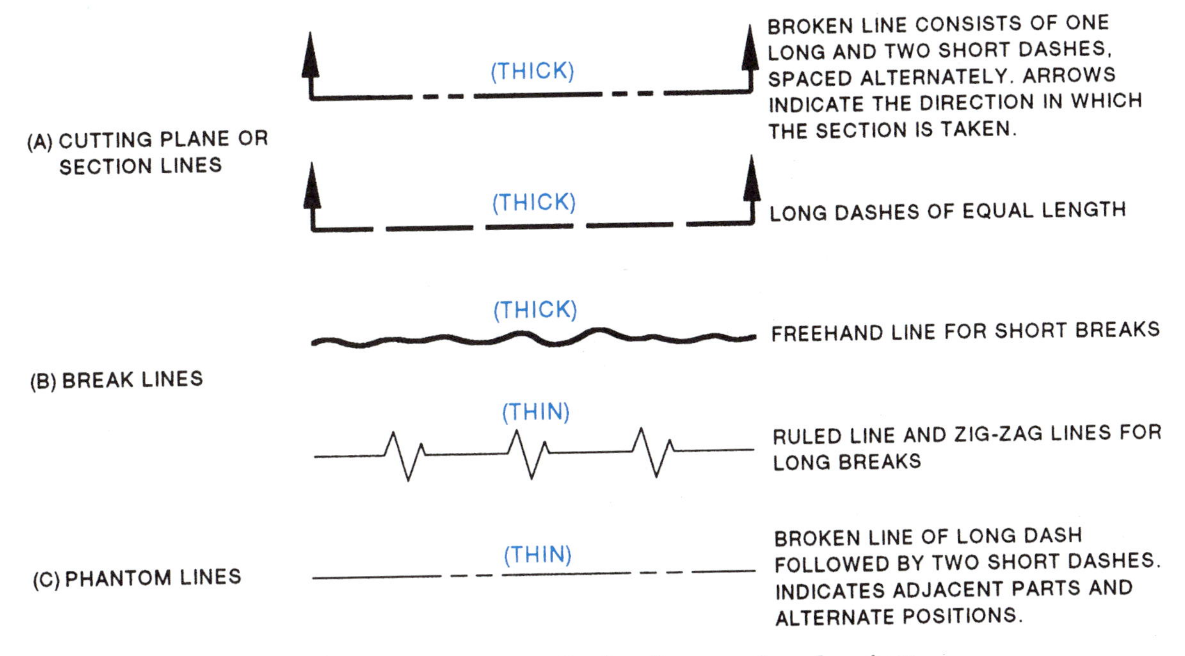

FIGURE 5-2 Samples of other lines used on drawings.

LINES USED IN COMBINATION

Most drawings consist of a series of object lines, hidden lines, center lines, extension lines, and dimension lines used in combination with each other to give a full description of a part or mechanism, figure 5-3.

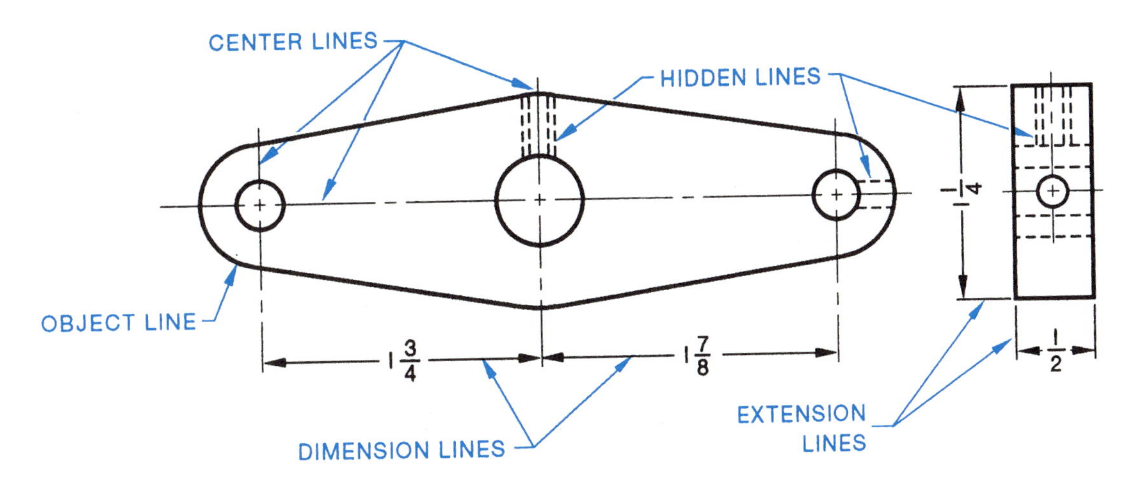

FIGURE 5-3 Lines used in combination.

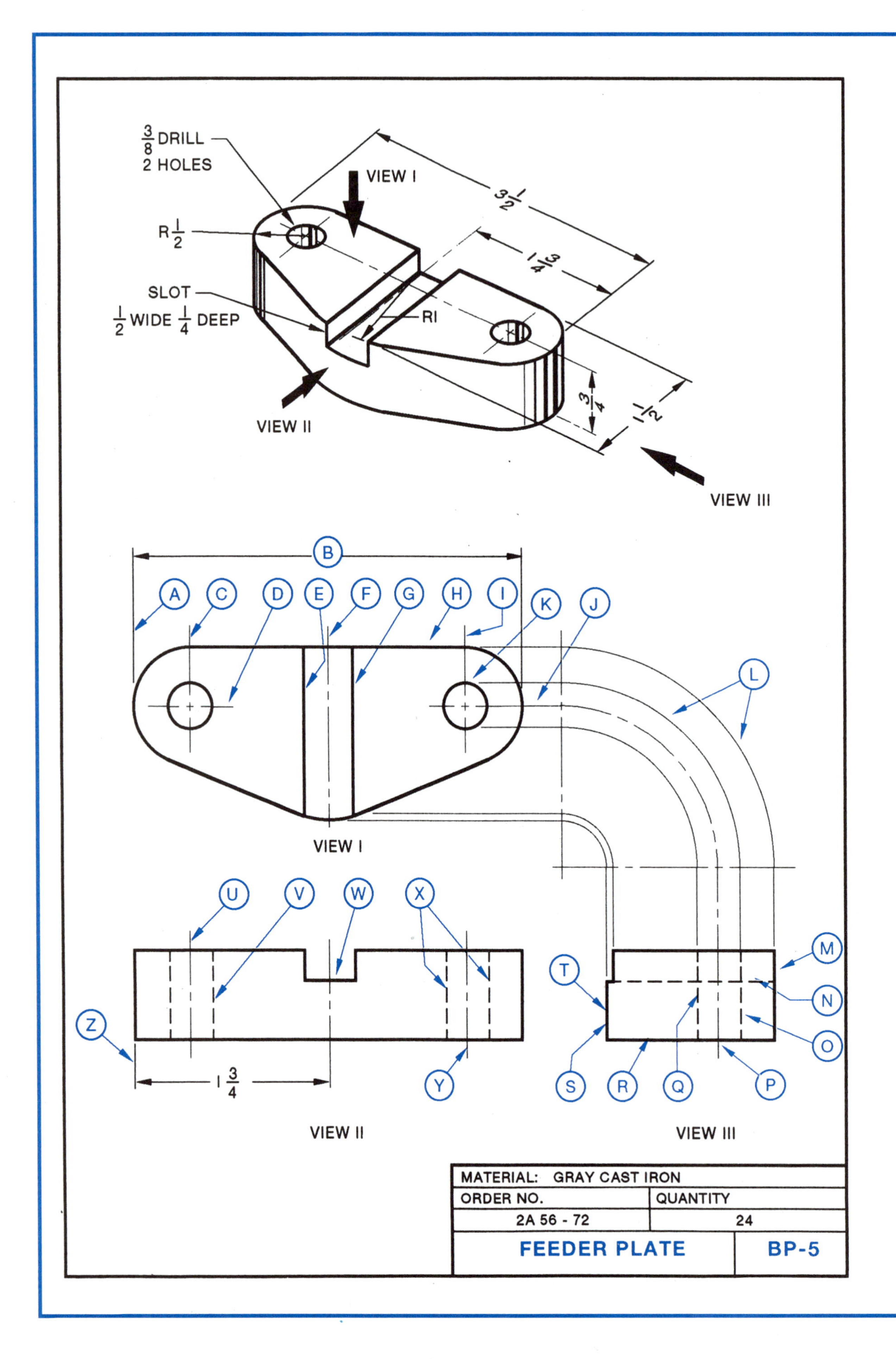

MATERIAL: GRAY CAST IRON	
ORDER NO.	QUANTITY
2A 56 - 72	24
FEEDER PLATE	**BP-5**

ASSIGNMENT—UNIT 5

FEEDER PLATE (BP-5)

Student's Name ______________________

1. What is the name of the part? 1. ______________________
2. What is the blueprint number? 2. ______________________
3. What is the Plate order number? 3. ______________________
4. How many parts are to be made? 4. ______________________
5. Name the material specified for the part. 5. ______________________
6. Study the Feeder Plate, BP-5.
   a. Locate and name each line from (A) to (Z) in the space provided for each in the table.
   b. Tell how each line from (A) to (L) is identified. (NOTE: Line (A) is filled in as a guide.)

Line on Drawing	(A) Name of Line	(B) How the Line is Identified
A	EXTENSION LINE	THIN, DARK, UNBROKEN LINE
B		
C		
D		
E		
F		
G		
H		
I		
J		
K		
L		
M		
N		
O		
P		
Q		
R		
S		
T		

Line	(A) Name of Line
U	
V	
W	
X	
Y	
Z	

SECTION 2 Views

Unit 6

THREE-VIEW DRAWINGS

Regularly-shaped flat objects which require only simple machining operations are often adequately described with notes on a one-view drawing (see Unit 9). However, when the shape of the object changes, portions are cut away or relieved, or complex machining or fabrication processes must be represented on a drawing, the one view may not be sufficient to describe the part accurately.

The number and selection of views is governed by the shape or complexity of the object. A view should not be drawn unless it makes a drawing easier to read or furnishes other information needed to describe the part clearly.

Throughout this text, as the student is required to interpret more complex drawings, the basic principles underlying the use of all additional views which are needed to describe the true shape of the object will be covered. Immediate application of these principles will then be made on typical industrial blueprints.

The combination of front, top, and right side views represents the method most commonly used by draftspersons to describe simple objects. The manner in which each view is obtained and the interpretation of each view is discussed in this section.

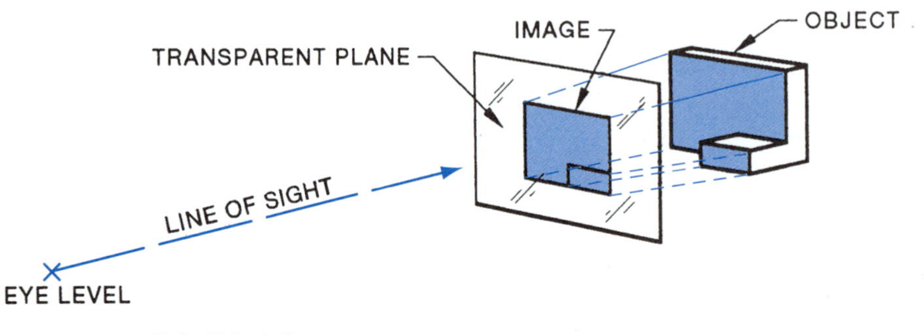

FIGURE 6-1 Projecting image for the front view.

THE FRONT VIEW

Before an object is drawn, it is examined to determine which views will best furnish the information required to manufacture the object. The surface which is to be shown as the observer looks at the object is called the *Front View.* To draw this view, the draftsperson goes through an imaginary process of raising the object to eye level and turning it so that only one side can be seen. If an imaginary transparent plane is placed between the eye and the face of the object, parallel to the object, the image projected on the plane is the same as that formed in the eye of the observer, figure 6-1.

Note in figure 6-1 that the rays converge as they approach the observer's eye. If, instead of converging, these rays are parallel as they leave the object, the image they form on the screen is equivalent to a Front View, as shown in figure 6-2.

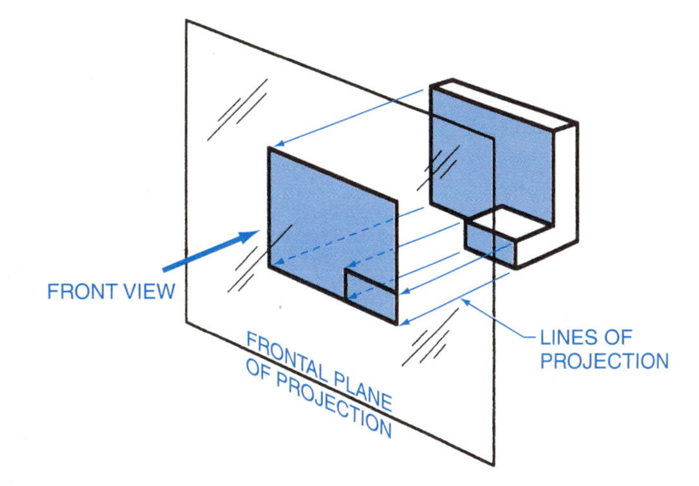

FIGURE 6-2 Front view of object.

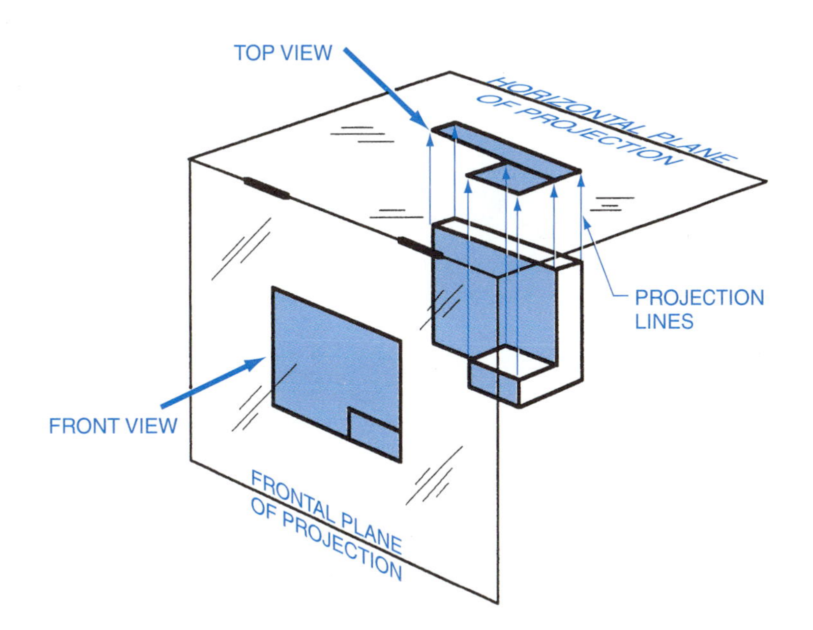

FIGURE 6-3 Projecting image to form the top view.

THE TOP VIEW

To draw a *Top View,* the draftsperson goes through a process similar to that required to obtain the Front View. However, instead of looking squarely at the front of the object, the view is seen from a point directly above it, figure 6-3.

When the horizontal plane on which the top view is projected is rotated so that it is in a vertical plane, as shown in figure 6-4, the front and top views are in their proper relationship. In other words, the top view is always placed immediately above and in line with the front view.

THE SIDE VIEW

A side view is developed in much the same way that the other two views were obtained. That is, the draftsperson imagines the view of the object from the side that is to be drawn. This person then proceeds to draw the object as it would appear if parallel rays were projected upon a vertical plane, figure 6-5.

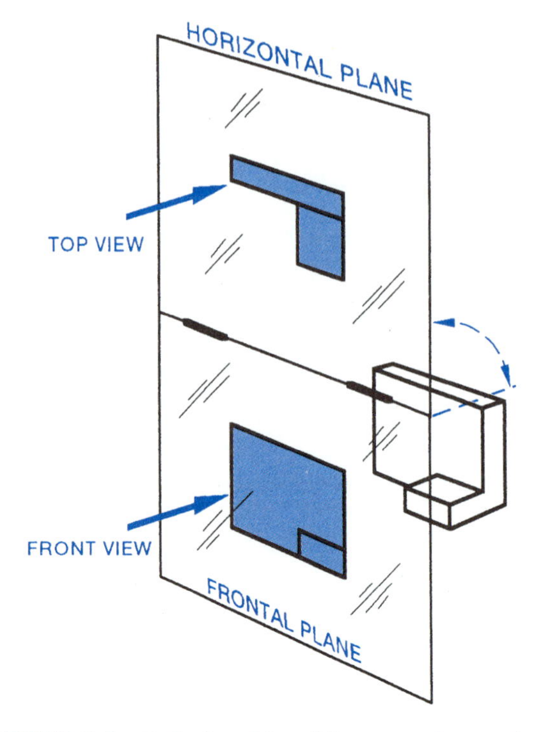

FIGURE 6-4 Relationship of front and top views.

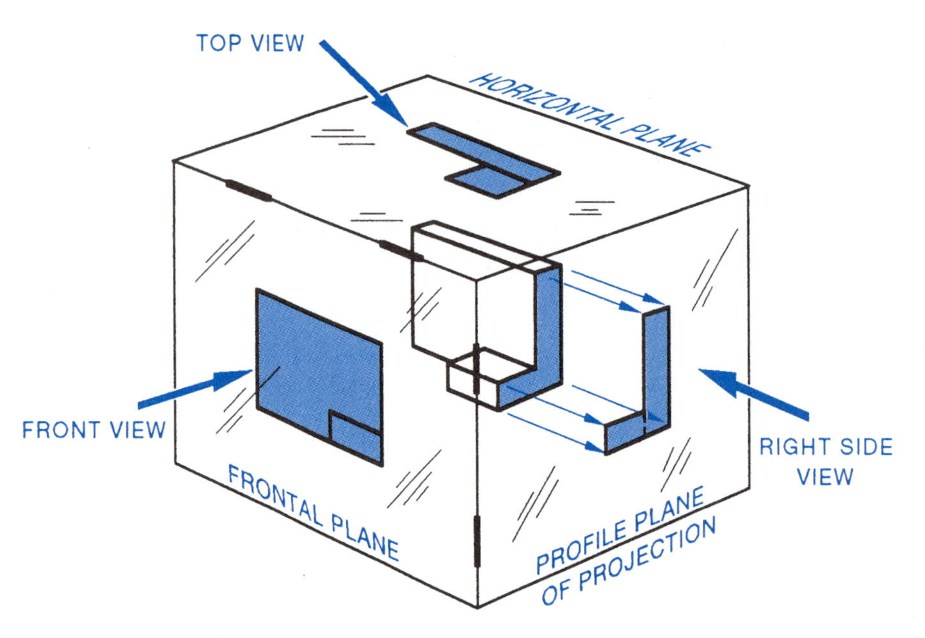

FIGURE 6-5 Projecting image to form the right side view.

FRONT, TOP, AND RIGHT SIDE VIEWS

By swinging the top of the imaginary projection box to a vertical position and the right side forward, the top view is directly above the front view, and the side view is to the right of the front view and in line with it. Figure 6-6 shows the front, top, and right side views in the positions they will occupy on a blueprint.

HEIGHT, WIDTH (LENGTH), AND DEPTH DIMENSIONS

The terms *height, width,* and *depth* refer to specific dimensions or part sizes. ANSI designations for these dimensions are shown in figure 6-7. *Height* is the vertical distance between two or more lines or surfaces (part features) which are in horizontal planes. *Width* refers to the horizontal distance between surfaces in profile planes. In the shop, the terms

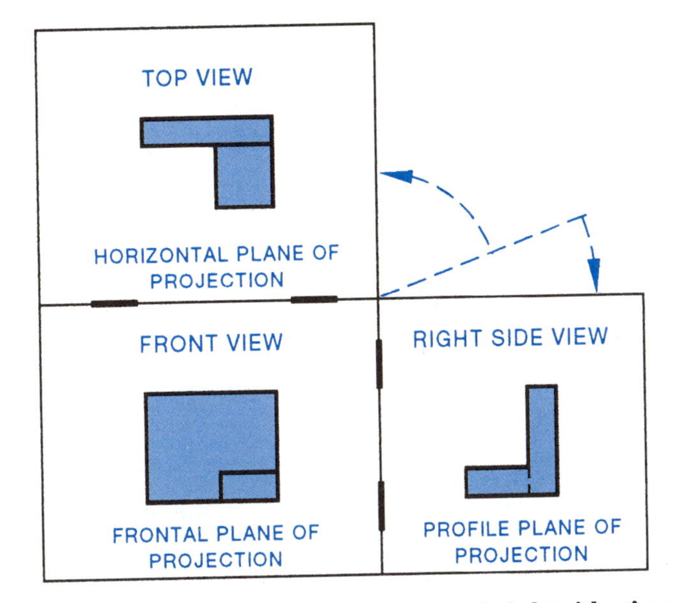

FIGURE 6-6 Positions of front, top, and right side views.

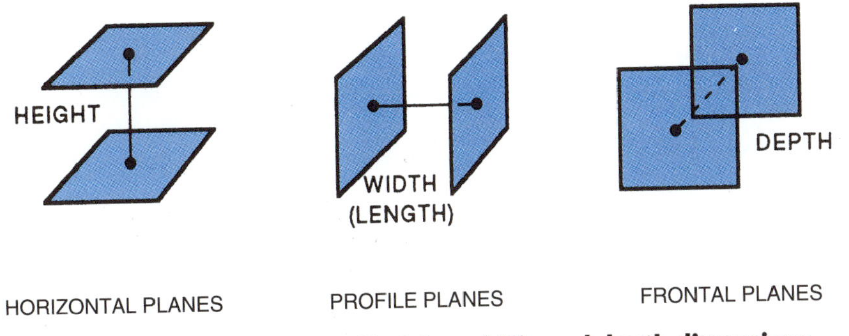

FIGURE 6-7 Designation of height, width, and depth dimensions.

length and *width* are used interchangeably. *Depth* is the horizontal (front to back) distance between two features in frontal planes. Depth is often identified in the shop as the *thickness* of a part or feature.

WORKING DRAWINGS

An actual drawing of a part shows only the top, front, and right side views without the imaginary transparent planes, figure 6-8. These views show the exact shape and size of the object, and define the relationship of one view to another.

A drawing, when completely dimensioned and with necessary notes added, is called a *working drawing* because it furnishes all the information required to construct the object, figure 6-9.

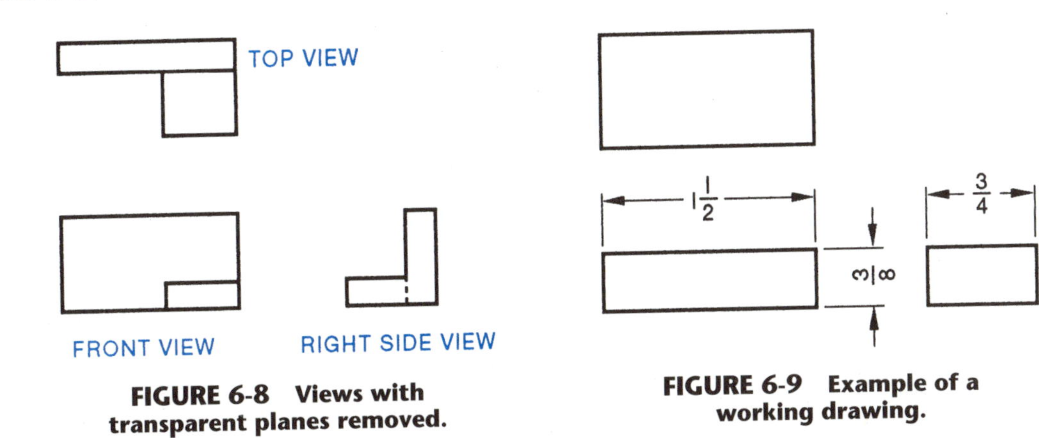

FIGURE 6-8 Views with transparent planes removed.

FIGURE 6-9 Example of a working drawing.

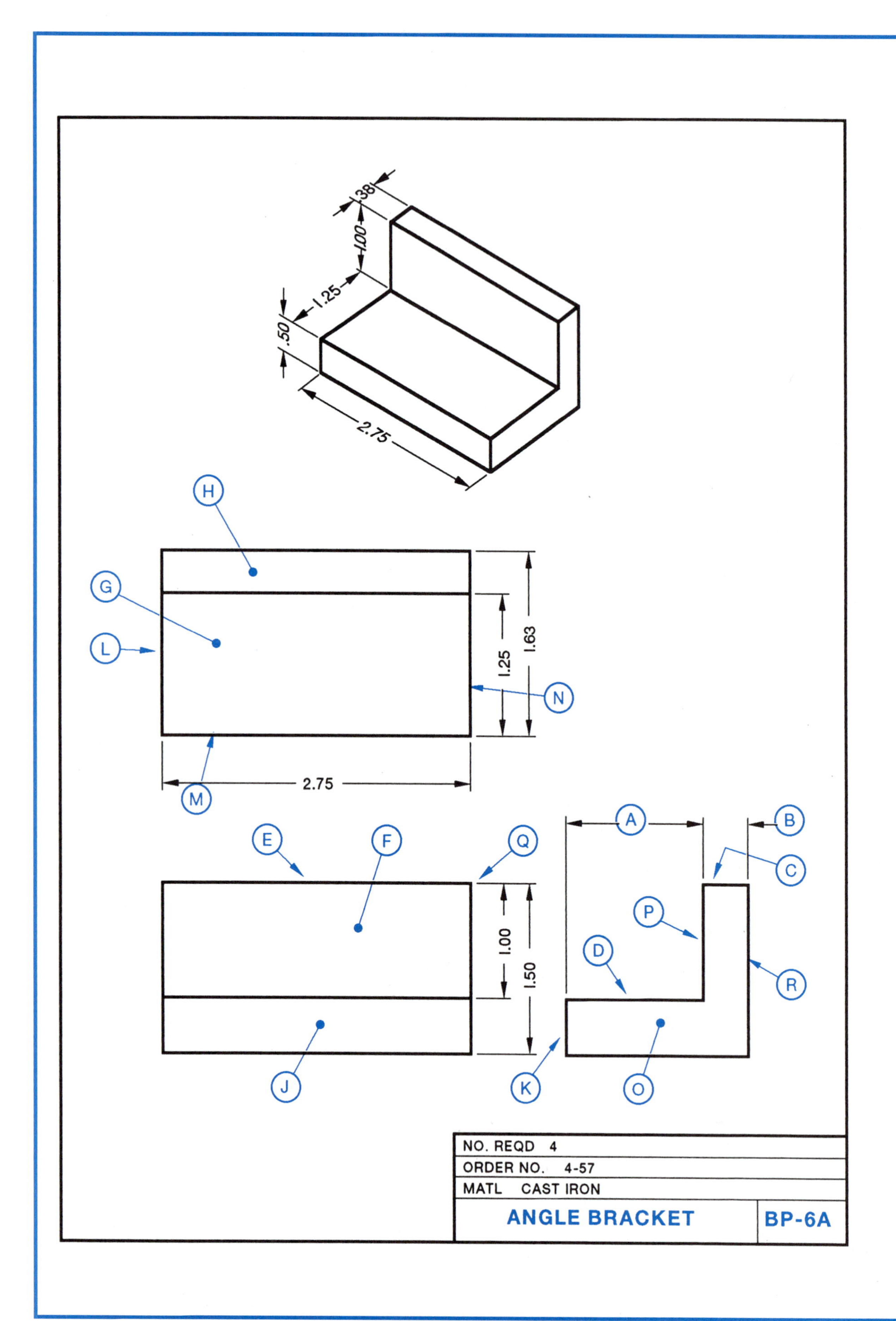
.38
1.00
1.25
.50
2.75
H
G
L
N
1.25
1.63
M
2.75
E
F
Q
A
B
C
P
D
R
1.00
1.50
J
K
O
NO. REQD 4
ORDER NO. 4-57
MATL CAST IRON
ANGLE BRACKET
BP-6A

ANGLE BRACKET (BP-6A)

1. How many Angle Brackets are required?
2. Name the material specified for the Angle Bracket.
3. State the order number of the Bracket.
4. What is the overall width (length) of the Bracket?
5. What is the overall height?
6. What is the overall depth?
7. What is dimension (A)?
8. What is dimension (B)?
9. What surface in the top view is represented by line (C) in the right side view?
10. Name the three views that are used to describe the shape and size of the part.
11. What surface in the top view is represented by line (D) in the right side view?
12. What line in the right side view represents surface (F) in the front view?
13. What line in the right side view represents surface (J) in the front view?
14. What line in the top view represents surface (O) in the right side view?
15. What line in the front view represents surface (H) in the top view?
16. What line in the right side view represents surface (H) in the top view?
17. What kind of lines are (E) (L) (C) (D) and (K)?
18. What kinds of lines are (A) and (B)?
19. What encircled letter denotes an extension line?
20. What encircled letter in the front view denotes an object line?

ASSIGNMENT A—UNIT 6

Student's Name ____________

1. ____________
2. ____________
3. ____________
4. ____________
5. ____________
6. ____________
7. ____________
8. ____________
9. ____________
10. ____________
11. ____________
12. ____________
13. ____________
14. ____________
15. ____________
16. ____________
17. ____________
18. ____________
19. ____________
20. ____________

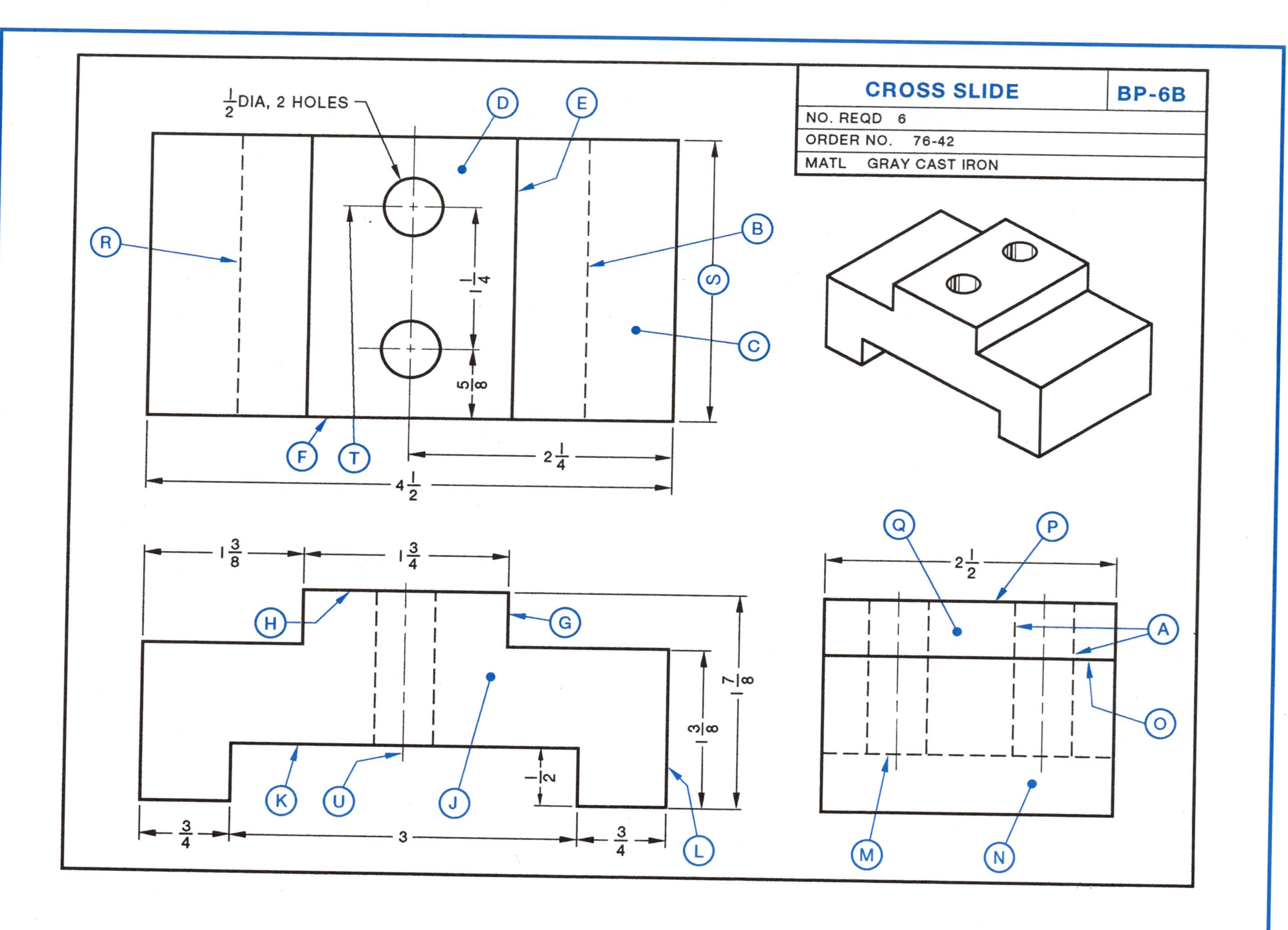
CROSS SLIDE
BP-6B
NO. REQD 6
ORDER NO. 76-42
MATL GRAY CAST IRON
1/2 DIA, 2 HOLES
D
E
R
B
S
C
1 1/4
5/8
F
T
2 1/4
4 1/2
1 3/8
1 3/4
H
G
J
K
U
L
1/2
1 3/8
1 7/8
3/4
3
3/4
Q
P
2 1/2
A
O
M
N

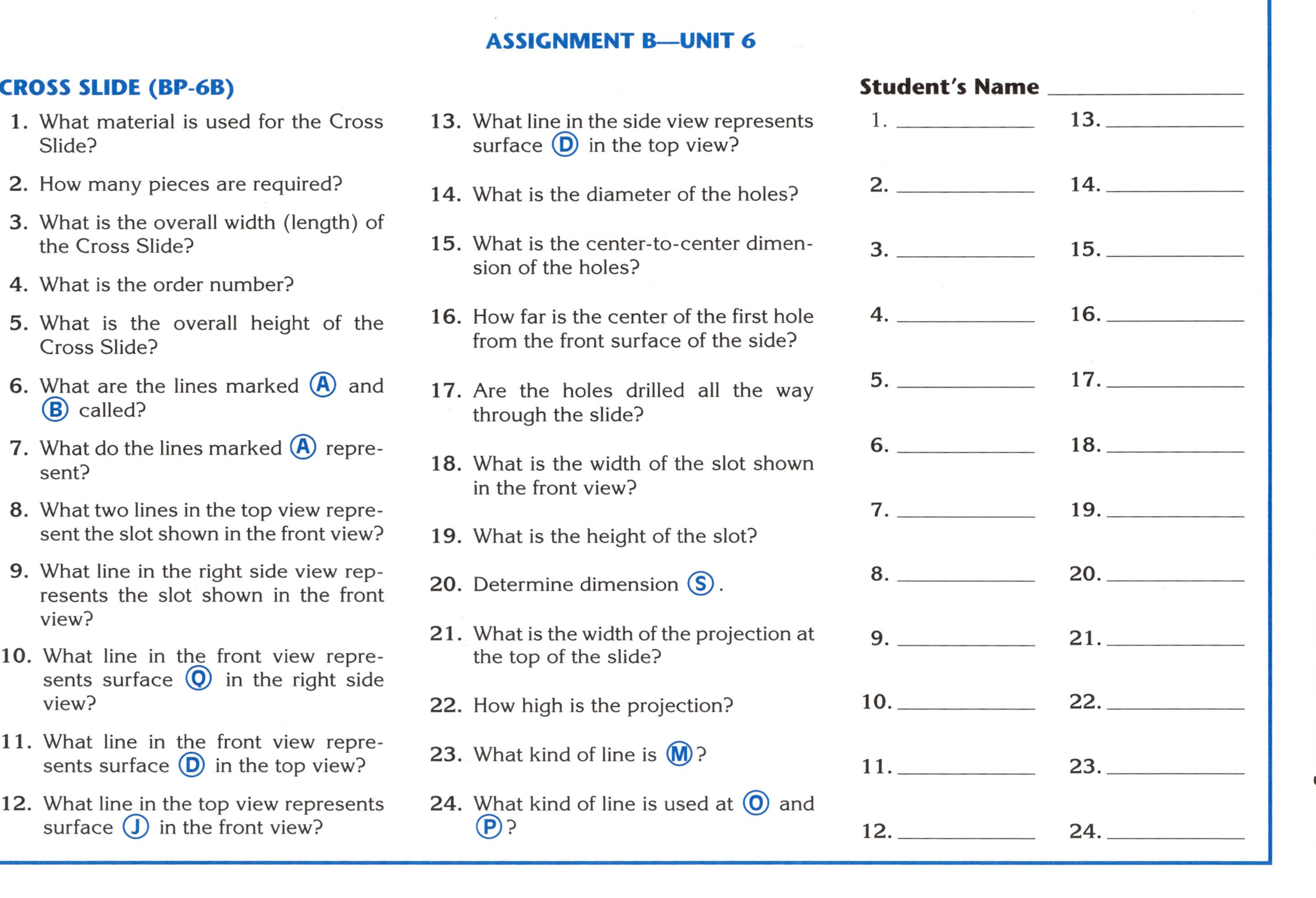

ASSIGNMENT B—UNIT 6

CROSS SLIDE (BP-6B)

1. What material is used for the Cross Slide?
2. How many pieces are required?
3. What is the overall width (length) of the Cross Slide?
4. What is the order number?
5. What is the overall height of the Cross Slide?
6. What are the lines marked (A) and (B) called?
7. What do the lines marked (A) represent?
8. What two lines in the top view represent the slot shown in the front view?
9. What line in the right side view represents the slot shown in the front view?
10. What line in the front view represents surface (Q) in the right side view?
11. What line in the front view represents surface (D) in the top view?
12. What line in the top view represents surface (J) in the front view?
13. What line in the side view represents surface (D) in the top view?
14. What is the diameter of the holes?
15. What is the center-to-center dimension of the holes?
16. How far is the center of the first hole from the front surface of the side?
17. Are the holes drilled all the way through the slide?
18. What is the width of the slot shown in the front view?
19. What is the height of the slot?
20. Determine dimension (S).
21. What is the width of the projection at the top of the slide?
22. How high is the projection?
23. What kind of line is (M)?
24. What kind of line is used at (O) and (P)?

Student's Name ____________

1. ____________	13. ____________
2. ____________	14. ____________
3. ____________	15. ____________
4. ____________	16. ____________
5. ____________	17. ____________
6. ____________	18. ____________
7. ____________	19. ____________
8. ____________	20. ____________
9. ____________	21. ____________
10. ____________	22. ____________
11. ____________	23. ____________
12. ____________	24. ____________

Unit 7

ARRANGEMENT OF VIEWS

The main purpose of a drawing is to give the technician sufficient information needed to build, inspect, or assemble a part or mechanism according to the specifications of the designer. Since the selection and arrangement of views depends upon the complexity of a part, only those views should be drawn which help in the interpretation of the drawing.

The average drawing which includes front, top, and side views is known as a three-view drawing. However, the designation of the views is not as important as the fact that the combination of views must give all the details of construction in the most understandable way.

The draftsperson usually selects as a front view of the object that view which best describes the general shape of the part. This front view may have no relationship to the actual front position of the part as it fits into a mechanism.

The names and positions of the different views that may be used to describe an object are illustrated in figure 7-1. Note that the back view may be placed in any one of three locations. The views which are easiest to read and, at the same time, furnish all the required information, should be the views selected for the drawing.

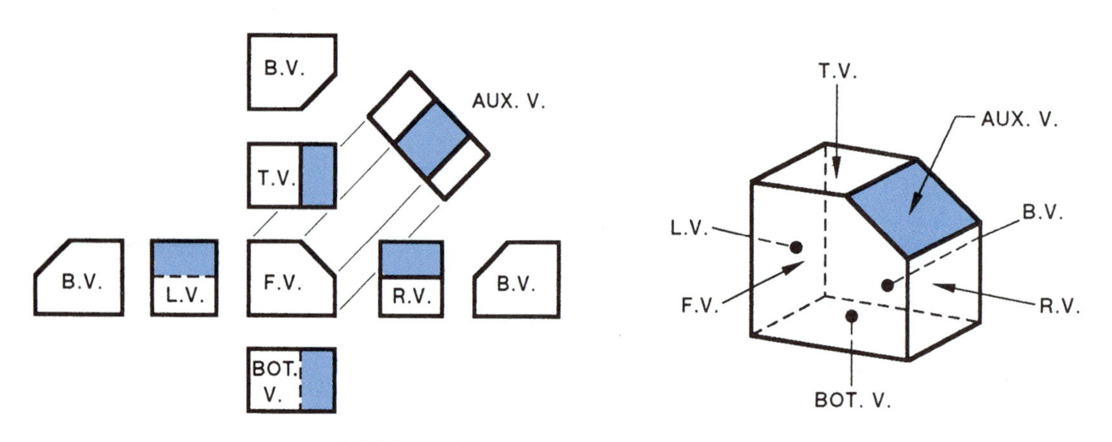

FIGURE 7-1 Identification of views.

The name and abbreviation for each view is identified throughout this text as shown in figure 7-2.

Name of View	Abbreviation
Front View	(F.V.)
Right Side View	(R.V.)
Left Side View	(L.V.)
Bottom View	(Bot. V.)
Back or Rear View	(B.V.)
Auxiliary View	(Aux. V.)
Top View	(T.V.)

FIGURE 7-2 Abbreviations of different views.

ASSIGNMENT—UNIT 7 BLOCK SLIDE (BP-7)

Student's Name ______________________________

Study the pictorial drawing of the SLIDE BASE (figure 7-3) with reference to the *front view.* Then, identify and place the name of each view in the spaces provided in figures 7-4 and 7-5.

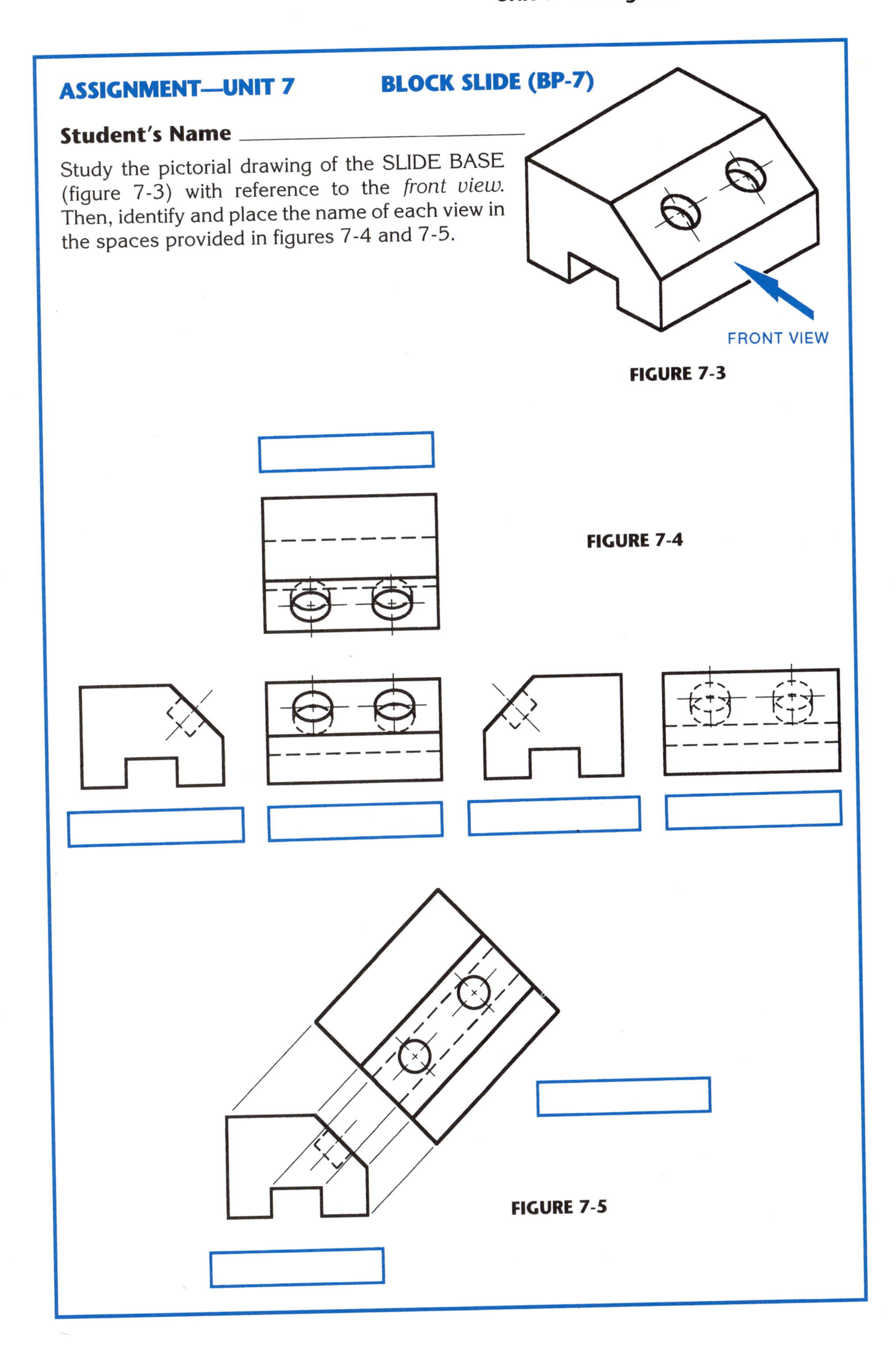

FIGURE 7-3

FIGURE 7-4

FIGURE 7-5

Unit 8

TWO-VIEW DRAWINGS

Simple, symmetrical flat objects and cylindrical parts, such as sleeves, shafts, rods, and studs, require only two views to give the full details of construction, figure 8-1. The two views usually include the front view and a right-side or left-side view, or a top or bottom view.

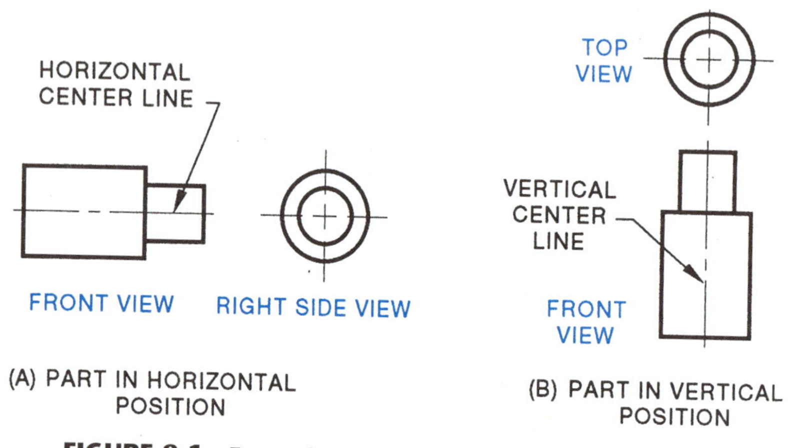

FIGURE 8-1 Examples of two-view drawings of a plug.

In the front view, figure 8-1, the center line runs through the axis of the part as a horizontal center line. If the plug is in a vertical position, the center line runs through the axis as a vertical center line.

The second view of the two-view drawing contains a horizontal and a vertical center line intersecting at the center of the circles which make up the part in this view.

The selection of views for a two-view drawing rests largely with the draftsperson or designer. Some of the combinations of views commonly used in industrial blueprints are shown in figure 8-2.

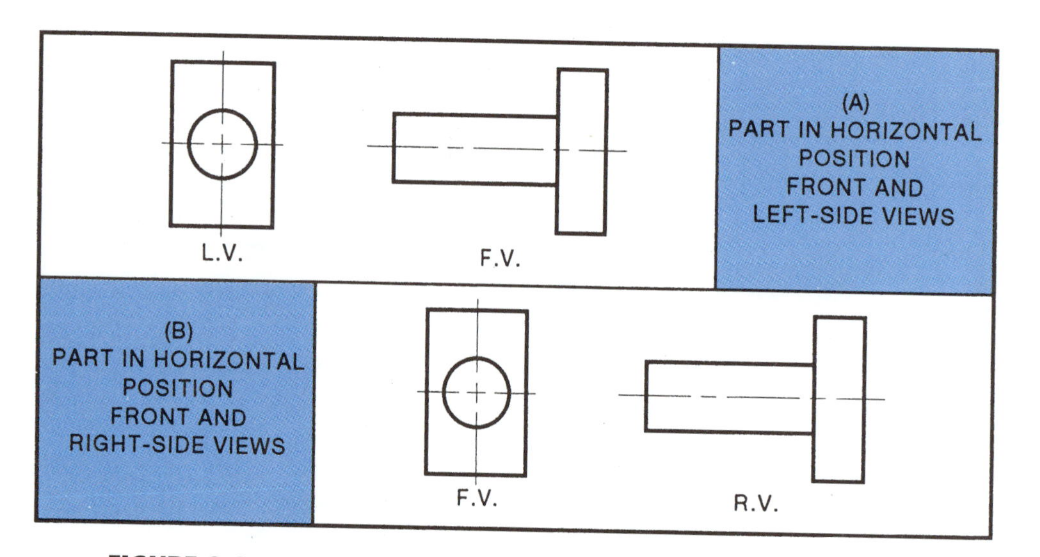

FIGURE 8-2 Combination of views for a two-view drawing of a pin.

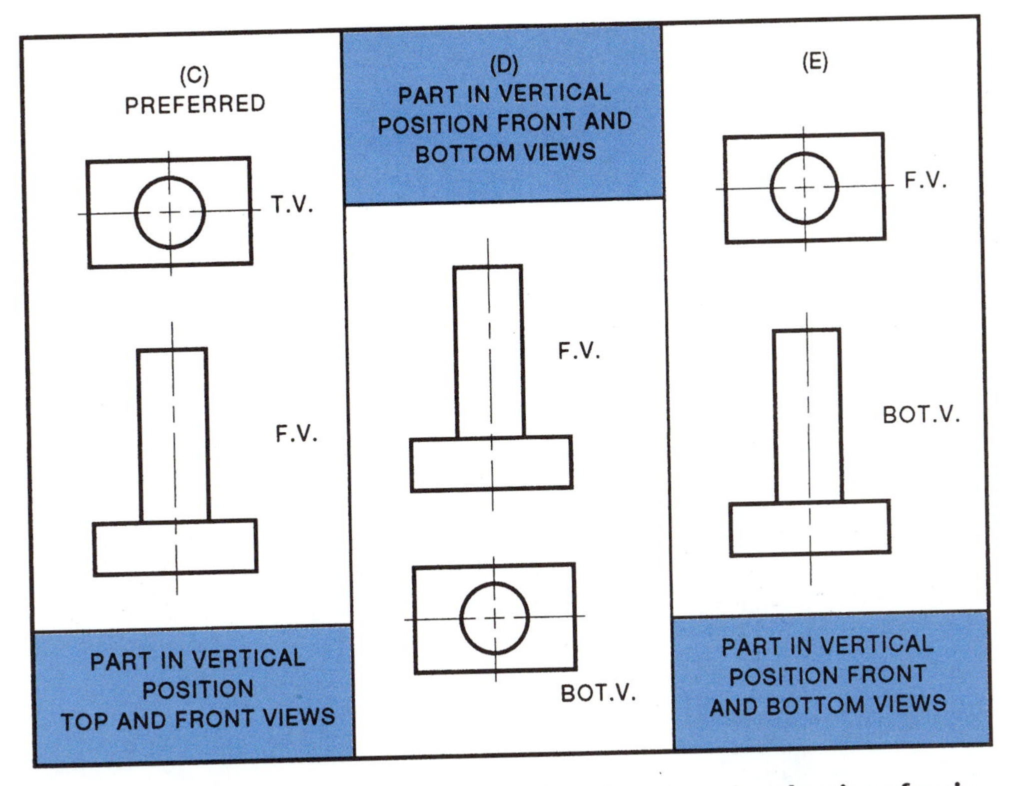

FIGURE 8-2 (CONT'D) **Combination of views for a two-view drawing of a pin.**

Note in figure 8-2 that different names are used to identify the same views. The name of each view depends on how the draftsperson views the object to obtain the front view.

A hidden detail may be straight, curved, or cylindrical. Whatever the shape of the detail, and regardless of the number or positions of views, the hidden detail is represented by a hidden edge or invisible edge line, figure 8-3.

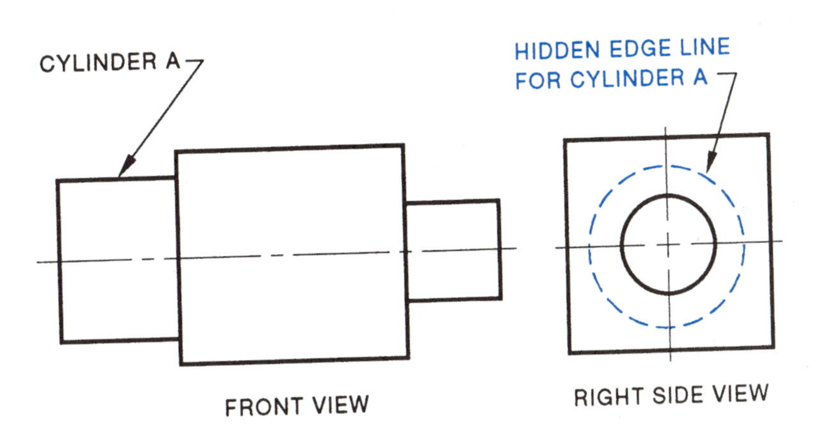

FIGURE 8-3 **Use of invisible edge lines.**

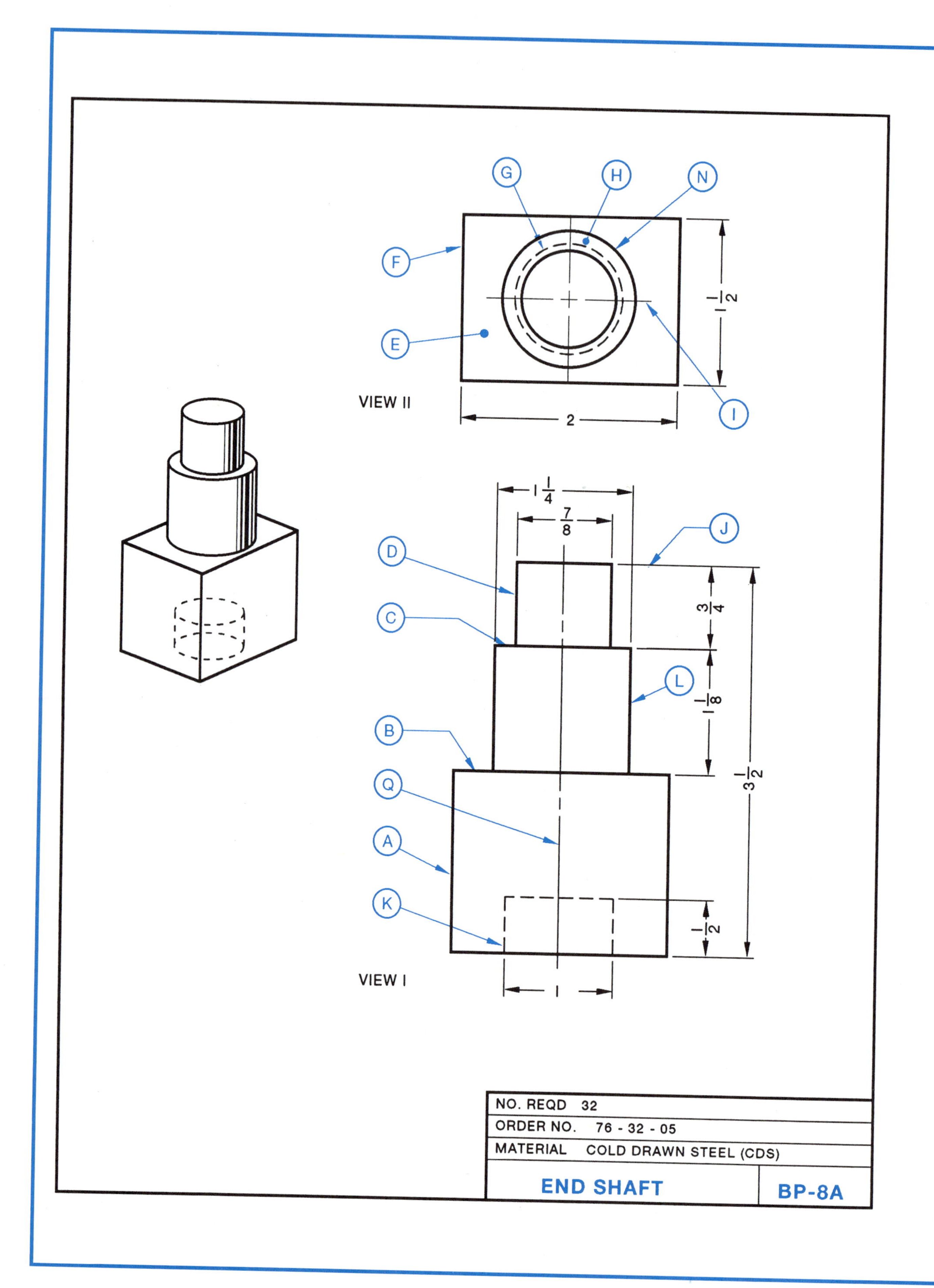

G
H
N
F
E
I
VIEW II
2
1 1/2
1 1/4
7/8
J
D
C
L
B
Q
A
K
3/4
1 1/8
3 1/2
1/2
VIEW I
NO. REQD 32
ORDER NO. 76 - 32 - 05
MATERIAL COLD DRAWN STEEL (CDS)
END SHAFT
BP-8A

END SHAFT (BP-8A)

1. Name the two views shown.
2. What line in View II represents surface Ⓐ?
3. What lettered surface in View II represents surface Ⓑ?
4. What circle in View II represents the 1″ hole?
5. What line in View 1 represents surface Ⓗ?
6. Name line Ⓘ.
7. What kind of line is Ⓓ?
8. Name line Ⓙ.
9. What kind of line is Ⓚ?
10. What circle in the top view represents diameter Ⓛ?
11. What letters in View I represent object lines?
12. What letters in Views I and II represent center lines?
13. Give the diameter of Ⓛ.
14. What is the smallest diameter of the shaft?
15. Determine the height (length) of the 1 1/4" diameter portion.
16. What is the height (length) of the rectangular part of the shaft?
17. Give the dimensions of the rectangular part.
18. Give the overall height (length) of the shaft.
19. What is the order number?
20. State the material from which the shaft is to be machined.

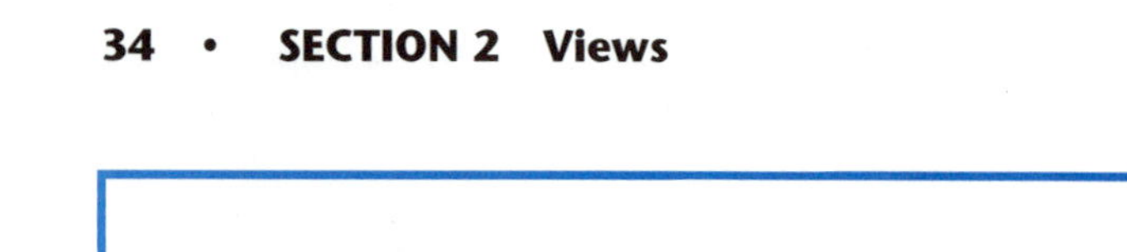

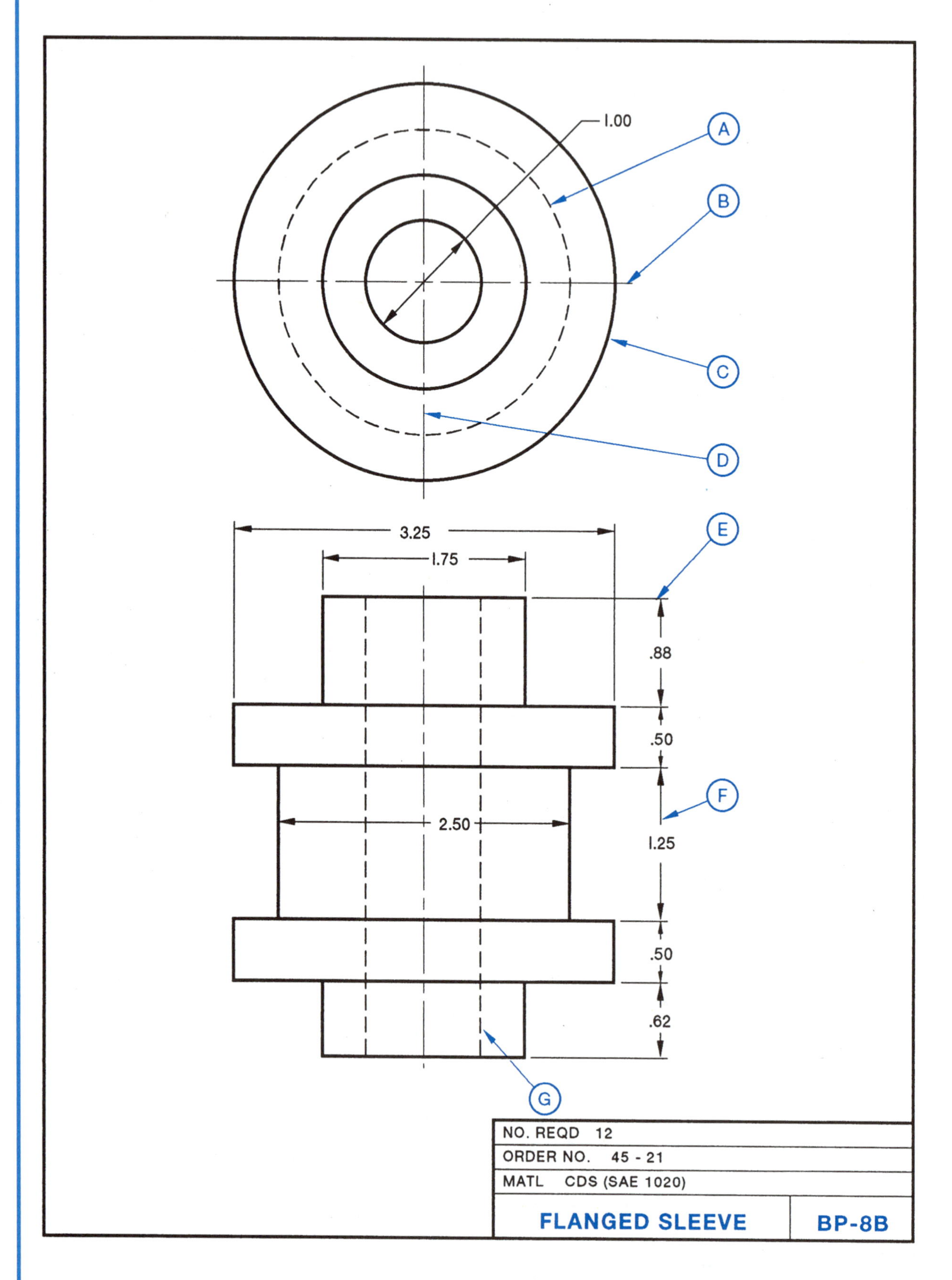

1.00
A
B
C
D
E
F
G
3.25
1.75
.88
.50
2.50
1.25
.50
.62
NO. REQD 12
ORDER NO. 45 - 21
MATL CDS (SAE 1020)
FLANGED SLEEVE
BP-8B

ASSIGNMENT B—UNIT 8

FLANGED SLEEVE (BP-8B)

1. What is the name of the part?
2. What is the order number?
3. How many pieces are required?
4. What material is used?
5. Name the two views which are used to represent the Flanged Sleeve.
6. Name the kind of line indicated by each of the following encircled letters.
   - (A)
   - (B)
   - (C)
   - (D)
   - (E)
   - (F)
   - (G)
7. What is the outside diameter of both flanges?
8. What is the height (thickness) of each flange?
9. What is the diameter of the center hole?
10. Does the hole go all the way through the center of the sleeve?
11. What is the diameter of the hidden circle?
12. Determine the total or overall height of the Flanged Sleeve.

Student's Name ____________________

1. ____________________
2. ____________________
3. ____________________
4. ____________________
5. ____________________
6. (A) ____________________
   (B) ____________________
   (C) ____________________
   (D) ____________________
   (E) ____________________
   (F) ____________________
   (G) ____________________
7. ____________________
8. ____________________
9. ____________________
10. ____________________
11. ____________________
12. ____________________

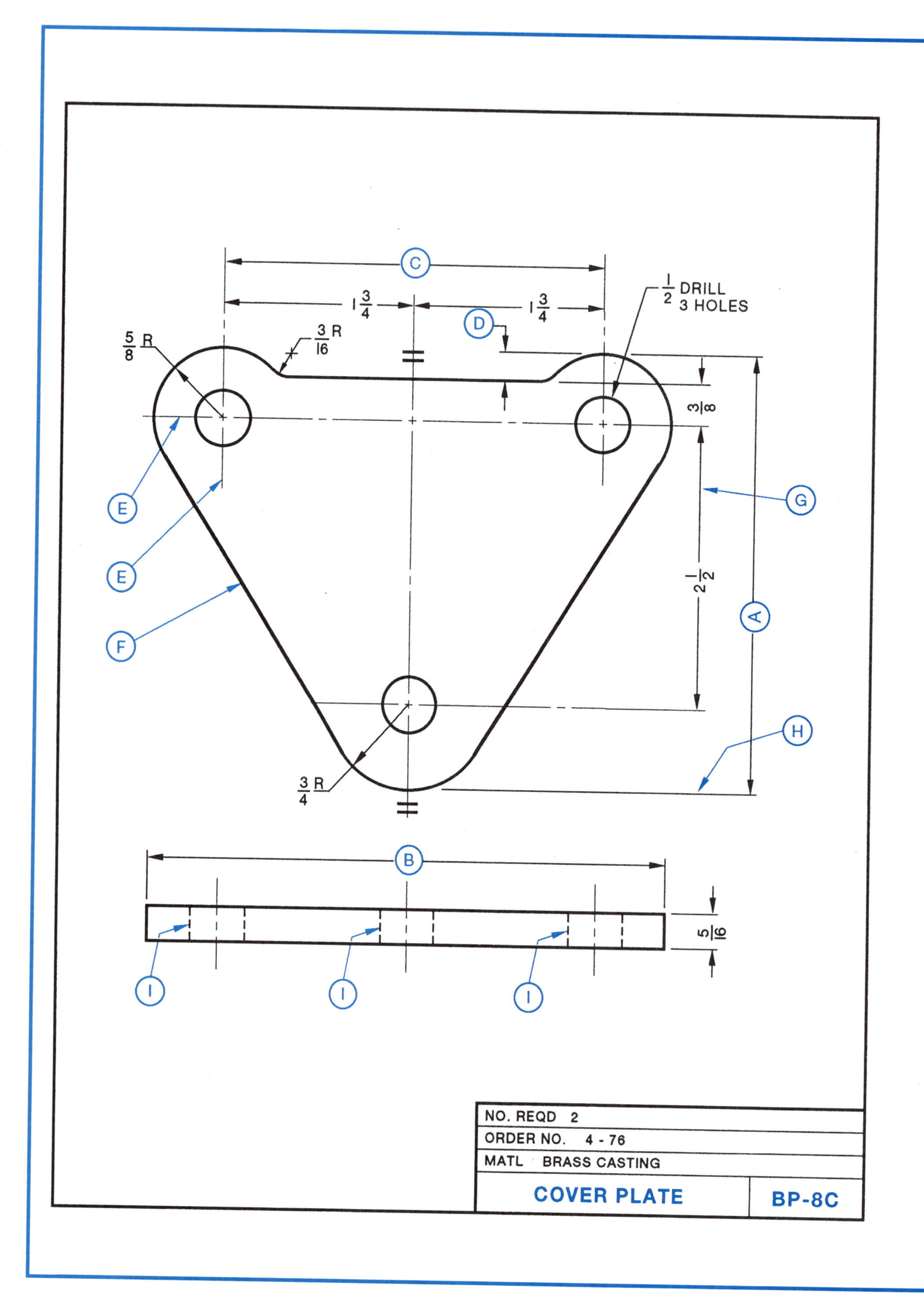
C
$1\frac{3}{4}$
$1\frac{3}{4}$
$\frac{1}{2}$ DRILL
3 HOLES
D
$\frac{5}{8}$ R
$\frac{3}{16}$ R
$\frac{3}{8}$
E
E
G
$2\frac{1}{2}$
A
F
H
$\frac{3}{4}$ R
B
$\frac{5}{16}$
I
I
I
NO. REQD 2
ORDER NO. 4 - 76
MATL BRASS CASTING
COVER PLATE
BP-8C

ASSIGNMENT C—UNIT 8

COVER PLATE (BP-8C)

1. Name the material specified for the part.
2. Name the two views used to describe the part.
3. Identify the kind of line indicated by each of the following encircled letters.
   Ⓔ
   Ⓕ
   Ⓖ
   Ⓗ
   Ⓘ
4. What is the overall depth Ⓐ?
5. What is the overall length Ⓑ?
6. How many holes are to be drilled?
7. What is the thickness (height) of the plate?
8. What is the diameter of the holes?
9. What is the distance between the center of one of the two upper holes and the center line of the Plate?
10. Give the center distance Ⓒ of the two upper holes.
11. What is the radius that forms the two upper rounds of the Plate?
12. What radius forms the lower part of the Plate?
13. What kind of line is drawn through the center of the Plate?
14. Determine distance Ⓓ.
15. How much stock is left between the edge of one of the upper holes and the outside of the piece?

Unit 9

ONE-VIEW DRAWINGS

Parts which are uniform in shape often require only one view to describe them adequately. This is particularly true of cylindrical objects where a one-view drawing saves drafting time and simplifies blueprint reading.

When a one-view drawing of a cylindrical part is used, the dimension for the diameter (according to ANSI standards) must be preceded by the symbol ∅, figure 9-1. The older, but still widely used, drafting room and shop practice for dimensioning diameters is to place the letters DIA after the dimension. In both applications the symbol ∅, or the letters **DIA**, and the use of a center line indicate that the part is cylindrical.

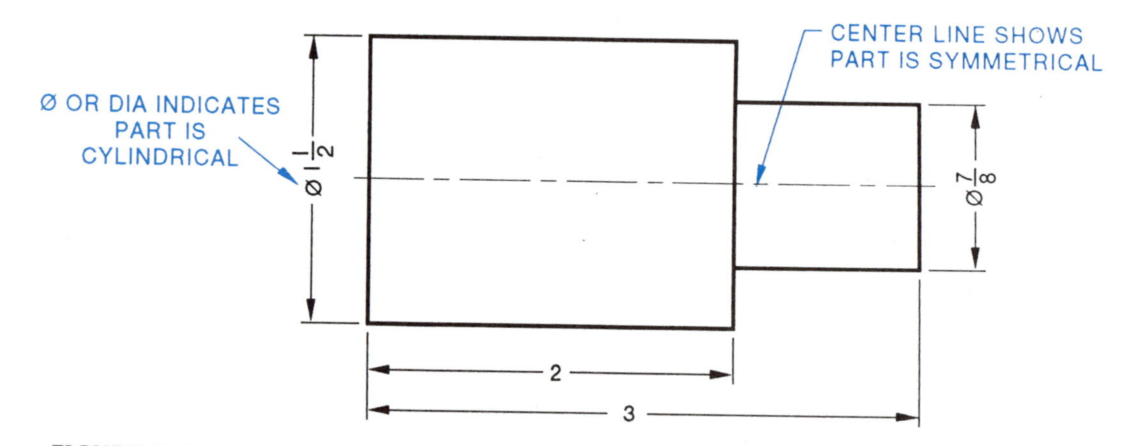

FIGURE 9-1 One-view drawing of a cylindrical shaft using the symbol ∅ for diameter.

Industrial drawings in this and succeeding units use either the recommended ANSI symbol ∅ or the letters **DIA**.

The one-view drawing is also used extensively for flat parts. With the addition of notes to supplement the dimensions on the view, the one view furnishes all the necessary information for accurately describing the part, figure 9-2.

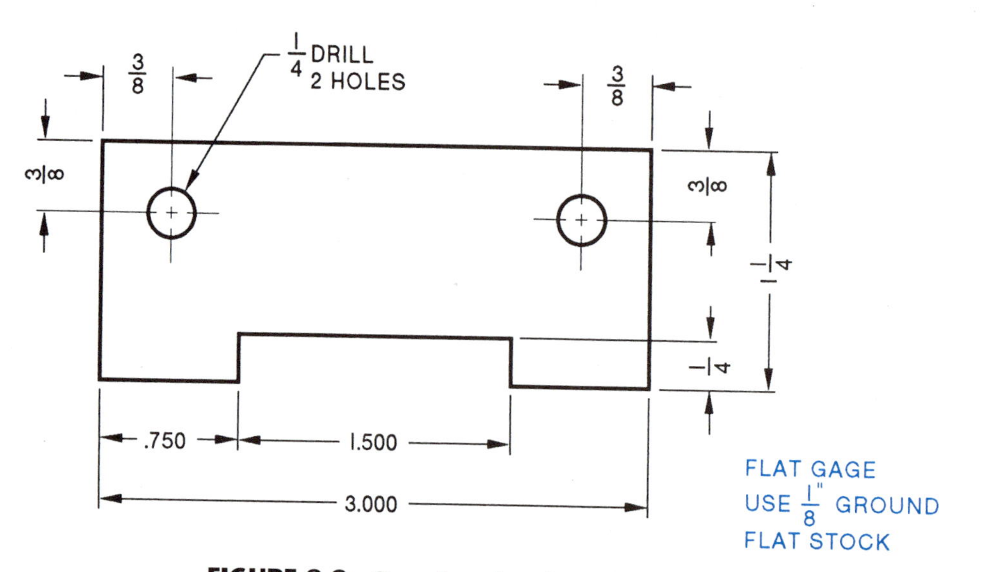

FIGURE 9-2 One-view drawing of a flat part.

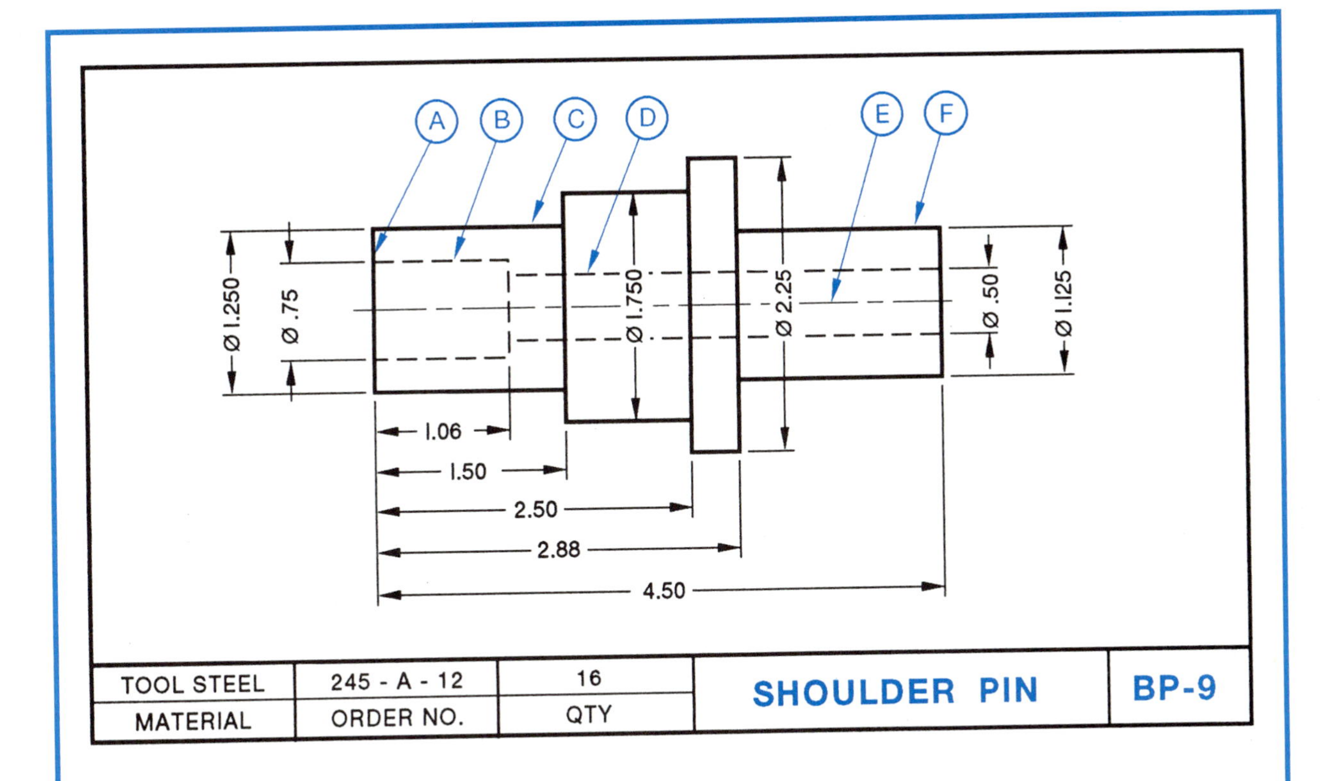

ASSIGNMENT—UNIT 9

SHOULDER PIN (BP-9)

1. Name the view represented on BP-9.
2. What is the shape of the Shoulder Pin?
3. How many outside diameters are shown?
4. What is the largest diameter?
5. What diameter is the smallest hole?
6. What is the overall length of the pin?
7. How deep is the .75" hole?
8. How wide is the ∅ 1.750" portion?
9. What letters represent object lines?
10. What kinds of lines are (B) and (D)?
11. What letter represents the center line?
12. What does the center line indicate about the holes and outside diameters?
13. Give the thickness of the ∅ 2.25" portion.
14. State the order number of the part.
15. What material is specified for the pins?

Student's Name ______________________

1. ______________________
2. ______________________
3. ______________________
4. ______________________
5. ______________________
6. ______________________
7. ______________________
8. ______________________
9. ______________________
10. ______________________
11. ______________________
12. ______________________
13. ______________________
14. ______________________
15. ______________________

Unit 10

AUXILIARY VIEWS

As long as all the surfaces of an object are parallel or at right angles to one another, they may be represented in one or more regular views, figure 10-1.

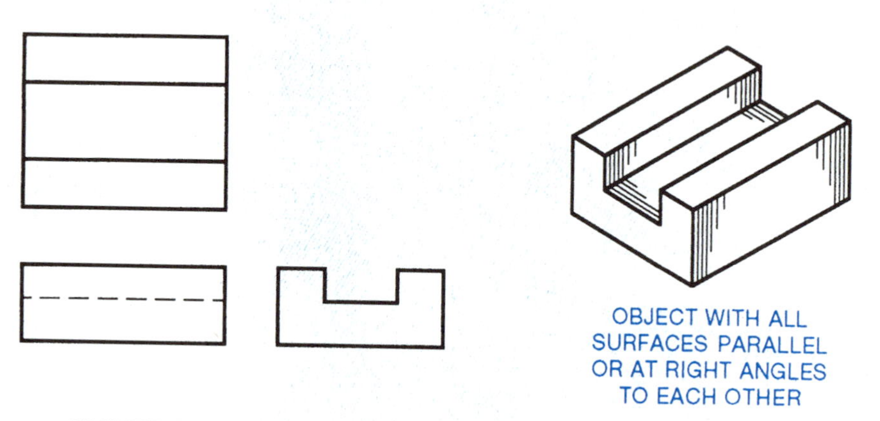

FIGURE 10-1 Representation of object using regular views.

The surfaces of such objects can be projected in their true sizes and shapes on either a horizontal or a vertical plane, or on any combination of these planes.

Some objects have one or more surfaces which slant and are inclined away from either a horizontal or vertical plane. In this situation the regular views will not show the true shape of the inclined surface, figure 10-2. If the true shape must be shown, the drawing must include an *auxiliary view* to represent the angular surface accurately. The auxiliary view is in addition to the regular views.

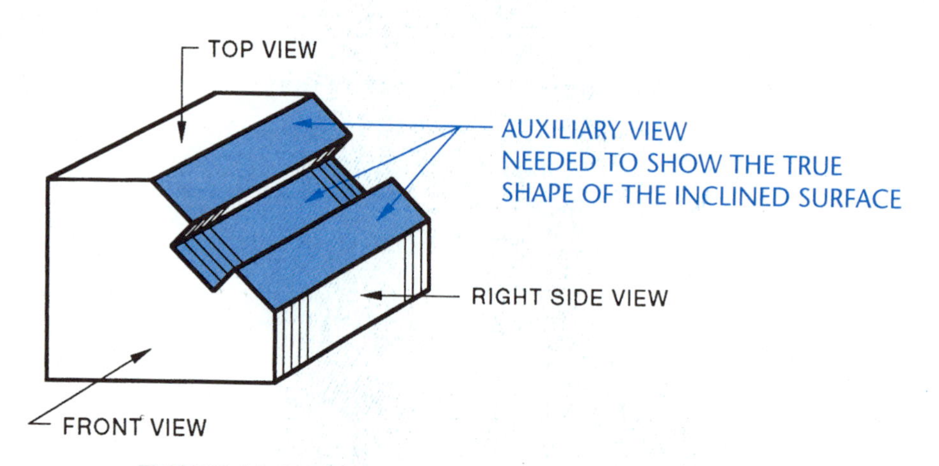

FIGURE 10-2 Object requires use of auxiliary view.

APPLICATION OF AUXILIARY VIEWS

Auxiliary views may be full views or partial views. Figure 10-3 shows an auxiliary view in which only the inclined surface and other required details are included. In an auxiliary view, the inclined surface is projected on an imaginary plane which is parallel to it. Rounded surfaces and circular holes, which are distorted and appear as ellipses in the regular views, will appear in their true shapes and sizes in an auxiliary view.

Auxiliary views are named according to the position from which the inclined face is seen. For example, the auxiliary view may be an auxiliary front, top, bottom, left side, or right side view. On drawings of complex parts involving compound angles, one auxiliary view may be developed from another auxiliary view. The first auxiliary view is called the *primary* view, and those views developed from it are called *secondary* auxiliary views. For the present, attention is focused on primary auxiliary views.

SUMMARY

Auxiliary views are usually partial views which show only the inclined surface of an object. In figure 10-3, the true size and shape of surface (A) is shown in the auxiliary view of the angular surface.

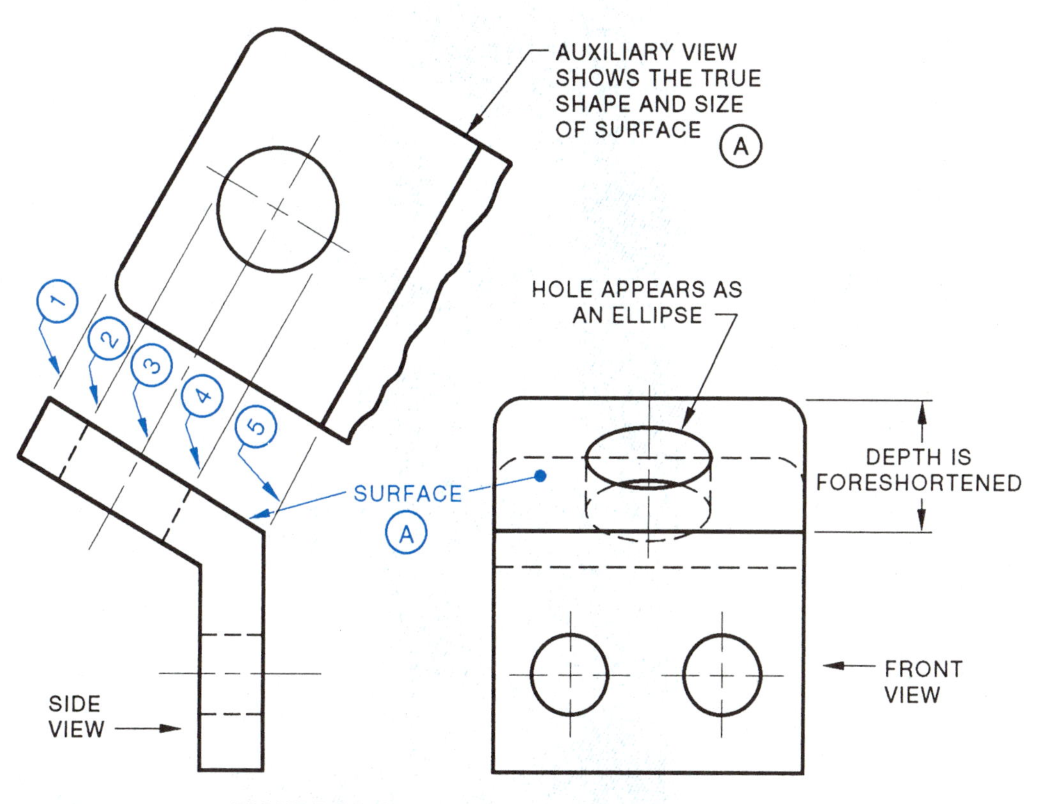

FIGURE 10-3 Application of an auxiliary view.

The draftsperson develops this view by projecting lines (1), (2), (3), (4), and (5) at right angles to surface (A). When the front view is compared with the auxiliary partial view, it can be seen that the hole in surface (A) appears as an ellipse in the front view. The depth of surface (A) is foreshortened in the regular view while the true shape and size are shown on the auxiliary view.

In figure 10-3, the combination of *left side view, auxiliary partial top view* and *front view,* when properly dimensioned, shows all surfaces in their true size and shape. As a result, these are the only views required to describe this part completely.

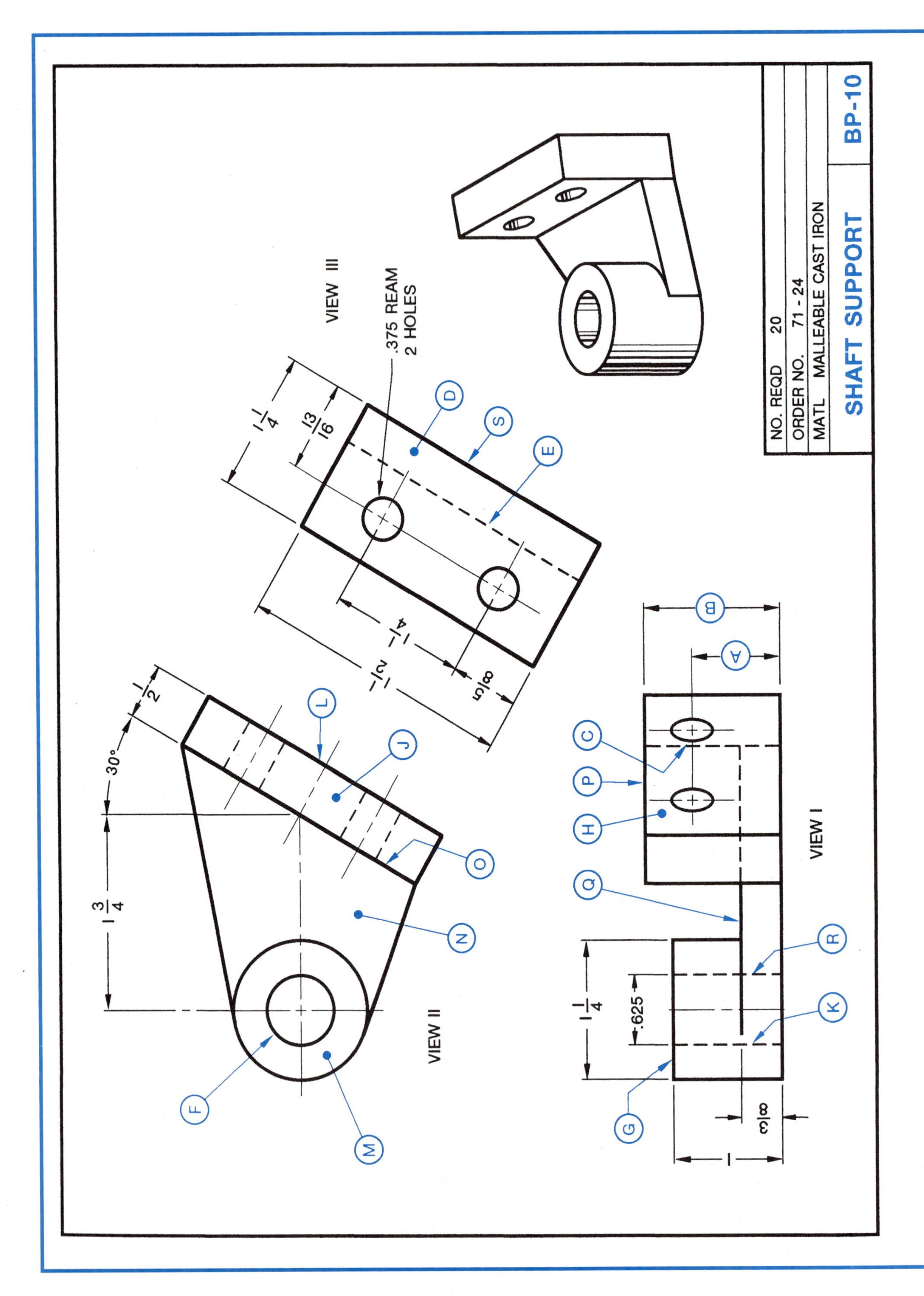
NO. REQD 20
ORDER NO. 71 - 24
MATL MALLEABLE CAST IRON
SHAFT SUPPORT
BP-10
VIEW I
VIEW II
VIEW III
.375 REAM
2 HOLES
30°
.625
A
B
C
D
E
F
G
H
J
K
L
M
N
O
P
Q
R
S

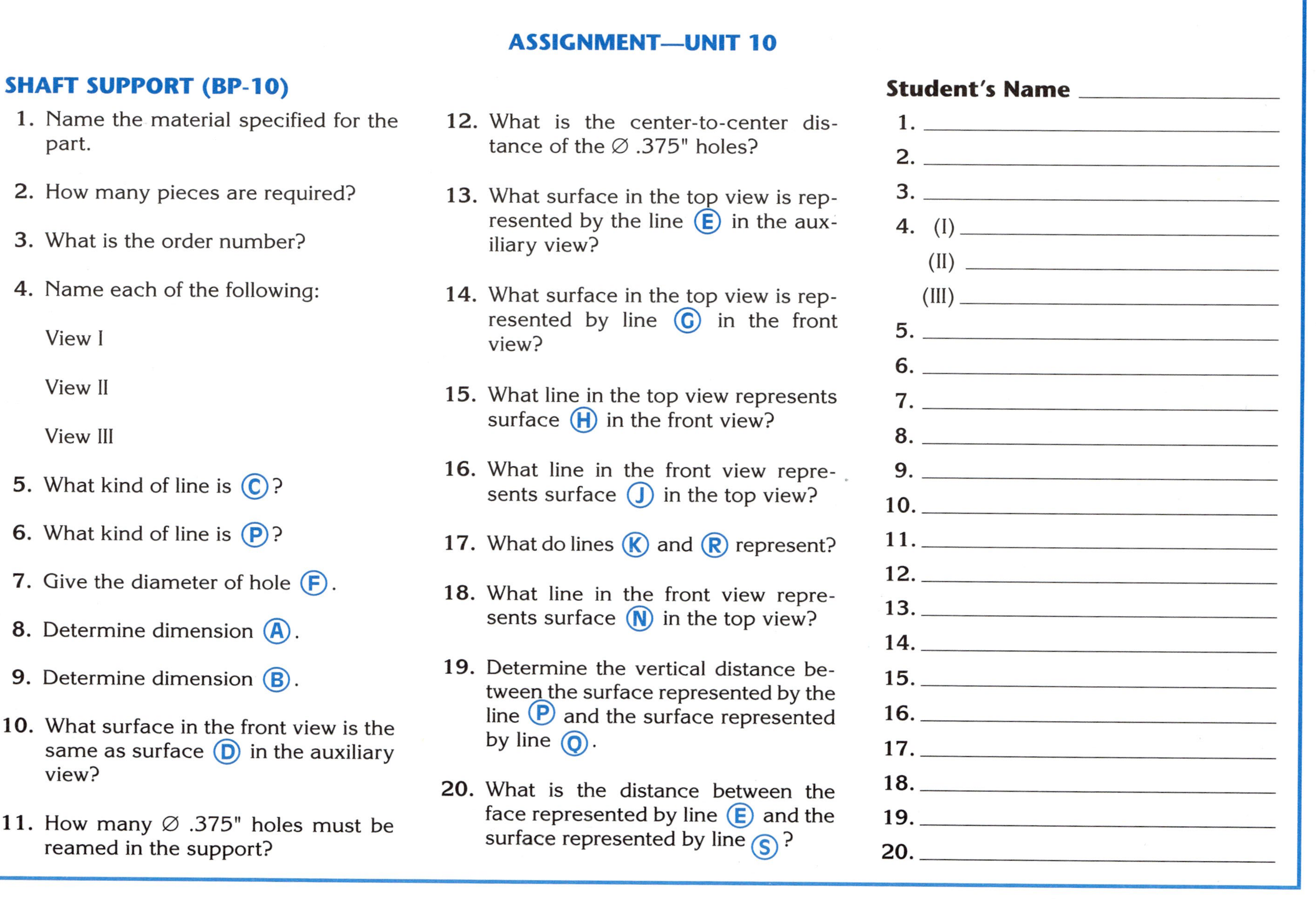

ASSIGNMENT—UNIT 10

SHAFT SUPPORT (BP-10)

1. Name the material specified for the part.
2. How many pieces are required?
3. What is the order number?
4. Name each of the following:

   View I

   View II

   View III
5. What kind of line is (C)?
6. What kind of line is (P)?
7. Give the diameter of hole (F).
8. Determine dimension (A).
9. Determine dimension (B).
10. What surface in the front view is the same as surface (D) in the auxiliary view?
11. How many ∅ .375" holes must be reamed in the support?
12. What is the center-to-center distance of the ∅ .375" holes?
13. What surface in the top view is represented by the line (E) in the auxiliary view?
14. What surface in the top view is represented by line (G) in the front view?
15. What line in the top view represents surface (H) in the front view?
16. What line in the front view represents surface (J) in the top view?
17. What do lines (K) and (R) represent?
18. What line in the front view represents surface (N) in the top view?
19. Determine the vertical distance between the surface represented by the line (P) and the surface represented by line (Q).
20. What is the distance between the face represented by line (E) and the surface represented by line (S)?

Student's Name ______________________

1. ______________________
2. ______________________
3. ______________________
4. (I) ______________________

   (II) ______________________

   (III) ______________________
5. ______________________
6. ______________________
7. ______________________
8. ______________________
9. ______________________
10. ______________________
11. ______________________
12. ______________________
13. ______________________
14. ______________________
15. ______________________
16. ______________________
17. ______________________
18. ______________________
19. ______________________
20. ______________________

SECTION 3 Dimensions and Notes

Unit 11

SIZE AND LOCATION DIMENSIONS

Drawings consist of several types of lines which are used singly or in combination with each other to describe the shape and internal construction of an object or mechanism. However, to construct or machine a part, the blueprint or drawing must include dimensions which indicate exact sizes and locations of surfaces, indentations, holes, and other details.

The lines and dimensions, in turn, are supplemented by notes which give additional information. This information includes the kind of material used, the degree of machining accuracy required, details regarding the assembly of parts, and any other data which the craftsperson needs to know to make and assemble the part.

THE LANGUAGE OF DRAFTING

To insure some measure of uniformity in industrial drawings, the American National Standards Institute (ANSI) has established drafting standards. These standards are called the language of drafting and are in general use throughout the United States. While these drafting standards or practices may vary in some respects between industries, the principles are basically the same. The practices recommended by ANSI for dimensioning and for making notes are followed in this section.

Standards for Dimensioning

All drawings should be dimensioned completely so that a minimum of computation is necessary, and the parts can be built without scaling the drawing. However, there should not be a duplication of dimensions unless such dimensions make the drawing clearer and easier to read.

Many parts cannot be drawn full size because they are too large to fit a standard drawing sheet, or too small to have all details shown clearly. The draftsperson can, however, still represent such objects either by reducing (in the case of large objects) or enlarging (for small objects) the size to which the drawing is made.

This practice does not affect any dimensions as the dimensions on a drawing give the actual sizes. If a drawing 6" long represents a part 12" long, a note should appear in the title box of the drawing to indicate the *scale* that is used. This scale is the ratio of the drawing size to the actual size of the object. In this case, the scale 6" = 12" is called *half scale* or 1/2" = 1". Other common scales include the one-quarter scale (1/4" = 1"), one-eighth scale (1/8" = 1"), and the double size scale (2" = 1").

CONSTRUCTION DIMENSIONS

Dimensions used in building a part are sometimes called *construction dimensions.* These dimensions serve two purposes: (1) they indicate size, and (2) they give exact locations.

For example, to drill a through hole in a part, the technician must know the diameter of the hole, and the exact location of the center of the hole, figure 11-1.

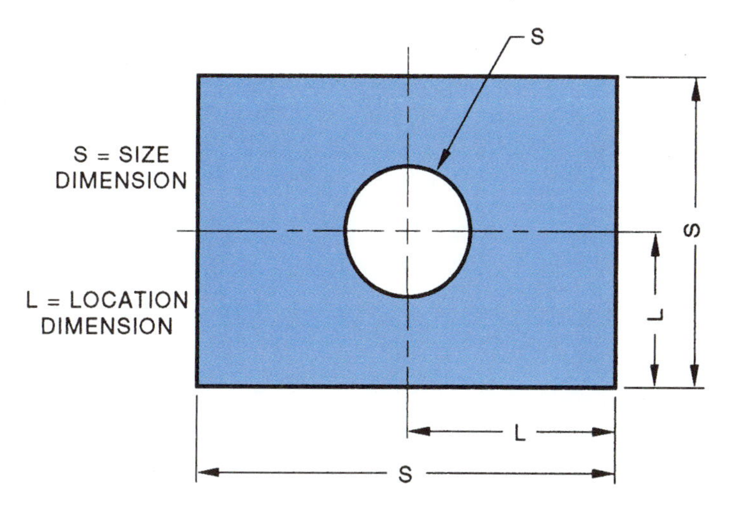

FIGURE 11-1 Dimensions indicating size and location.

TWO-PLACE DECIMAL DIMENSIONS

Dimensions may appear on drawings as *two-place decimals.* These are becoming widely used when the range of dimensional accuracy of a part is between 0.01″ larger or smaller than a specified dimension (nominal size). Where possible, two-place decimal dimensions are given in even hundredths of an inch.

Three- and four-place decimal dimensions continue to be used for more precise dimensions requiring machining accuracies in thousandths or ten-thousandths of an inch.

SIZE DIMENSIONS

Every solid has three size dimensions: depth or thickness, length or width, and height. In the case of a regular prism, two of the dimensions are usually placed on the principal view and the third dimension is placed on one of the other views, figure 11-2.

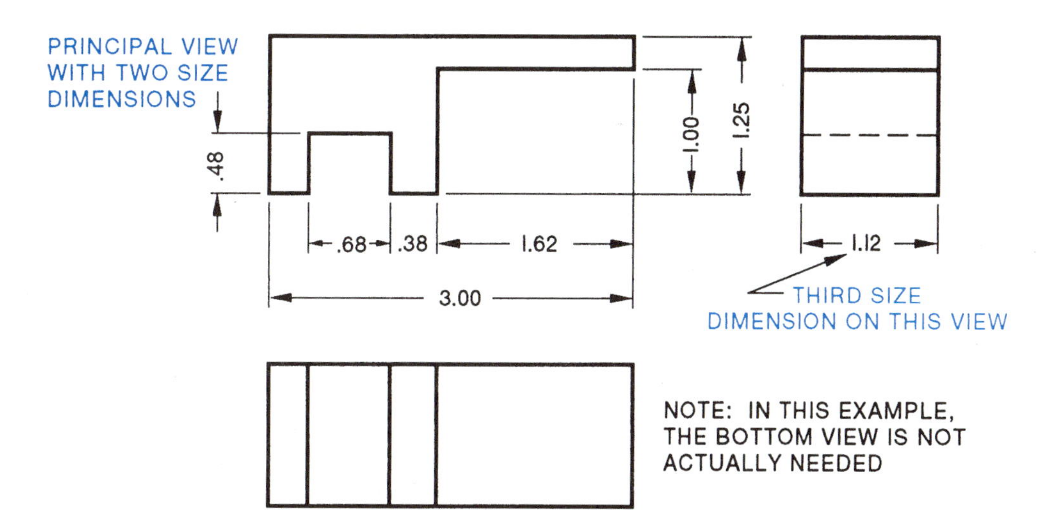

FIGURE 11-2 Placing size dimensions.

LOCATION DIMENSIONS

Location dimensions are usually made from either a center line or a finished surface. This practice is followed to overcome inaccuracies due to variations in measurement caused by surface irregularities.

Draftspersons and designers must know all the manufacturing processes and operations which a part must undergo in the shop. This technical information assists them in placing size and location dimensions, required for each operation, on the drawing. Figure 11-3 shows how size and location dimensions are indicated.

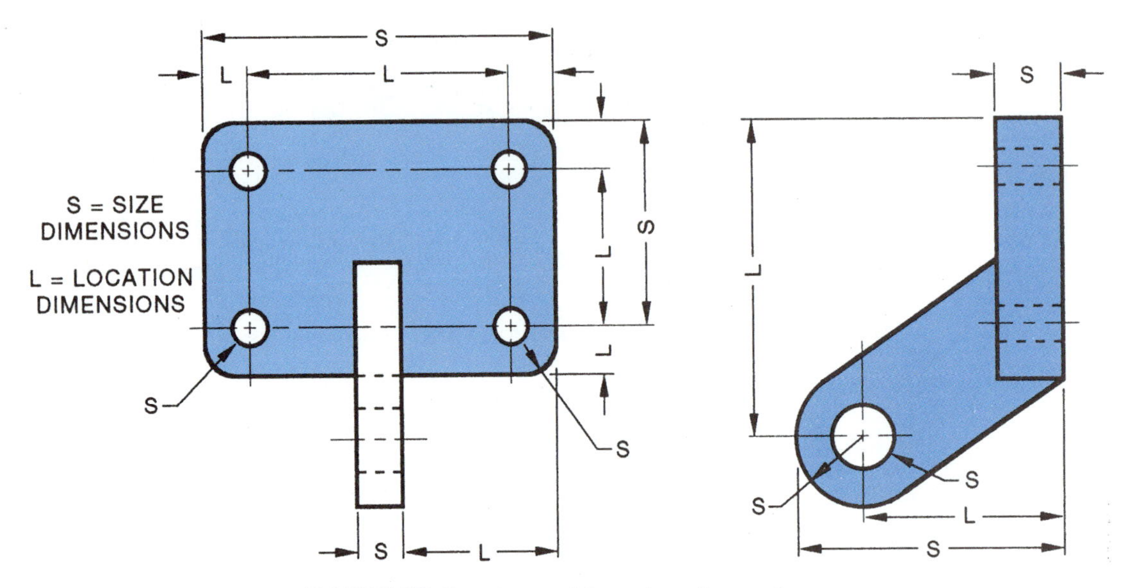

FIGURE 11-3 Size and location dimensions.

PLACING DIMENSIONS

In dimensioning a drawing, the first step is to place extension lines and external center lines where needed, figure 11-4.

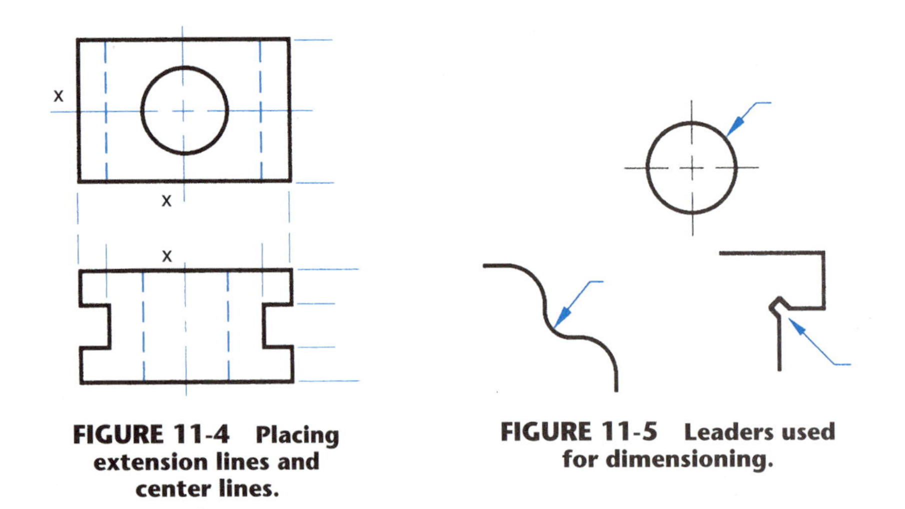

FIGURE 11-4 Placing extension lines and center lines.

FIGURE 11-5 Leaders used for dimensioning.

Dimension lines and leaders are added next. The term *leader* refers to a thin, inclined straight line terminating in an arrowhead. The leader directs attention to a dimension or note and the arrowhead identifies the feature to which the dimension or note refers. A few sample leaders are given in figure 11-5.

The common practice in placing dimensions is to keep them outside the outline of the object. The exception is where the drawing may be made clearer by inserting the dimensions within the object. When a dimension applies to two views, it should be placed between the two views as shown in figure 11-6.

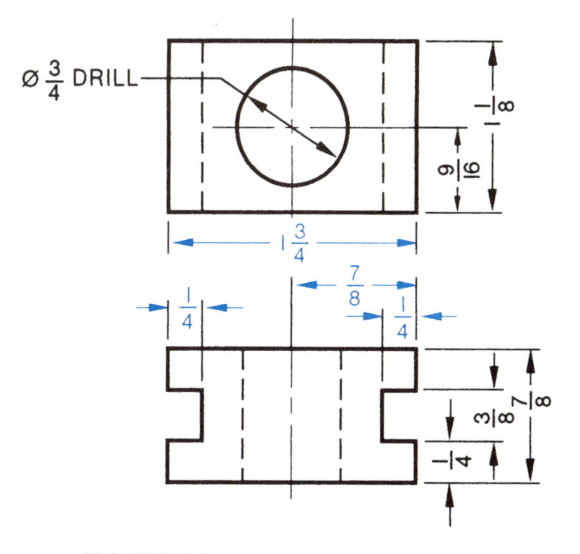

FIGURE 11-6 Placing dimensions between views.

Continuous Dimensions

Sets of dimension lines and numerals should be placed on drawings close enough so they may be read easily without any possibility of confusing one dimension with another. If a series of dimensions is required, the dimensions should be placed in a line as continuous dimensions, figure 11-7A. This method is preferred over the staggering of dimensions, figure 11-7B, because of ease in reading, appearance, and simplified dimensioning.

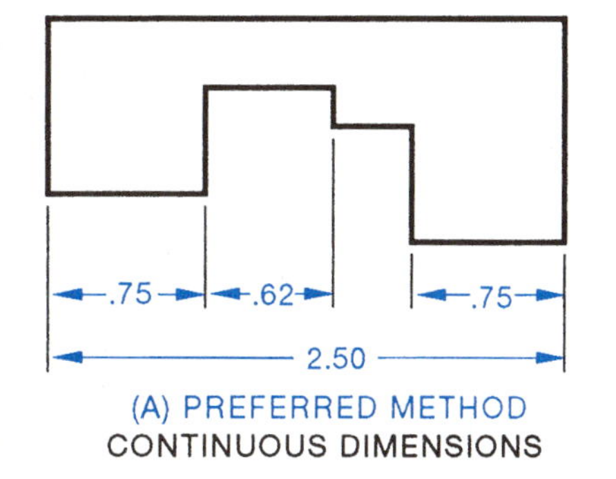

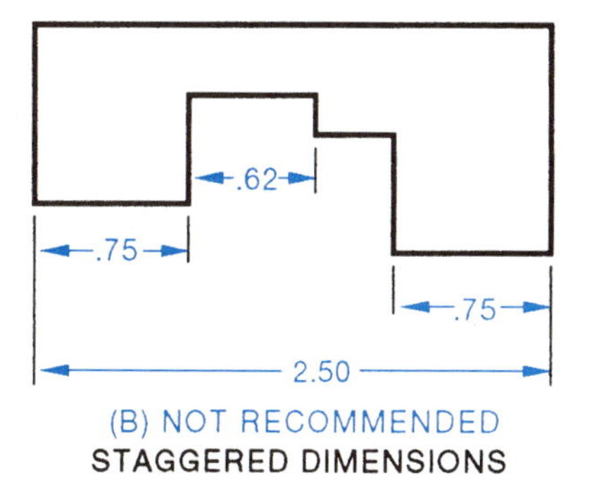

FIGURE 11-7 Placing series dimensions.

DIMENSIONS IN LIMITED SPACES

In the dimensioning of grooves or slots, the dimension, in many cases, extends beyond the width of the extension line. In such instances, the dimension is placed on either side of the extension line, or a leader is used, figure 11-8 (A, B, C, and D). Dots are also used on drawings for dimensioning when space is limited as shown in figure 11-8E.

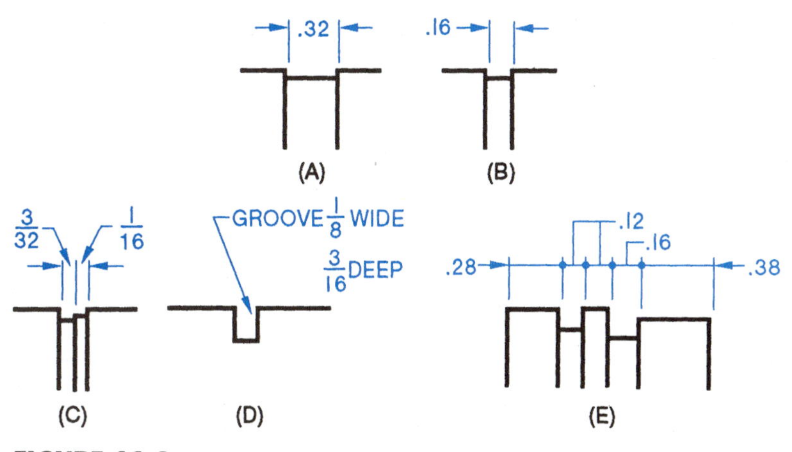

FIGURE 11-8 Dimensioning grooves and slots in limited spaces.

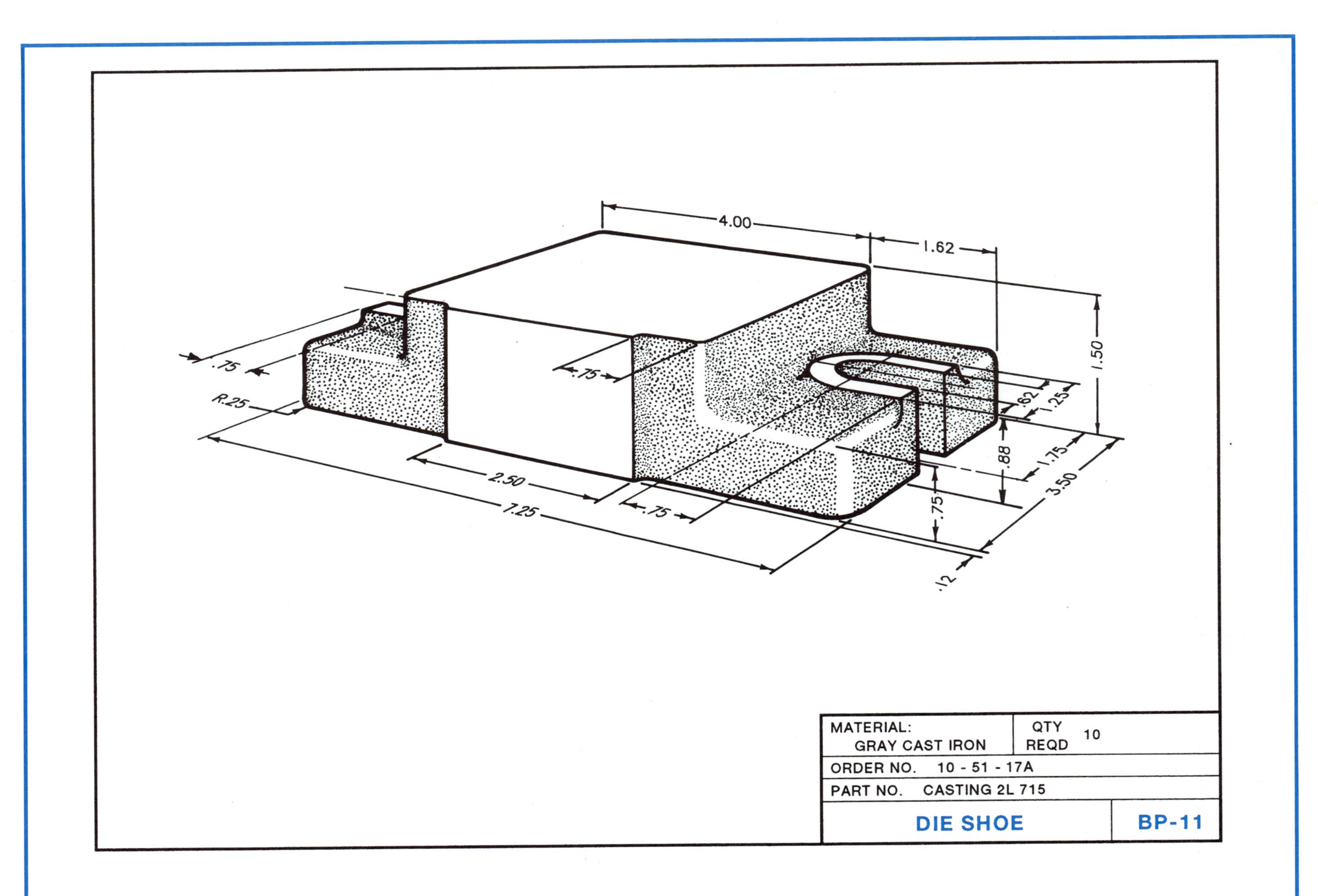
4.00
1.62
1.50
.75
R.25
.75
.62
1.25
.88
1.75
3.50
2.50
7.25
.75
.75
.12
MATERIAL:
GRAY CAST IRON
QTY REQD 10
ORDER NO. 10 - 51 - 17A
PART NO. CASTING 2L 715
DIE SHOE
BP-11

DIE SHOE (BP-11)

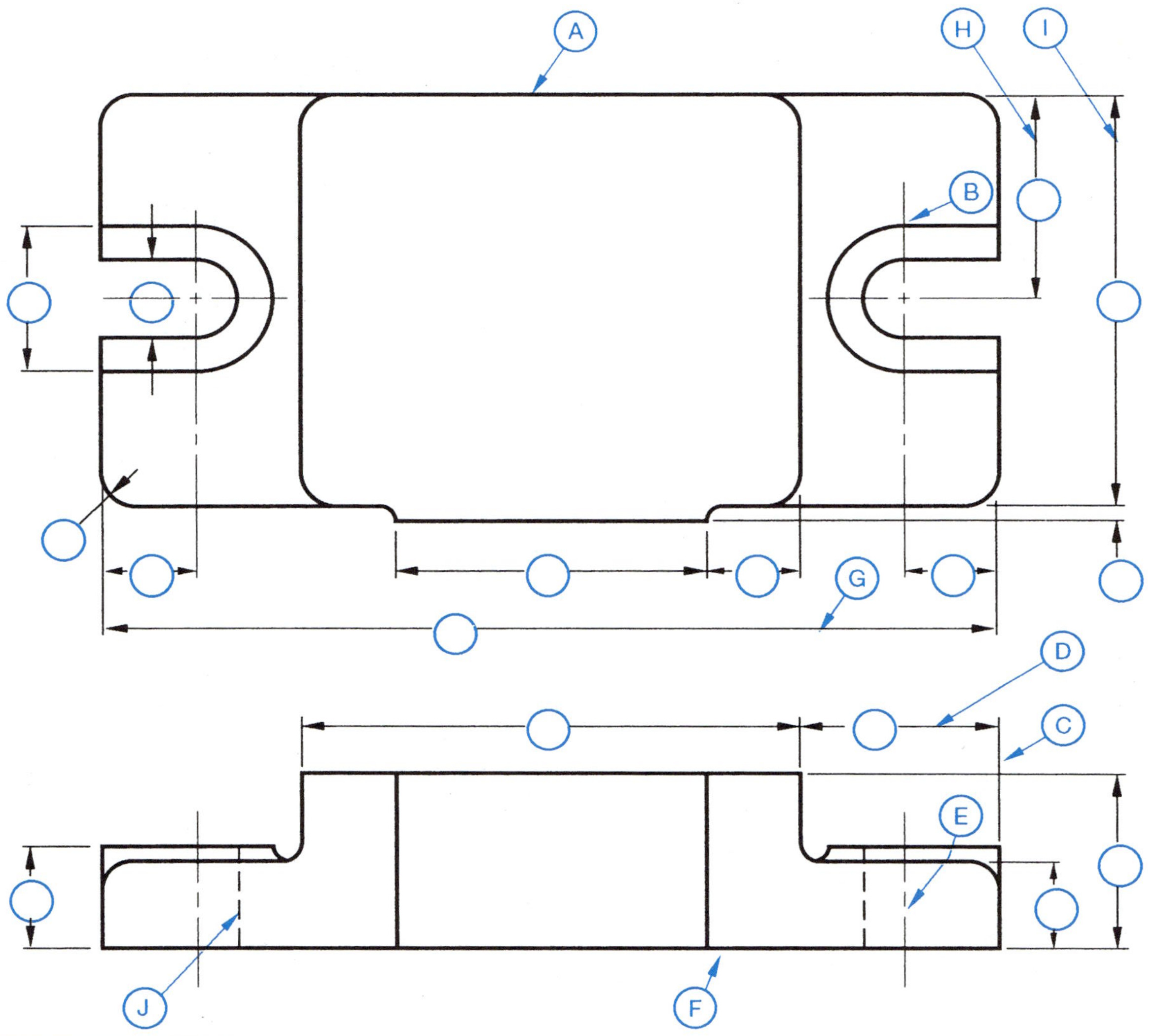

ASSIGNMENT—UNIT 11

Student's Name ______________

NOTE: STUDY THE SKETCH OF THE DIE SHOE

1. Give the dimensions in both views in the circles provided.
2. Indicate one letter on the two-view drawing which identifies each of the following types of lines:
   (a) Extension ______________ ◯
   (b) Dimension ______________ ◯
   (c) Object ______________ ◯
   (d) Center line ______________ ◯
   (e) Hidden edge ______________ ◯
3. Place an X in the correct block to identify the kind of dimension each letter refers to.

Dimension	Size Dimension	Location Dimension
G is a		
H is a		
I is a		

Unit 12

DIMENSIONING CYLINDERS, CIRCLES, AND ARCS

ALIGNED AND UNIDIRECTIONAL METHODS OF PLACING AND READING DIMENSIONS

There are two standard methods of placing dimensions. In the older (almost obsolete) method, called the *aligned* method, each dimension is placed in line with the dimension to which it refers, figure 12-1A. The second method, the *unidirectional* method (recommended by ANSI), has all numbers or values placed horizontally (one direction), regardless of the direction of the dimension line. All values are read from the bottom, figure 12-1B.

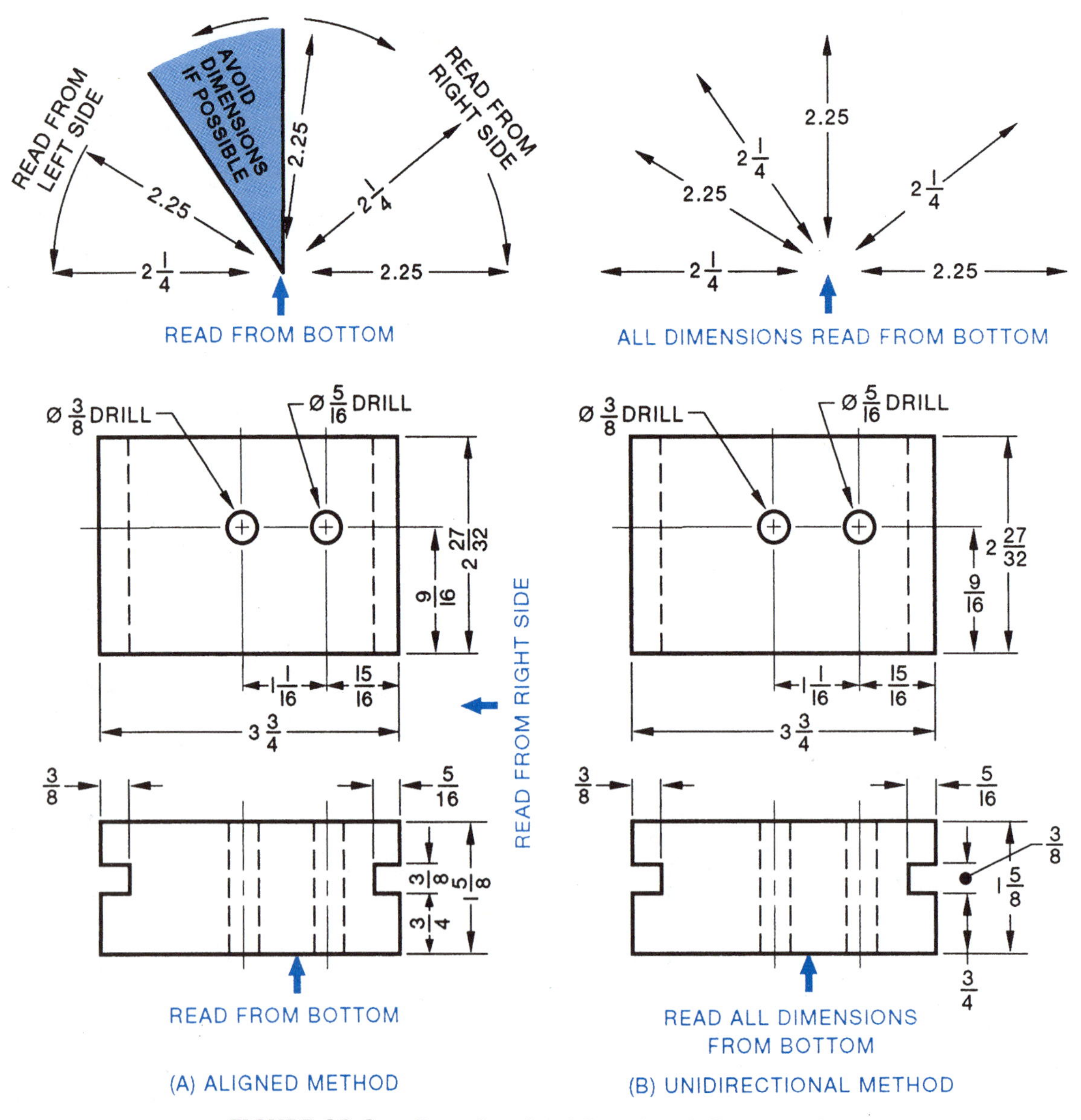

FIGURE 12-1 Aligned and unidirectional dimensioning.

The aligned and unidirectional methods are illustrated in figures 12-1A and 12-1B on similar drawings of the same part. Note at (A) that the aligned dimensions are read from both the bottom and the right side. By contrast, the unidirectional dimensions, figure 12-1B, are read from the bottom (one direction only).

DIMENSIONING CYLINDERS

The length and diameter of a cylinder are usually placed in the view which shows the cylinder as a rectangle, figure 12-2. This method of dimensioning is preferred because on small diameter cylinders and holes a dimension placed in the hole is confusing.

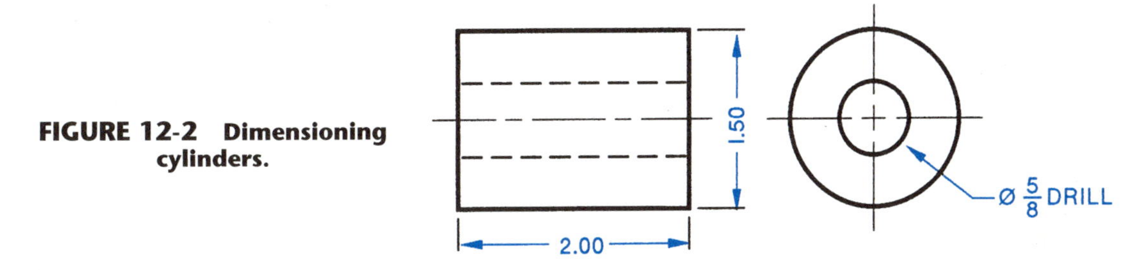

FIGURE 12-2 Dimensioning cylinders.

Many round parts, with cylindrical surfaces symmetrical about the axis, can be represented on one-view drawings. The abbreviation for diameter, DIA, is used with the dimension in such instances because no other view is needed to show the shape of the surface, figure 12-3A. A diameter may also be identified according to ANSI standards by using the symbol Ø preceding the dimension. In shop practice, the symbol Ø often appears after the dimension. On two-view drawings, DIA may be omitted, figure 12-3B. In other words, when a cylinder is dimensioned, DIA should follow the dimension unless it is evident that the dimension refers to a diameter.

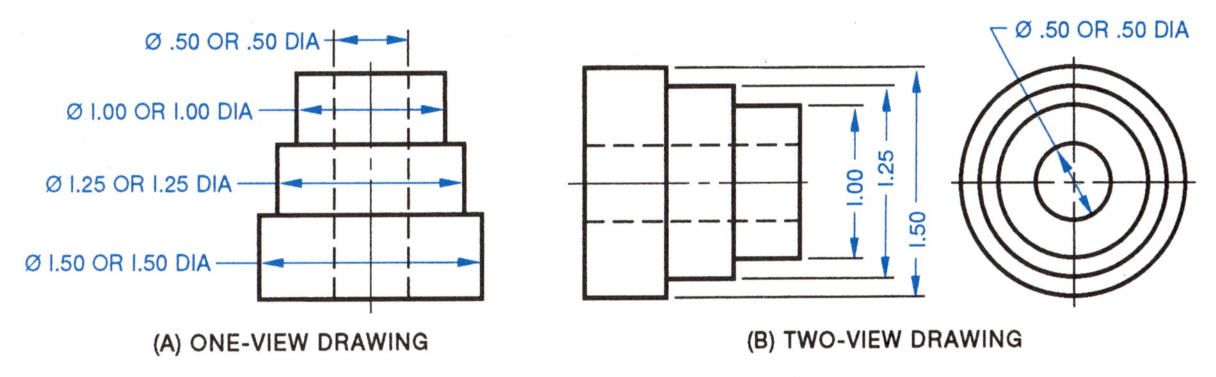

FIGURE 12-3 Dimensioning circles.

DIMENSIONING ARCS

An arc is always dimensioned by giving the radius. ANSI standards require a radius dimension to be preceded by the letter (symbol) **R** as shown in figure 12-4(A). However, the older (but still widely used) dimensioning practice is to place the symbol R after the dimension (figure 12-4(B)).

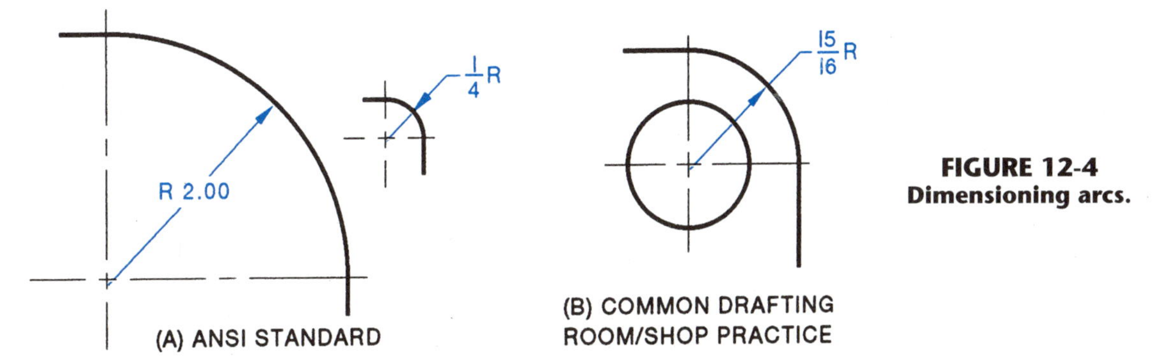

FIGURE 12-4 Dimensioning arcs.

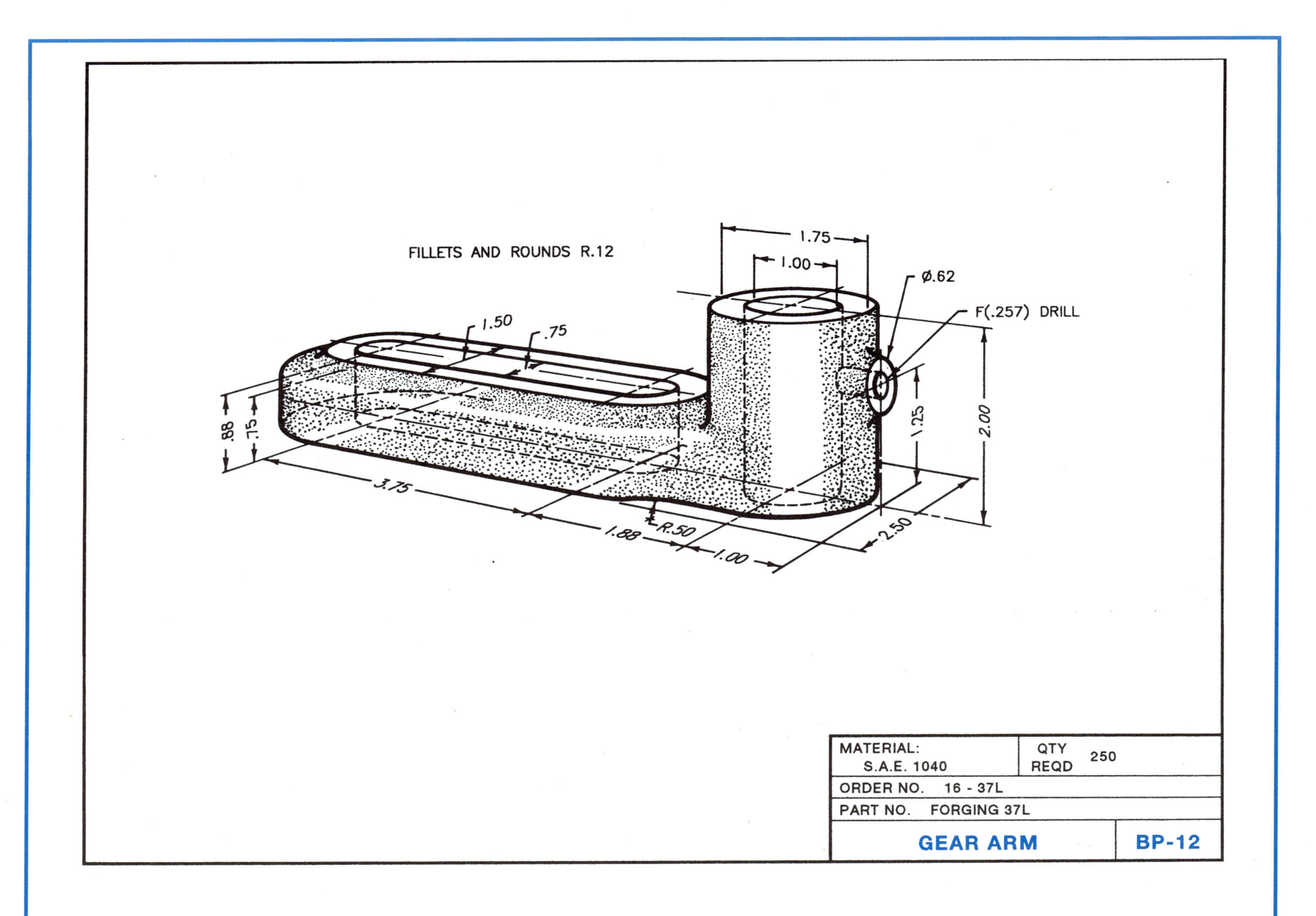
FILLETS AND ROUNDS R.12
1.75
1.00
Ø.62
F(.257) DRILL
1.50
.75
.88
.75
1.25
2.00
3.75
1.88
R.50
1.00
2.50
MATERIAL:
S.A.E. 1040
QTY REQD 250
ORDER NO. 16 - 37L
PART NO. FORGING 37L
GEAR ARM
BP-12

GEAR ARM (BP-12)

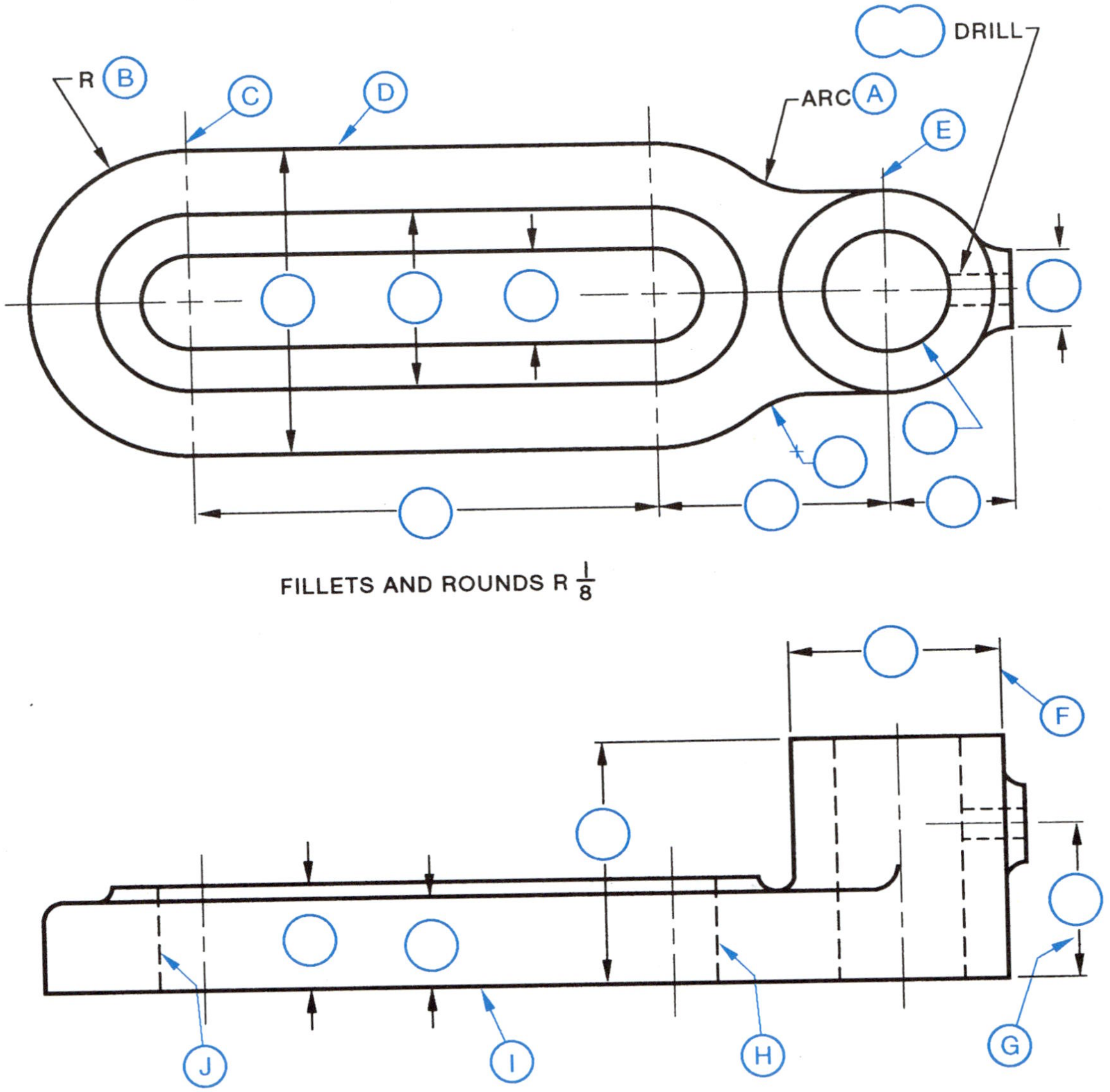

ASSIGNMENT—UNIT 12

Student's Name ________________

NOTE: STUDY THE SKETCH OF THE GEAR ARM

1. Give the dimensions required in both views in the circles provided, using the unidirectional method.
2. Dimension arc Ⓐ and radius Ⓑ.
3. What outside diameter is the upright portion?

   ____

4. Give the overall length of the elongated slot.

   ____

5. Determine the overall length of the Gear Arm.

   ____

6. Name the two views.

   ____________ and ____________ views.

7. Refer to the two-view drawing. Give one letter which identifies each type of line.
   - (a) Center line ________
   - (b) Object line ________
   - (c) Extension line ________
   - (d) Hidden edge ________
   - (e) Dimension line ________

Unit 13

SIZE DIMENSIONS FOR HOLES AND ANGLES

DIMENSIONING HOLES

The diameters of holes which are to be formed by drilling, reaming, or punching should have the diameter, preferably on a leader, followed by a note indicating (1) the operation to be performed, and (2) the number of holes to be produced, figure 13-1.

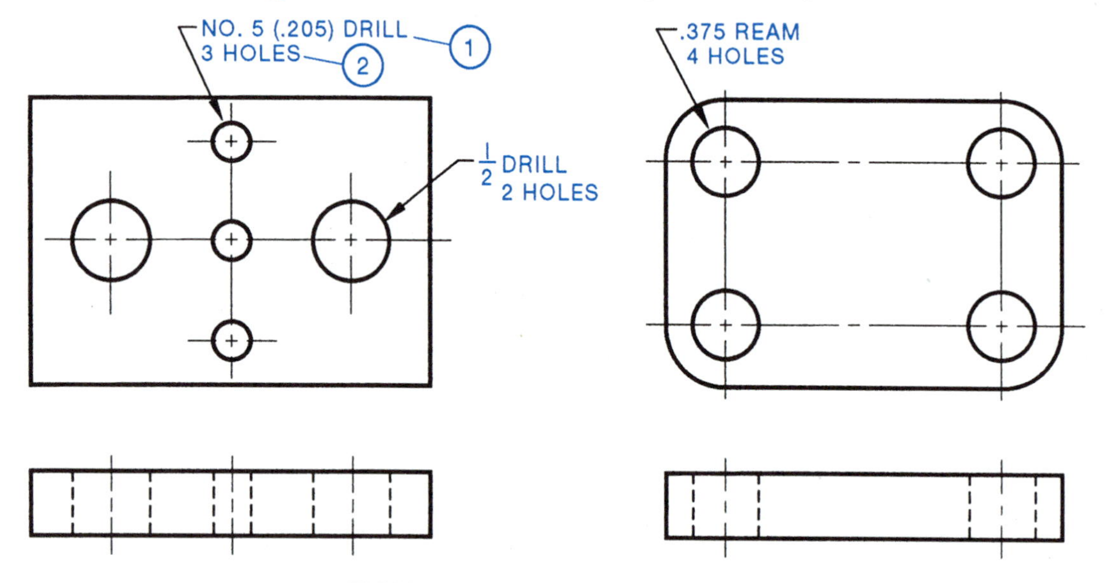

FIGURE 13-1 Dimensioning holes.

DIMENSIONING COUNTERBORED HOLES

A *counterbored hole,* figure 13-2, is one that has been machined to a larger diameter for a specified depth so that a bolt or pin will fit into this recessed hole. The counterbored hole provides a flat surface for the bolt or pin to seat against.

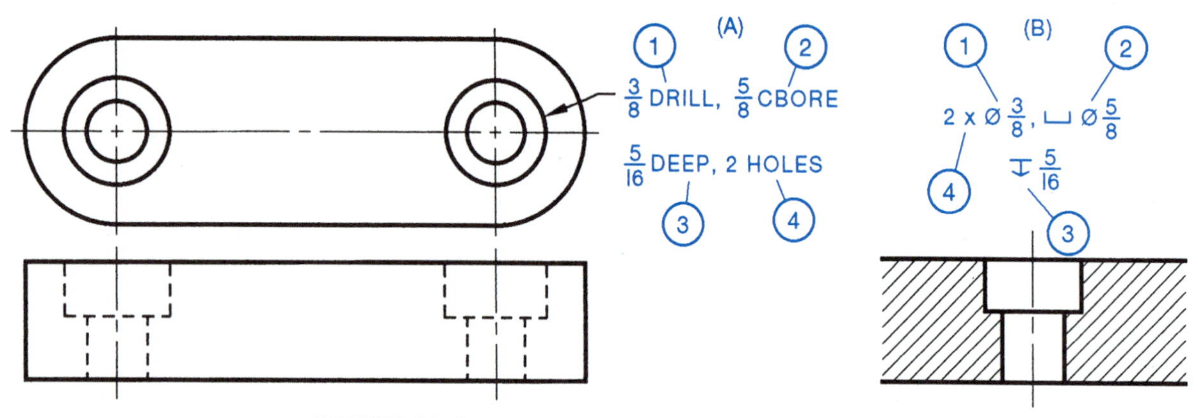

FIGURE 13-2 Dimensioning counterbored holes.

Counterbored holes are dimensioned by giving (1) the diameter of the drill, (2) the diameter of the counterbore, (3) the depth, and (4) the number of holes, figure 13-2(A). The dimensioning at (B) shows the use of the new ANSI symbol ⌴ to represent a counterbored hole and ↧ to indicate depth. In the example, the number of holes (2X) is followed by the counterbore symbol ⌴, the diameter symbol ∅, and the diameter $\frac{5}{8}$. The depth symbol ↧ is followed by the dimension $\frac{5}{16}$.

DIMENSIONING COUNTERSUNK HOLES

A *countersunk hole,* figure 13-3, is a cone-shaped recess machined in a part to receive a cone-shaped flat head screw or bolt.

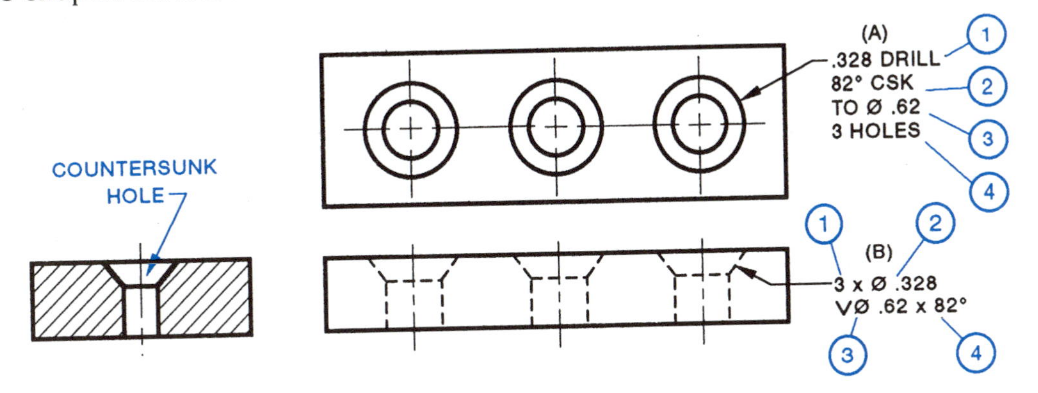

FIGURE 13-3 Dimensioning countersunk holes.

Countersunk holes are dimensioned by giving ① the diameter of the hole, ② the angle at which the hole is to be countersunk, ③ the diameter at the large end of the hole, and ④ the number of holes to be countersunk, (figure 13-3A). When the new ANSI symbol for countersunk ∨ is used, the countersunk holes are dimensioned as shown in figure 13-3B; ① number of holes (**3X**); ② diametral symbol and diameter (∅ .328); ③ countersunk hole symbol ∨, diametral symbol (∅), and diameter (.62); and the included angle of the countersink (82°).

DIMENSIONING ANGLES

The design of a part may require some lines to be drawn at an angle. The amount of the divergence (the amount the lines move away from each other) is indicated by an angle measured in degrees or fractional parts of a degree. The degree is indicated by the symbol ° placed after the numerical value of the angle. For example, in figure 13-4B, 45° indicates that the angle measures 45 degrees. Two common methods of dimensioning angles show (1) linear dimensions or (2) the angular measurement as illustrated in figure 13-4.

FIGURE 13-4 Methods of dimensioning angles.

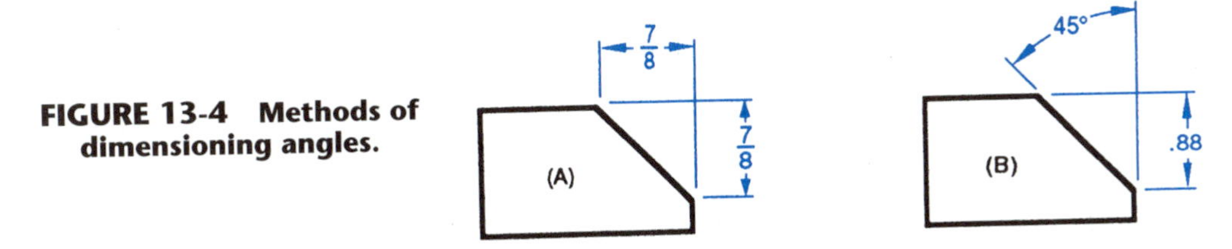

The dimension line for an angle should be an arc whose ends terminate in arrowheads. The numeral indicating the degrees in the angle is read in a horizontal position, except where the angle is large enough to permit the numerals to be placed along the arc, figure 13-5.

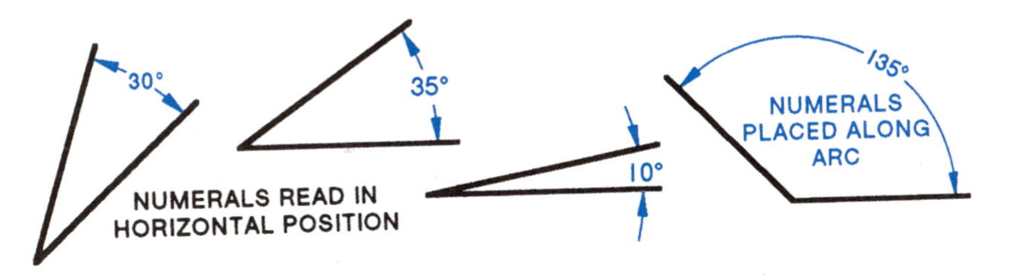

FIGURE 13-5 Placing angular dimensions.

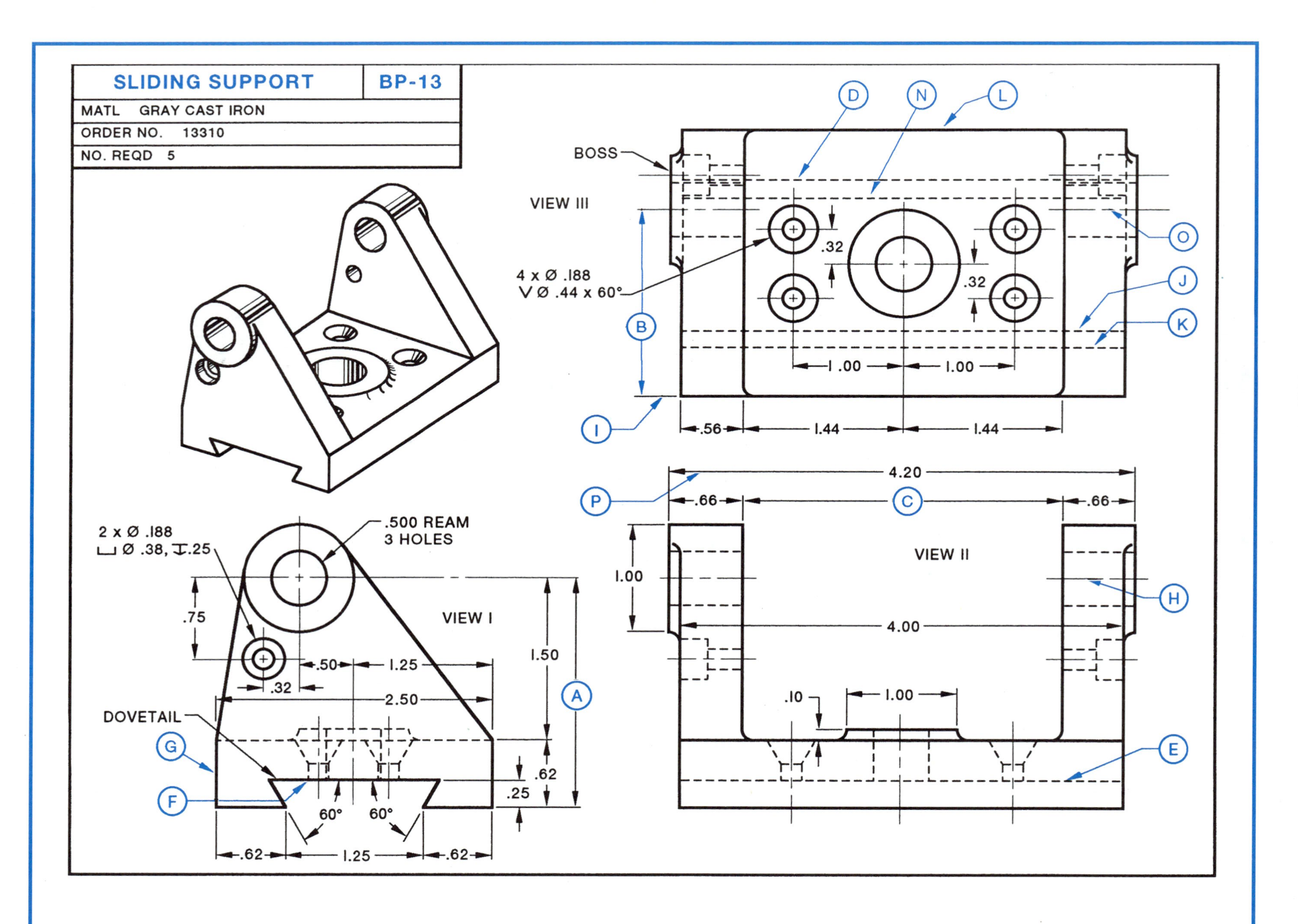
SLIDING SUPPORT
BP-13
MATL GRAY CAST IRON
ORDER NO. 13310
NO. REQD 5
BOSS
VIEW III
4 x Ø .188
V Ø .44 x 60°
.32
.32
1.00
1.00
.56
1.44
1.44
4.20
.66
.66
VIEW II
1.00
4.00
.10
1.00
2 x Ø .188
⌴ Ø .38, ⊥ .25
.500 REAM
3 HOLES
VIEW I
.75
.50
1.25
.32
2.50
1.50
DOVETAIL
.62
.25
60°
60°
.62
1.25
.62
A
B
C
D
E
F
G
H
I
J
K
L
N
O
P

ASSIGNMENT—UNIT 13

SLIDING SUPPORT (BP-13)

1. Name view I which shows the shape of the dovetail.
2. Name view III in which the bottom pad appears as a circle.
3. Name view II.
4. Name the kind of line shown at (E).
5. What surface in view I is represented by line (E)?
6. Name the kind of line shown at (G).
7. What line in view III represents surfaces (G)?
8. Name the kind of line shown at (H).
9. What line in view III represents the line (H)?
10. Name the kind of line shown at (I).
11. Name the kind of line shown at (J).
12. What lines in the top view represent the dovetail?
13. What does the line (E) in the front view represent?
14. Determine height (A).
15. How many bosses are shown on the uprights?
16. What is the outside diameter of the boss?
17. Determine dimension (B).
18. How far off from the center of the support is the center of the two holes in the bosses of the uprights?
19. Give the dimensions for the counterbored holes.
20. What dimensions are given for the countersunk holes?
21. Give the dimensions of the reamed holes.
22. What is the dimension (C)?
23. How wide is the opening in the dovetail?
24. How deep is the dovetail machined?
25. What is the angle to the horizontal at which the dovetail is cut?

Student's Name ______________

1. ______	14. ______
2. ______	15. ______
3. ______	16. ______
4. ______	17. ______
5. ______	18. ______
6. ______	19. ______
7. ______	20. ______
8. ______	21. ______
9. ______	22. ______
10. ______	23. ______
11. ______	24. ______
12. ______	25. ______
13. ______	

Unit 14

LOCATION DIMENSIONS FOR POINTS, CENTERS, AND HOLES

DIMENSIONING A POINT OR A CENTER

A point or a center of an arc or circle is generally measured from two finished surfaces. This method of locating the center is preferred to making an angular measurement.

In figure 14-1, the center of the circle and arc may be found easily by scribing the vertical and horizontal center lines from the machined surfaces.

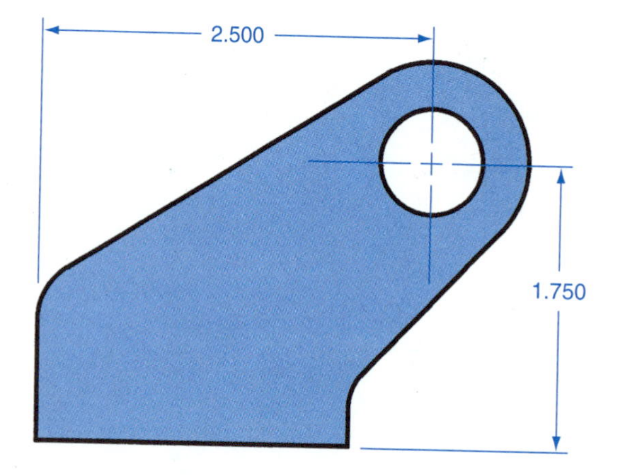

FIGURE 14-1 Dimensioning the center of a circle.

DIMENSIONING EQUALLY SPACED HOLES ON A CIRCLE

If a number of holes are to be equally spaced on a circle, the exact location of the first hole is given by location dimensions. To locate the remaining holes, the location dimensions are followed by ① the diameter of the holes, ② the number of holes, and ③ the notation **EQUALLY SPACED** or **EQ SP**, figure 14-2.

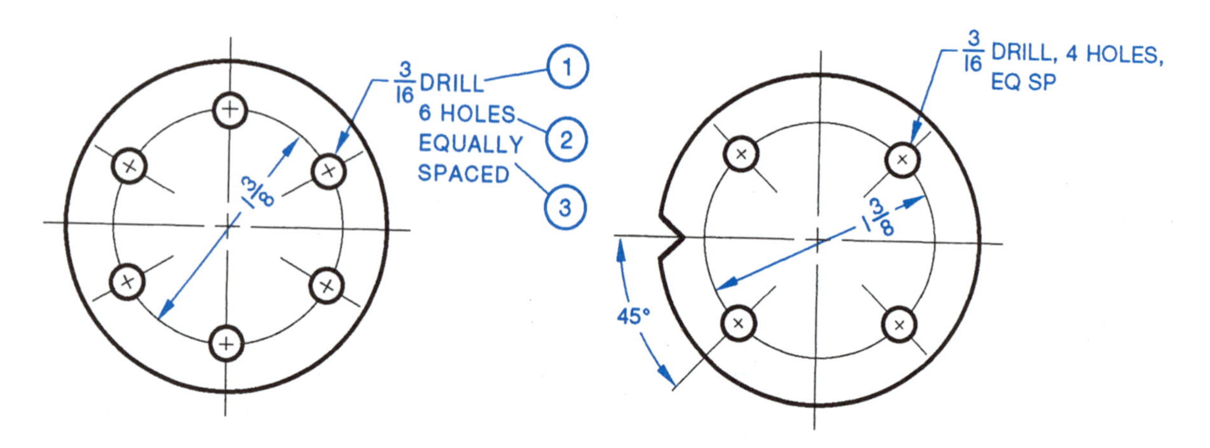

FIGURE 14-2 Dimensioning holes equally spaced on a circle.

DIMENSIONING UNEQUALLY SPACED HOLES ON A CIRCLE

When holes are to be located on a circle, the diameter of the circle should be given so as to fix the exact center of each hole. The size and position of each hole are noted on the drawing, figure 14-3A. If more than one hole is the same diameter, then a notation may be used to indicate this fact, figure 14-3B.

For example, the notation **.500 REAM–5 HOLES** means that the five holes on the drawing are reamed 1/2" in diameter.

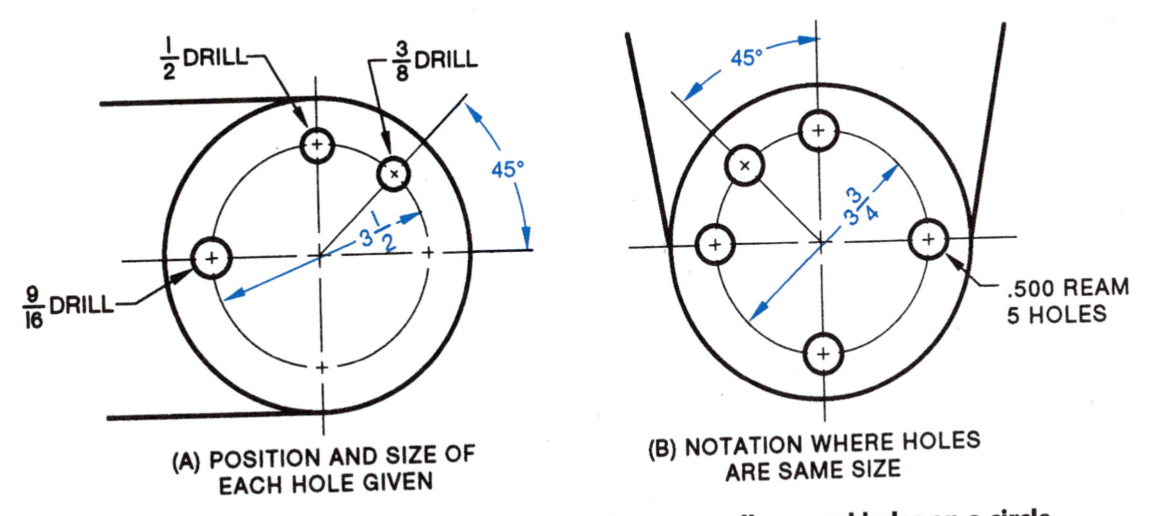

FIGURE 14-3 Two methods of dimensioning unequally spaced holes on a circle.

The machining process to be used in producing holes is often determined by the manufacturer of the part. When this is the case, the hole may be dimensioned by using the symbol for diameter (∅) followed by the hole size. For example, the $\frac{1}{2}$ **DRILL**, $\frac{3}{8}$ **DRILL**, and $\frac{9}{16}$ **DRILL** sizes shown in figure 14-3(A) may also be dimensioned ∅ $\frac{1}{2}$, ∅ $\frac{3}{8}$, and ∅ $\frac{9}{16}$, respectively.

DIMENSIONING HOLES NOT ON A CIRCLE

Holes are often dimensioned in relation to one another and to a finished surface. Dimensions are usually given, in such cases, in the view which shows the shape of the holes, that is, square, round, or elongated. The preferred method of placing these dimensions is shown in figure 14-4A.

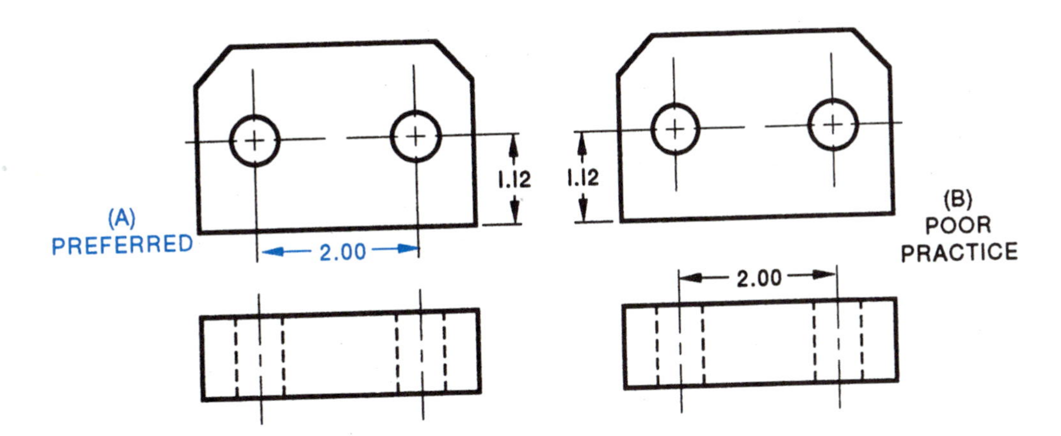

FIGURE 14-4 Placing dimensions for holes.

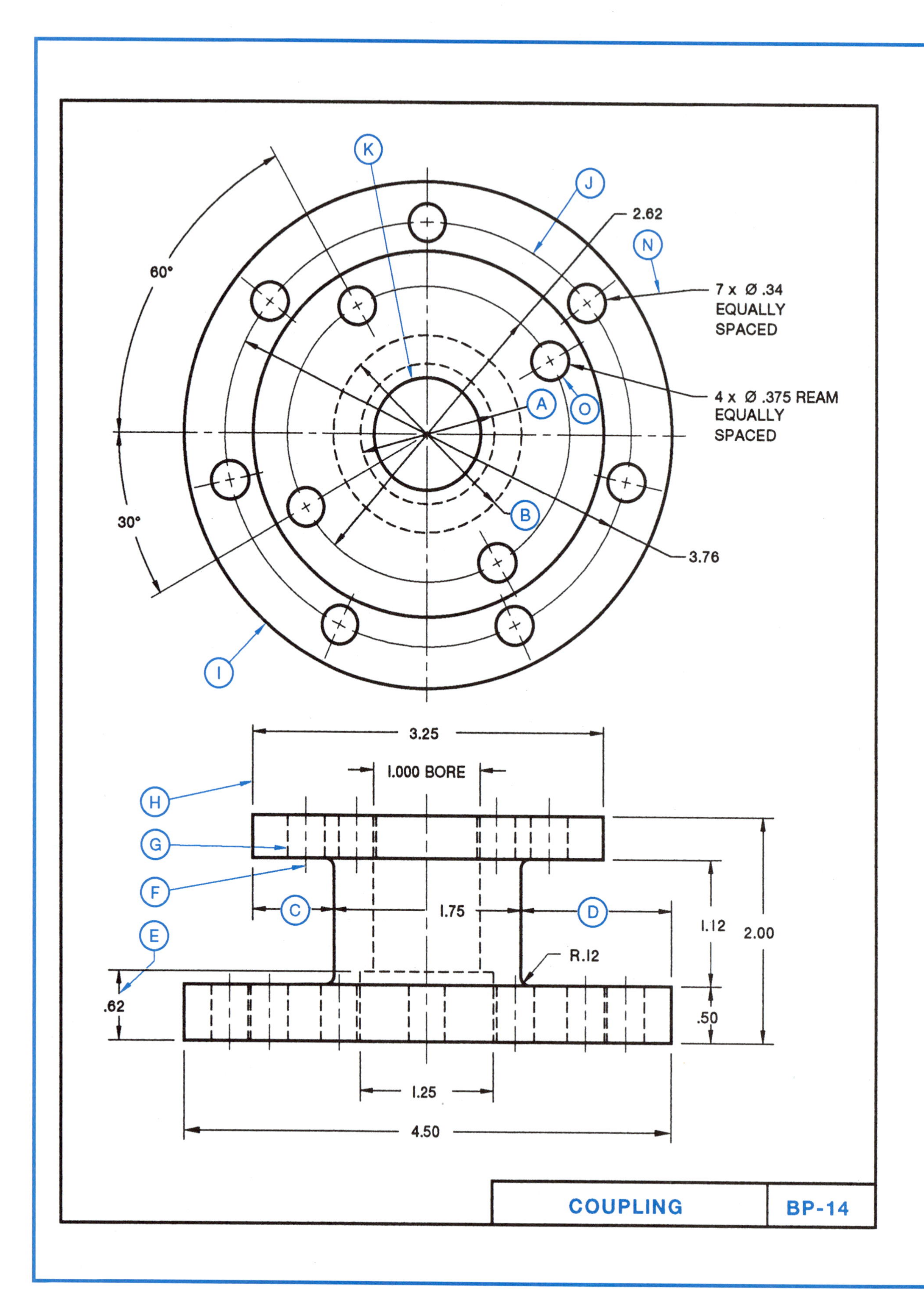
K
J
2.62
N
60°
7 x Ø .34
EQUALLY
SPACED
A
O
4 x Ø .375 REAM
EQUALLY
SPACED
B
30°
3.76
I
3.25
1.000 BORE
H
G
F
C
1.75
D
E
1.12
2.00
R.12
.62
.50
1.25
4.50
COUPLING
BP-14

COUPLING (BP-14)

1. Name the view which shows the width (height) of the Coupling.
2. Name the view in which the holes are shown as circles.
3. Name the kind of line shown at Ⓔ Ⓕ Ⓖ Ⓗ Ⓘ Ⓙ.
4. What circle represents the 4.50" diameter flange?
5. What circle represents the 1.000" bore diameter?
6. Name the kind of line shown at Ⓝ.
7. How many holes are to be drilled in the larger flange?
8. Indicate the drill size to be used.
9. Give the diameter circle on which the equally spaced holes are drilled in the larger flange.
10. How many holes are to be reamed in the smaller flange?
11. How deep is the 1.25" diameter hole bored?
12. Give the diameter of the reamed holes.
13. State the angle with the horizontal center line used for locating reamed hole Ⓞ.
14. What is the overall width (height) of the Coupling?
15. What is the diameter of the smaller flange?
16. What is the diameter of the circle Ⓐ? Ⓑ?
17. What is the depth of the 1.000″ bored hole?
18. What is the thickness of the larger flange?
19. If a 5/16" bolt is used in the drilled holes, what will be the clearance between the hole and the bolt?
20. Determine distances Ⓒ and Ⓓ.

ASSIGNMENT—UNIT 14

Student's Name ____________________

1. ____________________
2. ____________________
3. Ⓔ ____________________
   Ⓕ ____________________
   Ⓖ ____________________
   Ⓗ ____________________
   Ⓘ ____________________
   Ⓙ ____________________
4. ____________________
5. ____________________
6. ____________________
7. ____________________
8. ____________________
9. ____________________
10. ____________________
11. ____________________
12. ____________________
13. ____________________
14. ____________________
15. ____________________
16. Ⓐ ____________________
    Ⓑ ____________________
17. ____________________
18. ____________________
19. ____________________
20. Ⓒ ____________________
    Ⓓ ____________________

Unit 15

DIMENSIONING LARGE ARCS AND BASE LINE DIMENSIONS

DIMENSIONING ARCS WITH CENTERS OUTSIDE THE DRAWING

When the center of an arc falls outside the limits of the drawing, a broken dimension line is used as illustrated in figure 15-1. This dimension line gives the size of the arc and indicates that the arc center lies on a center line outside the drawing. This technique is used, also, when a dimension line interferes with other parts of a drawing.

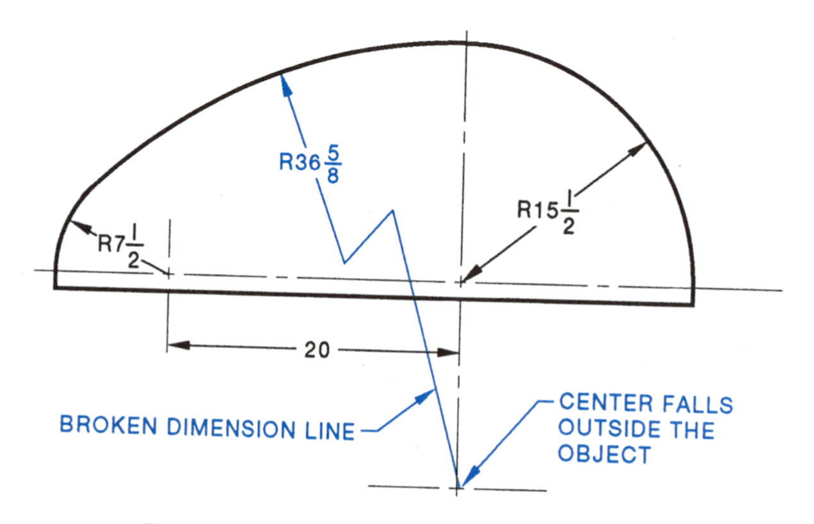

FIGURE 15-1 Using broken dimension line.

BASE LINE DIMENSIONING

In base line dimensioning, all measurements are made from common finished surfaces called *base lines* or *reference lines,* figure 15-2. Base line dimensioning is used where

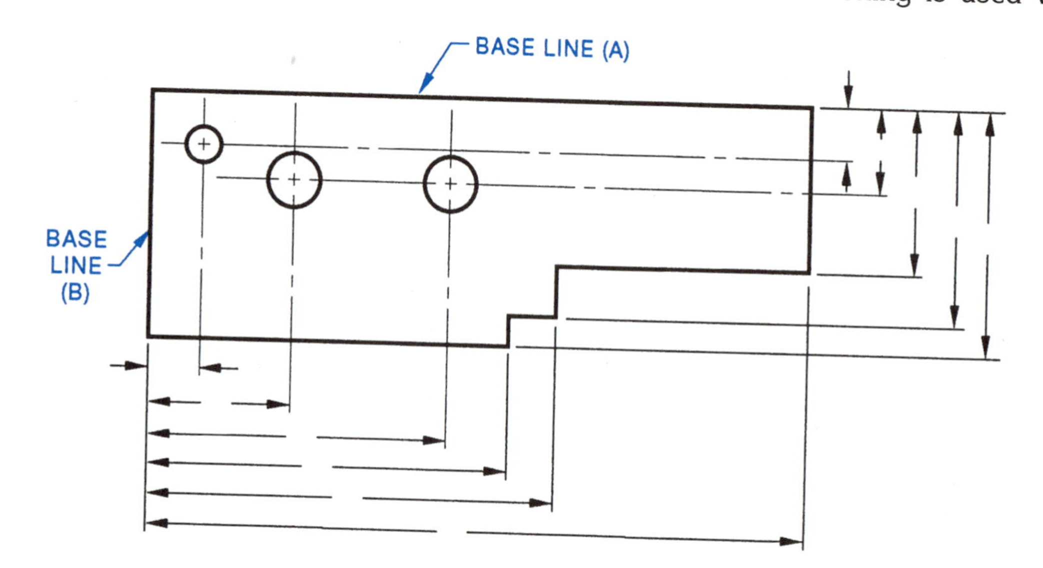

FIGURE 15-2 Base line dimensioning from two machined edges.

accurate layout work to precision limits is required. Errors are not cumulative with this type of dimensioning because all measurements are taken from the base lines.

Dimensions and measurements may be taken from one or more base lines. In figure 15-2, the two base lines are at right angles to each other. The horizontal dimensions are measured from base line (B) which is a machined edge. The vertical dimensions are measured from surface (A) which is at a right angle to surface (B) and is also a machined surface.

An application of base line dimensioning, where a center line is used as the reference line, is shown in figure 15-3A. Base line dimensioning may also be applied to irregular shapes such as the template shown in figure 15-3B.

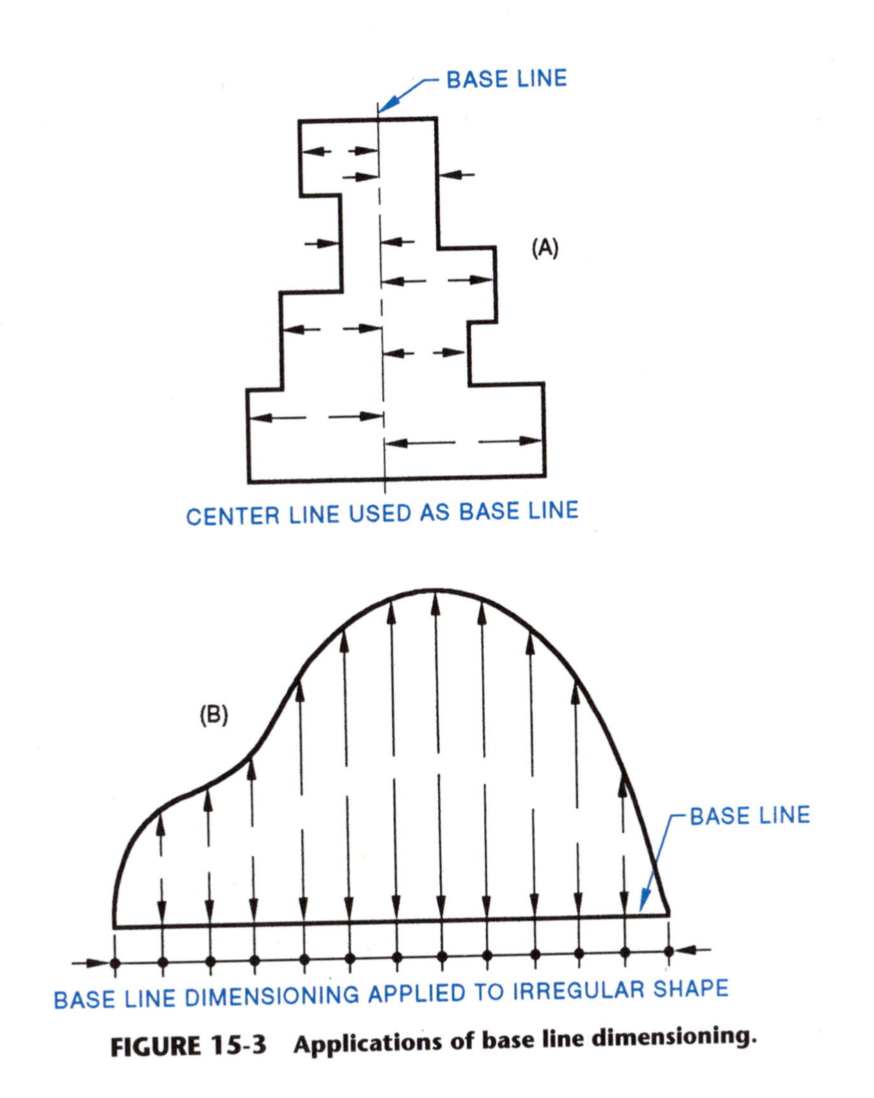

FIGURE 15-3 Applications of base line dimensioning.

Base line dimensions simplify the reading of a drawing and also permit greater accuracy in making the part.

STATING INCHES

Dimensions under 72 inches are usually stated in inches so the inch (") symbol may be omitted on a drawing. This applies primarily to manufacturing industries. By comparison, measurements for structural work and the building industry are usually given in feet and inches.

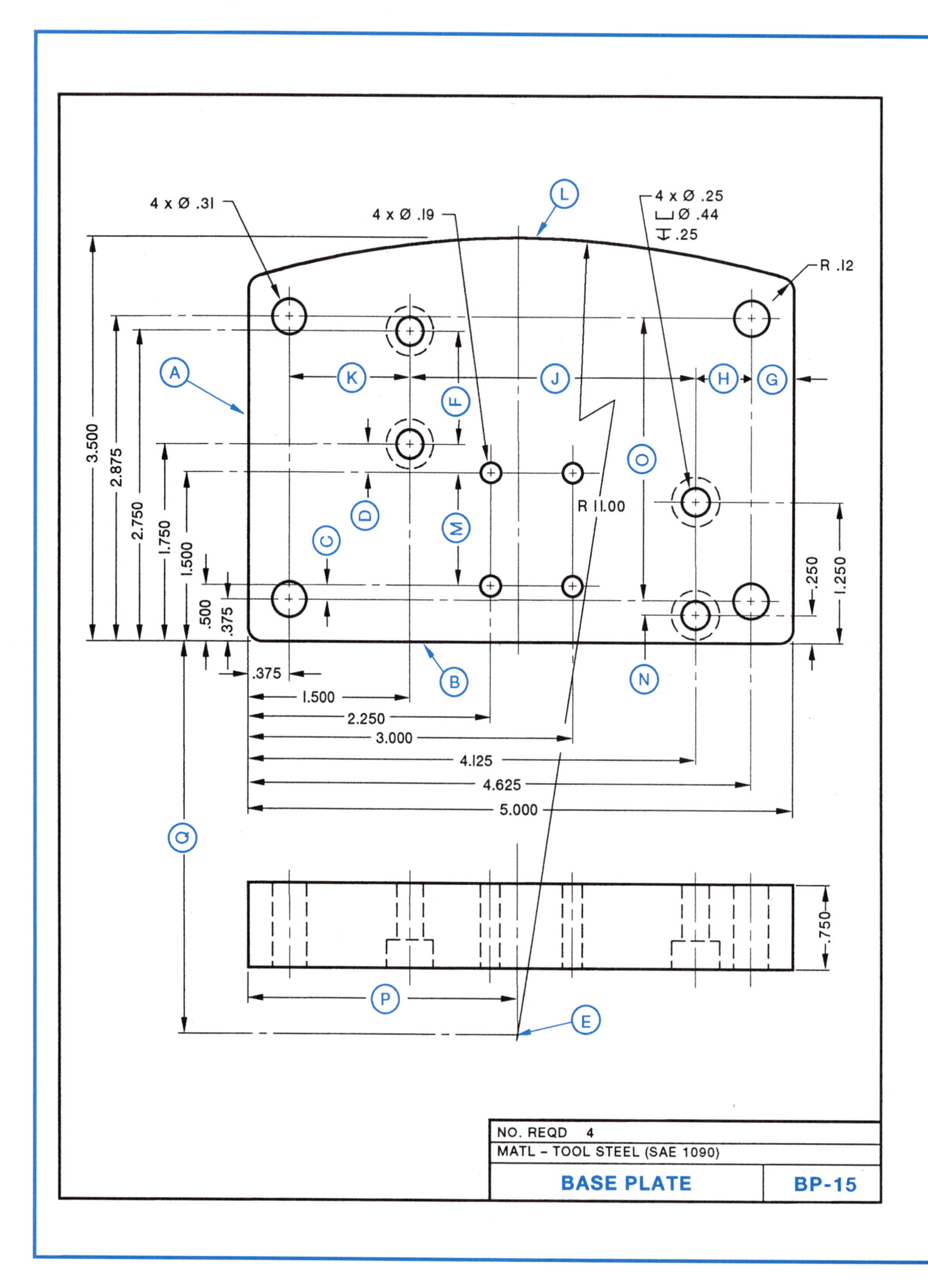
4 x Ø .31
4 x Ø .19
4 x Ø .25
⌴ Ø .44
↧ .25
R .12
R 11.00
3.500
2.875
2.750
1.750
1.500
.500
.375
.375
1.500
2.250
3.000
4.125
4.625
5.000
.250
1.250
.750
A
B
C
D
E
F
G
H
J
K
L
M
N
O
P
Q
NO. REQD 4
MATL – TOOL STEEL (SAE 1090)
BASE PLATE
BP-15

ASSIGNMENT—UNIT 15

BASE PLATE (BP-15)

1. Give the name of the part.
2. What material is used for the Base Part?
3. How many parts are required?
4. What is the length of the Base Plate?
5. What is the height of the Base Plate?
6. What is the thickness of the Plate?
7. How many 5/16" holes are to be drilled?
8. How many 3/16" holes are to be drilled?
9. How many 1/4" holes are to be drilled?
10. Give (a) the diameter and (b) the depth of counterbore for the .25" holes.
11. What system of dimensioning is used on this drawing?
12. Give the letter of the base line in the Top View from which all depth dimensions are taken.
13. Give the letter of the base line in the Top View from which all horizontal dimensions are taken.
14. Compute the following depth dimensions: Ⓒ Ⓓ Ⓜ Ⓕ.
15. Compute the following horizontal dimensions: Ⓖ Ⓗ Ⓙ Ⓚ.
16. Compute dimensions Ⓝ and Ⓞ.
17. Give the radius to which the corners are rounded.
18. What is the radius of arc Ⓛ?
19. What letter indicates the center for arc Ⓛ?
20. Compute dimensions Ⓟ and Ⓠ.

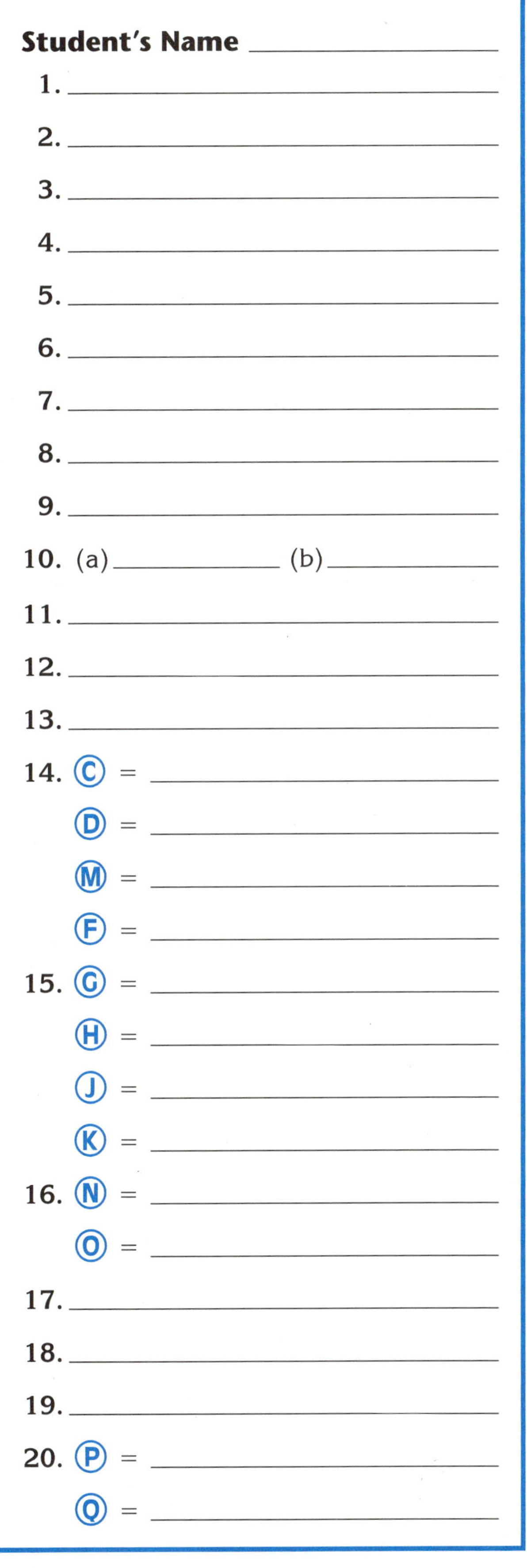

Student's Name ______________

1. ______________
2. ______________
3. ______________
4. ______________
5. ______________
6. ______________
7. ______________
8. ______________
9. ______________
10. (a) ________ (b) ________
11. ______________
12. ______________
13. ______________
14. Ⓒ = ______________
    Ⓓ = ______________
    Ⓜ = ______________
    Ⓕ = ______________
15. Ⓖ = ______________
    Ⓗ = ______________
    Ⓙ = ______________
    Ⓚ = ______________
16. Ⓝ = ______________
    Ⓞ = ______________
17. ______________
18. ______________
19. ______________
20. Ⓟ = ______________
    Ⓠ = ______________

Unit 16

TOLERANCES: FRACTIONAL AND ANGULAR DIMENSIONS

TOLERANCES

As a part is planned, the designer must consider (1) its function either as a separate unit or as a part which must move in a fixed position in relation to other parts, (2) the operations required to produce the part, (3) the material to be used, (4) the quantity to be produced, and (5) the cost. Each of these factors influences the degree of accuracy to which a part is machined.

The dimensions given on a drawing are an indication of what the limits of accuracy are. These limits are called *tolerances.* Parts may have a tolerance given in fractions or decimal inches or decimal millimeters. CAD drawings can dimension in any of these systems. For CNC-manufactured parts, decimal dimensions are required.

SPECIFYING FRACTIONAL TOLERANCES

The note in figure 16-1, **TOLERANCES ON FRACTIONAL DIMENSIONS ARE ± 1/64"**, indicates that the dimension given in fractions on the drawing may be any size between a 64th of an inch larger to a 64th of an inch smaller than the specified size.

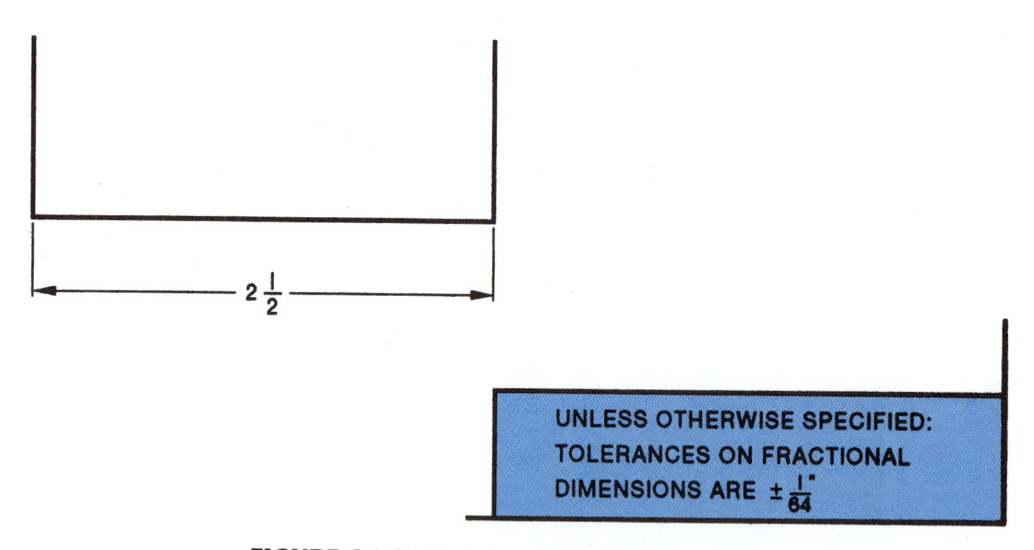

FIGURE 16-1 Fractional tolerances applied.

For example, on the 2 1/2" dimension in figure 16-1:

1. The tolerance given on the drawing is ± 1/64".
2. The largest size to which the part may be machined is 2 1/2" + 1/64" = 2 33/64".
3. The smallest size to which the part may be machined is 2 1/2" − 1/64" = 2 31/64".

The larger size is called the *upper limit;* the smaller size is called the *lower limit.*

ANGULAR DIMENSIONS

Angles may be dimensioned in degrees or parts of a degree.

1. Each degree is one three hundred sixtieth of a circle (1/360).
2. The degree may be divided into smaller units called *minutes*. There are 60 minutes in each degree.
3. Each minute may be divided into smaller units called *seconds*. There are 60 seconds in each minute.

To simplify the dimensioning of angles, symbols are used to indicate degrees, minutes and seconds, figure 16-2. For example, twelve degrees, fifteen minutes and forty-five seconds can also be written 12° 15' 45".

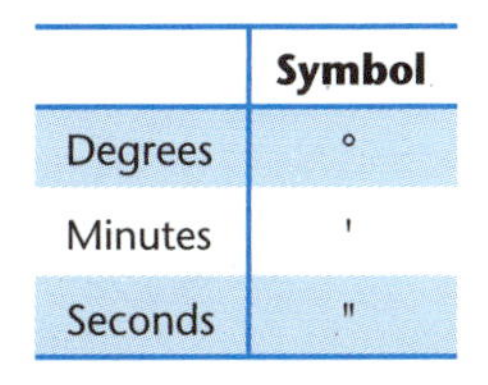

	Symbol
Degrees	°
Minutes	'
Seconds	"

FIGURE 16-2 Symbols used for dimensioning angles.

Decimalized angles are now preferred. To convert angles given in whole degrees, minutes, and seconds, the following steps should be followed.

Example: Convert 12° 15' 45" into decimal degrees

1. Convert minutes into degrees by dividing by 60 (60' = 1°)

   15 ÷ 60 = .25°

2. Convert seconds into degrees by dividing seconds by 3600 (3600" = 1)

   45" ÷ 3600 = .01°

3. Add whole degrees plus decimal degrees

   12° + .25° + .01° = 12.26°

therefore 12° 15' 45" = 12.26° decimal degrees

SPECIFYING ANGULAR TOLERANCES

The tolerance on an angular dimension may be given in a note on the drawing as shown in figure 16-3. The tolerance may also be shown on the angular dimension itself, figure 16-4.

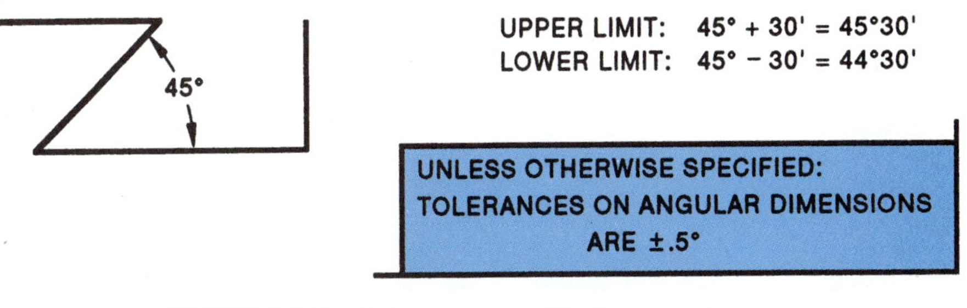

FIGURE 16-3 Tolerance specified as a note.

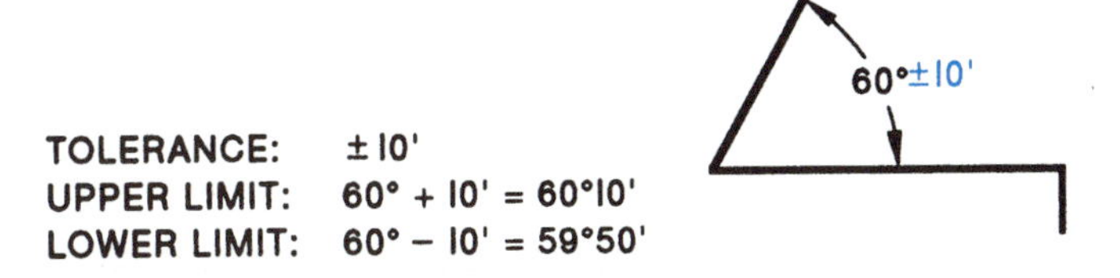

FIGURE 16-4 Tolerance specified on angular dimension.

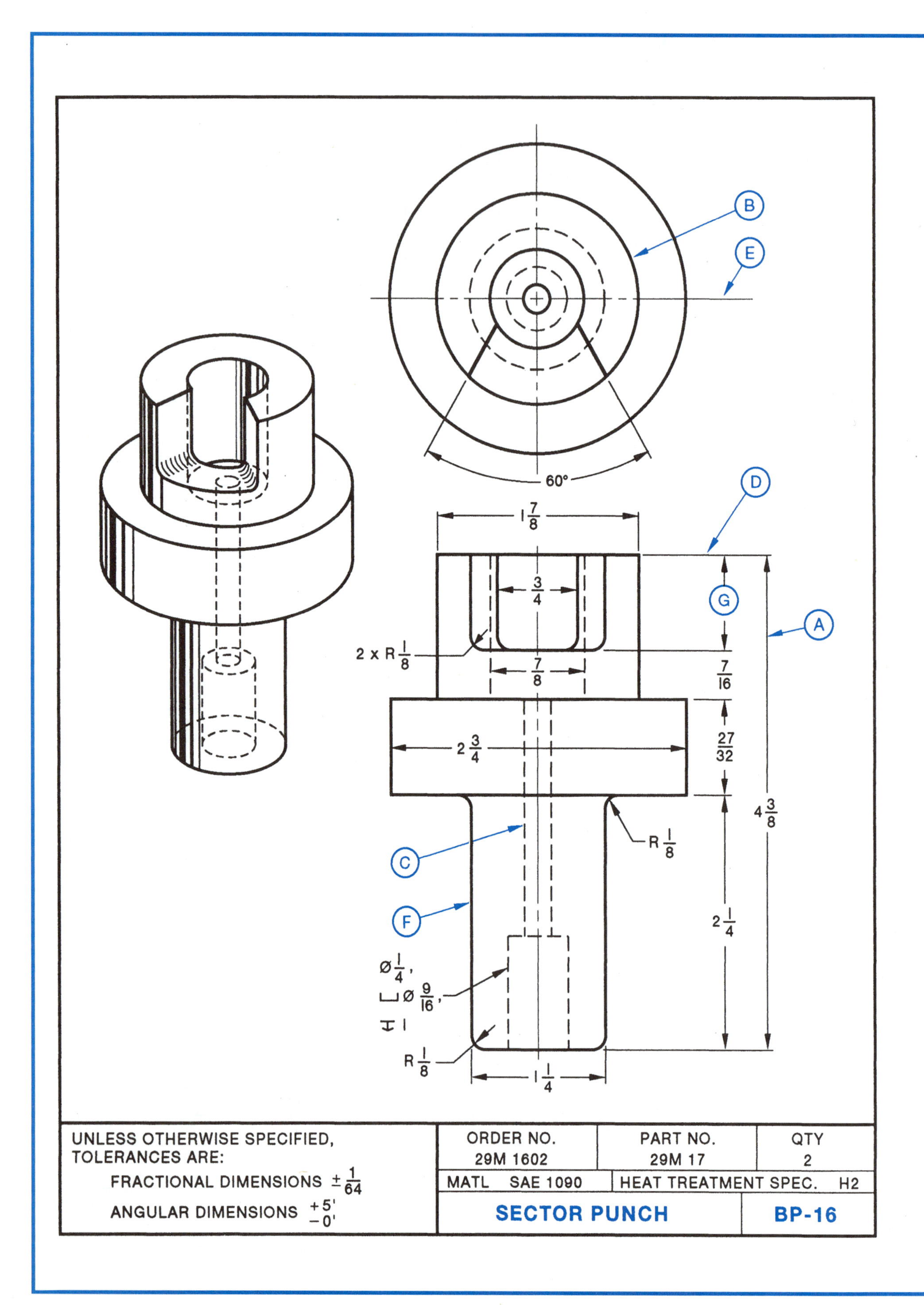

B
E
60°
D
1 7/8
3/4
G
A
2 x R 1/8
7/8
7/16
27/32
2 3/4
4 3/8
R 1/8
C
2 1/4
F
Ø 1/4, ⌴ Ø 9/16, ↧ 1
R 1/8
1 1/4
UNLESS OTHERWISE SPECIFIED, TOLERANCES ARE:
FRACTIONAL DIMENSIONS ± 1/64
ANGULAR DIMENSIONS +5' −0'
ORDER NO. 29M 1602
PART NO. 29M 17
QTY 2
MATL SAE 1090
HEAT TREATMENT SPEC. H2
SECTOR PUNCH
BP-16

SECTOR PUNCH (BP-16)

1. What letter is used to denote a
   (a) Hidden edge line
   (b) Dimension line
   (c) Extension line
   (d) Center line
2. Determine dimension Ⓒ.
3. What is the diameter of the drilled pilot hole for the counterbore?
4. Give the diameter and depth of the counterbore.
5. What are the upper and lower limits of tolerance for fractional dimensions?
6. What limit of tolerance is specified for angular dimensions?
7. What is the largest size to which the 2 3/4" diameter can be turned?
8. What is the lower limit to which the 2 3/4" diameter can be machined?
9. Give the upper limit to which diameter Ⓑ may be machined.
10. Give the upper and lower limit on the diameter for shank Ⓕ.
11. If shank Ⓕ is machined to the upper limit length, how long will it be?
12. If shank Ⓕ is machined 2 3/16" long, how much under the lower limit size will it be?
13. How much over the upper limit will the shank be if it is 2 5/16" long?
14. What is the upper limit of accuracy for the 60° angle?
15. If the height of the 1 7/8" diameter punch measures 1 9/32", is it over, under, or within the specified limits of accuracy?

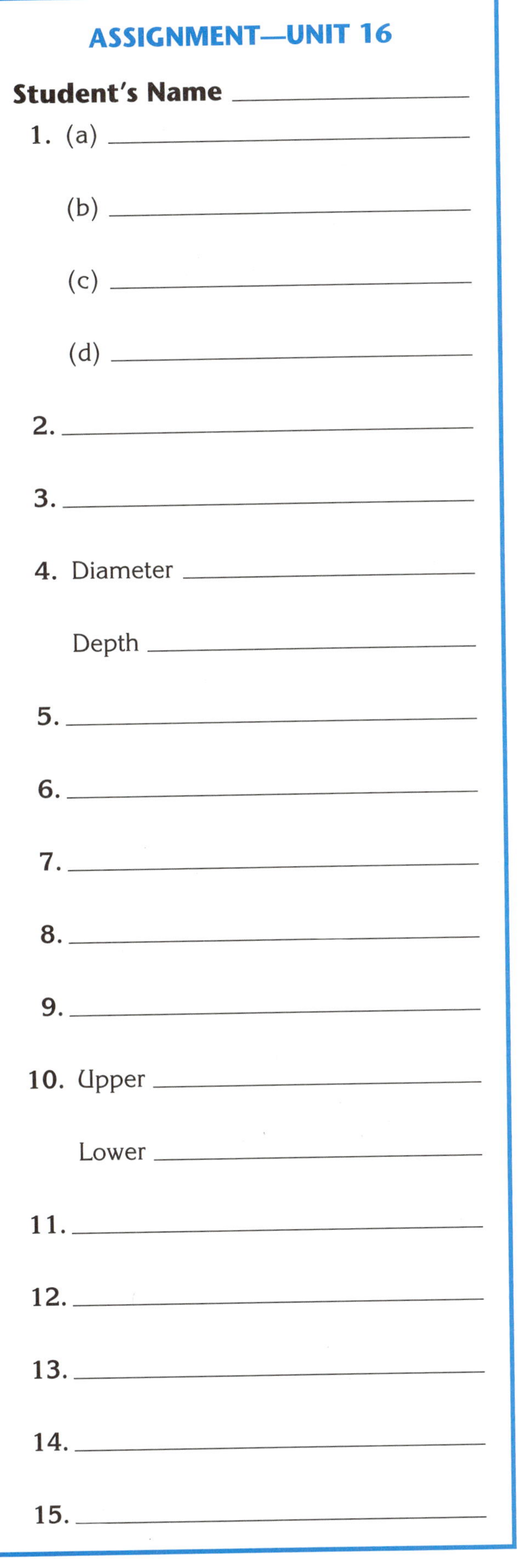

ASSIGNMENT—UNIT 16

Student's Name ____________

1. (a) ____________
   (b) ____________
   (c) ____________
   (d) ____________
2. ____________
3. ____________
4. Diameter ____________
   Depth ____________
5. ____________
6. ____________
7. ____________
8. ____________
9. ____________
10. Upper ____________
    Lower ____________
11. ____________
12. ____________
13. ____________
14. ____________
15. ____________

Unit 17

TOLERANCES: UNILATERAL AND BILATERAL TOLERANCES, AND DECIMAL DIMENSIONS

UNILATERAL AND BILATERAL TOLERANCES

The limits of accuracy to which a part is to be produced may fall within one of two classifications of tolerances. A dimension is said to have a *unilateral* (single) tolerance when the total tolerance is in one direction only, either (+) or (−). The examples of unilateral tolerances shown in figure 17-1 indicate that part (A) meets standards of accuracy when the basic dimension varies in one direction only and is between 3" and 3 1/64"; part (B) may vary from 2.44" to 2.43"; and angle (C) may vary between 60° and 59° 45'.

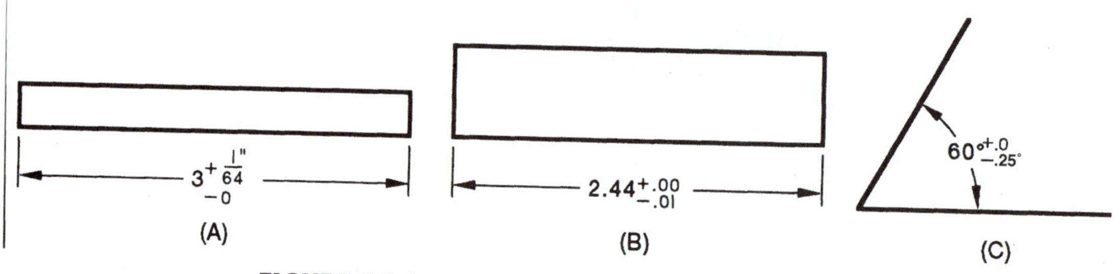

FIGURE 17-1 Examples of unilateral tolerances.

Bilateral tolerances applied to dimensions mean that the dimensions may vary from a larger size (+) to a smaller size (−) than the basic dimension (nominal size). In other words, the basic dimension may vary in both directions. The basic 3" dimension in figure 17-2A with a ± 1/64" tolerance may vary between 3 1/64" and 2 63/64". The basic 2.44" dimension in figure 17-2B with a bilateral tolerance of ± .01" is acceptable within a range of 2.45" and 2.43". The 60° angle in figure 17-2C with a tolerance of ± .25° may range between 60.25° and 59.75°. These bilateral tolerances are shown in figure 17-2.

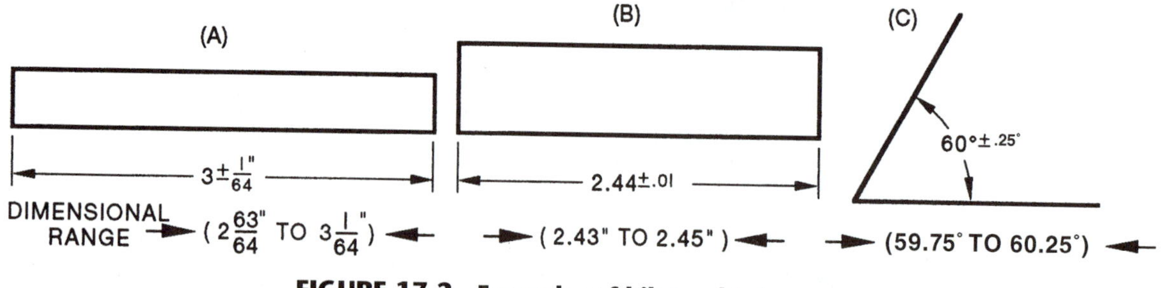

FIGURE 17-2 Examples of bilateral tolerances.

When the dimensions within the tolerance limits appear on a drawing, they are expressed as a range from the smaller to the larger dimension as indicated in figure 17-2.

DECIMAL DIMENSIONS

The decimal system of dimensioning is used almost exclusively in industry because of the ease with which computations can be made. In addition, the dimension can be measured with precision instruments to a high degree of accuracy and is required for CNC applications.

Dimensions in the decimal system can be read quickly and accurately. Dimensions are read in thousandths, 1/1000" = (.001"), in ten thousandths, 1/10,000" = (.0001"), and in even finer divisions if necessary.

SPECIFYING DECIMAL TOLERANCES

Decimal tolerances can be applied in both the English and the metric systems of measurement. Tolerances on decimal dimensions which are expressed in terms of two, three, four, or more decimal places may be given on a drawing in several ways. One of the common methods of specifying a tolerance that applies on all dimensions is to use a note, figure 17-3.

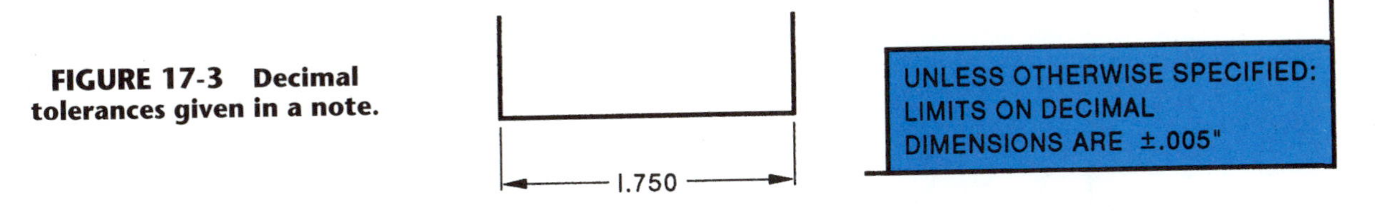

FIGURE 17-3 Decimal tolerances given in a note.

For example, the 1.750" dimension in the figure may be machined to a size ranging from

1.750''' + .005" = 1.755"

TO

1.750" – .005" = 1.745"

The larger size (1.755") is called the *upper limit*. The smaller size (1.745") is called the *lower limit*.

A tolerance on a decimal dimension also may be included as part of the dimension, as shown in figure 17-4.

1.750 ±.002

UPPER LIMIT 1.750" + .002" = 1.752"

LOWER LIMIT 1.750" –.002" = 1.748"

FIGURE 17-4 Decimal tolerance included with dimension.

Bilateral tolerances are not always equal in both directions. It is common practice for a drawing to include either a (+) tolerance or a (–) tolerance that is greater than the other.

In cases where the plus and minus tolerances are not the same, such as plus .001" and minus .002", the dimension may be shown on the drawing as in figure 17-5.

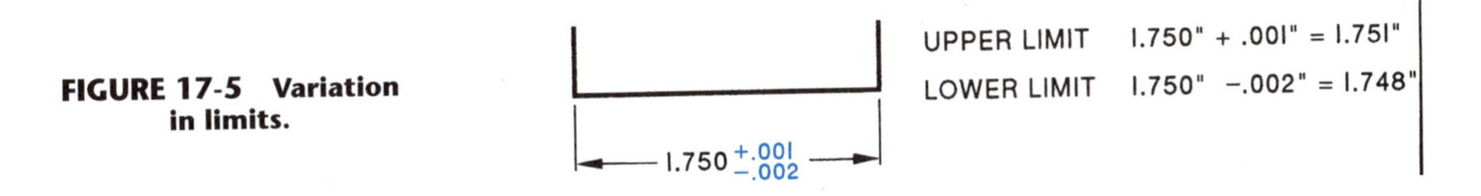

FIGURE 17-5 Variation in limits.

The same variation between the upper and lower limit can be given as in figure 17-6. The dimension above the line is the upper limit; the dimension below the line is the lower limit.

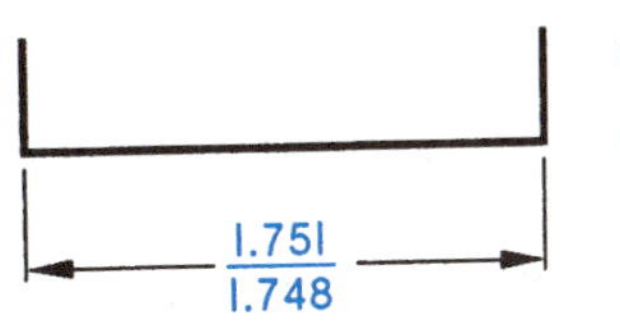

UPPER LIMIT 1.751"

LOWER LIMIT 1.748"

FIGURE 17-6 Upper and lower limit dimension.

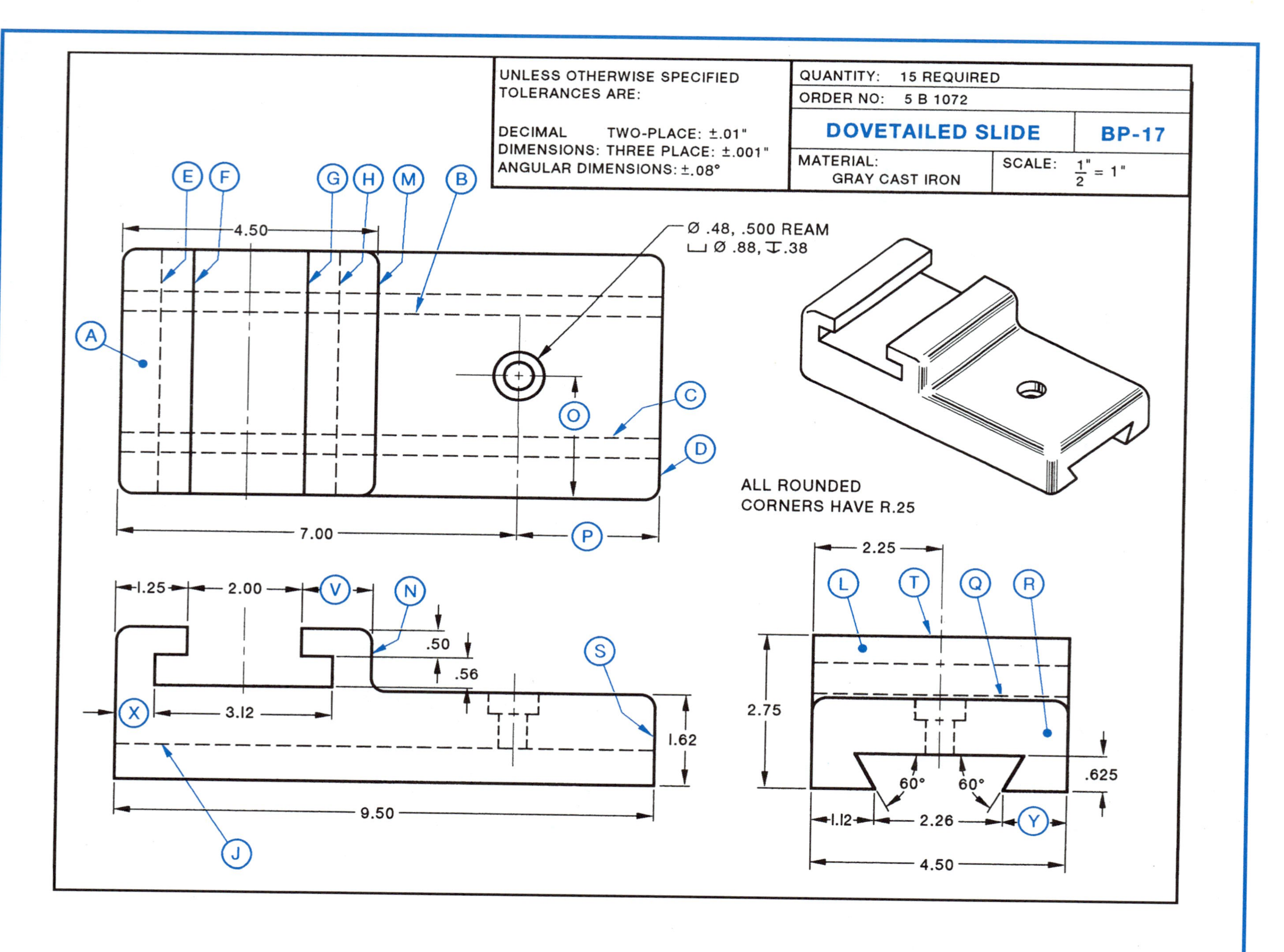
UNLESS OTHERWISE SPECIFIED
TOLERANCES ARE:
DECIMAL TWO-PLACE: ±.01"
DIMENSIONS: THREE PLACE: ±.001"
ANGULAR DIMENSIONS: ±.08°
QUANTITY: 15 REQUIRED
ORDER NO: 5 B 1072
DOVETAILED SLIDE
BP-17
MATERIAL:
GRAY CAST IRON
SCALE: 1/2" = 1"
Ø .48, .500 REAM
⌴ Ø .88, ↧.38
ALL ROUNDED
CORNERS HAVE R.25
4.50
7.00
1.25
2.00
.50
.56
3.12
1.62
9.50
2.25
2.75
60°
60°
.625
1.12
2.26
4.50
A
B
C
D
E
F
G
H
J
L
M
N
O
P
Q
R
S
T
V
X
Y

ASSIGNMENT—UNIT 17

DOVETAILED SLIDE (BP-17)

1. What type of lines are Ⓑ, Ⓒ, Ⓗ, Ⓔ and Ⓙ?
2. What tolerance is allowed on
   (a) Decimal dimensions?
   (b) Angular dimensions?
3. What is the minimum overall height of the Dovetailed Slide?
4. Give the upper limit dimension for the 60° angle.
5. Give the upper and lower limit dimensions for Ⓟ.
6. What is the maximum depth to which the counterbored hole can be bored?
7. What line in the top view represents surface Ⓡ of the side view?
8. What line in the front view represents surface Ⓛ?
9. What line in the side view represents surface Ⓐ of the top view?
10. What dimension indicates how far line Ⓙ is from the base of the slide?
11. What two lines in the top view indicate the opening of the dovetail?
12. How wide is the opening in the dovetail?
13. At what angle to the horizontal is the dovetail cut?
14. Give dimension Ⓨ.
15. To what depth into the piece is the dovetail cut?
16. What is the vertical distance from surface Ⓠ to surface Ⓣ?
17. What is the upper limit dimension between surfaces Ⓕ and Ⓖ?
18. What is the full depth of the tee slot?
19. Compute dimensions Ⓥ and Ⓧ.
20. What is the horizontal distance from line Ⓝ to line Ⓢ?
21. What classification of tolerances applies to all dimensions?
22. Change the decimal and angular tolerances so that only the (−) tolerances apply. Then, determine the upper and lower limit dimensions for:
    (a) The angular dimension
    (b) The distance between surfaces Ⓕ and Ⓖ.
    (c) Distance Ⓟ.

Student's Name ______________

1. ______________
2. (a) ______________
   (b) ______________
   (c) ______________
3. ______________
4. ______________
5. ______________
   ____
6. ______________
7. ______________
8. ______________
9. ______________
10. ______________
11. ______________
12. ______________
13. ______________
14. ______________
15. ______________
16. ______________
17. ______________
18. ______________
19. ______________
    ____
20. ______________
21. ______________
22. (a) ______________
    (b) ______________
    (c) ______________

Unit 18

REPRESENTING AND DIMENSIONING EXTERNAL SCREW THREADS

Screw threads are used widely (1) to fasten two or more parts securely in position, (2) to transmit power such as a feed screw on a machine, and (3) to move a scale on an instrument used for precision measurements.

PROFILE OF UNIFIED AND THE AMERICAN NATIONAL THREADS

The shape or profile of the thread is referred to as the *thread form*. One of the most common thread forms resembles a **V.** Threads with an included angle of 60° originally were called Sharp **V.** Later, these threads were called U.S. Standard, American National, and Unified threads. Changes and improvements in the thread forms have been made over the years through standards which have been established by professional organizations. These include the National Screw Thread Commission, the American Standards Association, the U.S. Standards Institute and (currently) the American National Standards Institute (ANSI). The newer Unified (**UN**) thread form and sizes are designed to correct production problems related to tolerances and classes of fit. The *Unified Thread System* was agreed upon by Canada, the United States, and the United Kingdom.

The most commonly used thread forms are the *Unified* and the *American National.* **UN** threads are mechanically interchangeable with the former American National threads of the same size and pitch. The symbol **UN** is found on drawings to designate the Unified form, and **N** designates the American National form. The only difference between these two forms is in the shape of the top (crest) and the bottom (root) of the thread. Both the crest and the root of the American National thread form are flat. In contrast, the crest and root of the Unified thread form may be either flat or rounded. The root is always rounded on **UNR** external threads. The characteristics of the basic profile for American National and Unified Screw Threads are shown in figure 18-1.

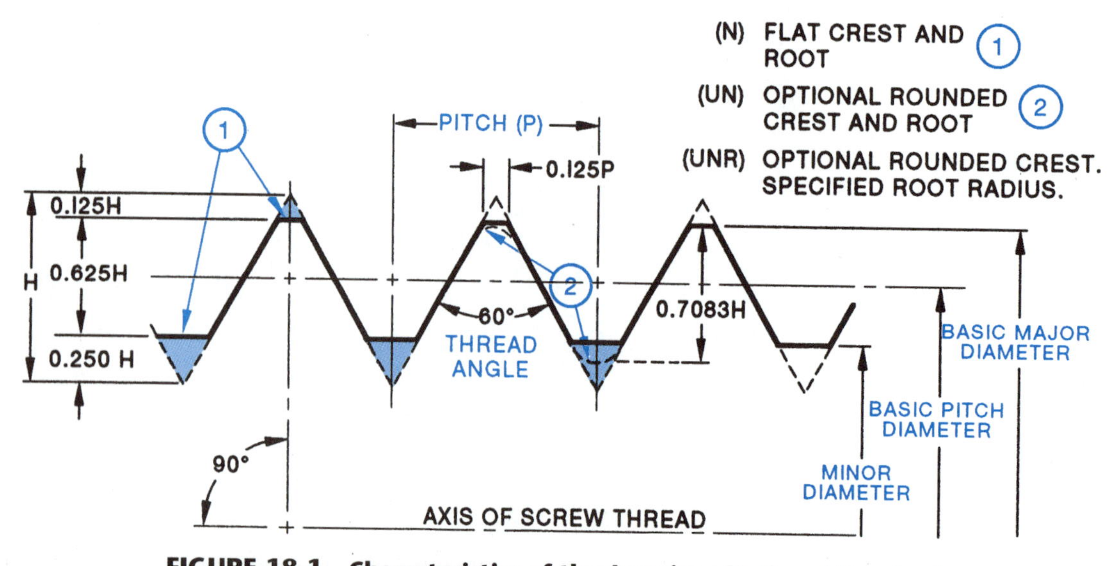

FIGURE 18-1 Characteristics of the American National Standard profile for Unified Screw Threads (N, UN, and UNR).

AMERICAN NATIONAL STANDARD AND UNIFIED SCREW THREADS

There are five common thread forms which use the 60° included angle. These forms are represented in figure 18-2 as (A) the Sharp **V**, (B) Unified, **UN**, (C) American Standard, **N**, (D) 60° Stub, and (E) ANSI Straight, **NPS,** and Taper, **NPT**, Pipe Threads. While flat crests and roots are shown for the pipe threads, some rounding off occurs in manufacturing.

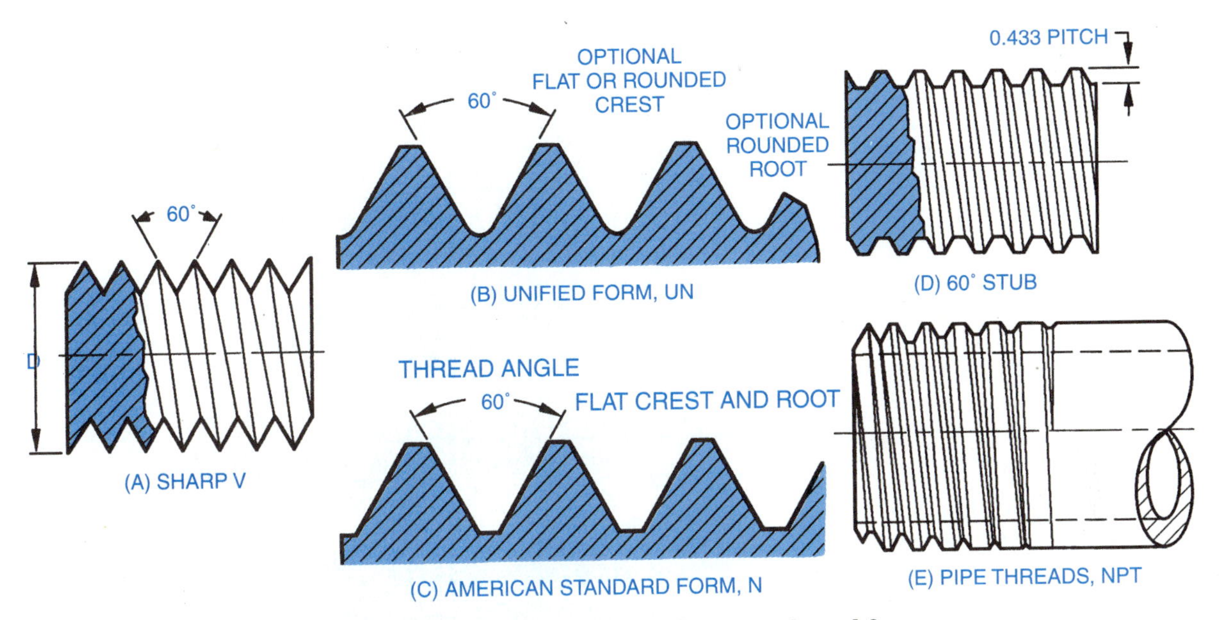

FIGURE 18-2 Common 60° angle screw thread forms.

OTHER COMMON THREAD FORMS

A square form of thread, or a modified form, is used when great power or force is needed. The square form shown in figure 18-3A is called a *Square thread.* The more popular modified thread form has an included angle of 29° and is known as the *Acme thread,* figure 18-3B. The *Buttress thread,* figure 18-3C, is used where power is to be transmitted in one direction. The load-resisting flank may be inclined from 1° to 5°.

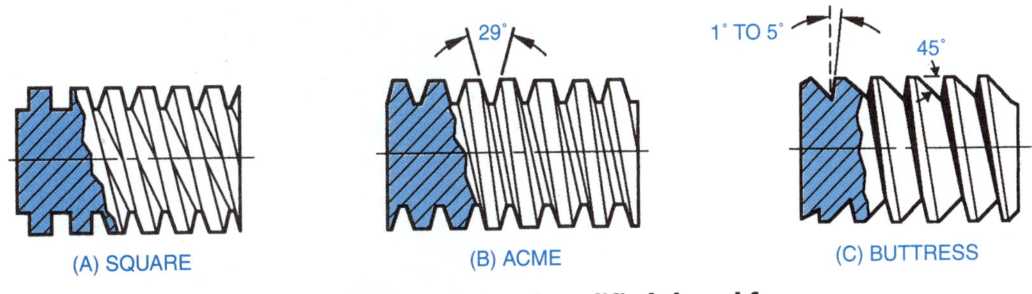

FIGURE 18-3 Square and modified thread forms.

THREAD SERIES

Each thread form has a standard number of threads per inch for a given diameter. The thread series designates the fineness or coarseness of the threads. For example, there are five basic series of 60° thread form screw threads, as follows.

① The *coarse thread series* is designated as **UNC** for Unified Coarse; **UNR** for Unified Coarse having a special rounded root but optional rounded crest; and **NC** for the American National Coarse Series threads.

② The *fine thread series* is similarly identified by the symbols **UNF, UNRF** for external threads, and **NF**.

③ *Extra fine series threads* are designated by the symbols **UNEF**, **UNREF** (for external threads only), and **NEF.**

④ *Nonstandard* or *special threads* are designated by the symbol **UNS** for Unified Special and **NS** for American National Standard Special Series threads.

⑤ A *constant pitch series* provides a fixed number of threads (pitch) for a range of diameters. The **8**, **12**, and **16** constant pitch series are preferred. Threads in each series are designated as **UN**, **UNR** (for external threads), and **N**. These letters (depending on the thread form) are preceded by the number of threads in the constant pitch series.

Figure 18-4 provides a summary of the five thread series and the designations as they appear on drawings.

	Designations on Drawings		
Thread Series	**Unified System**		**American Standard National System**
Coarse	∠1 UNC	∠2 UNRC	NC
Fine	UNF	UNRF	NF
Extra Fine	UNEF	UNREF	NEF
Special	UNS	UNRS	NS
Constant Pitch	∠3 * UN	* UNR	* N

∠1 Optional rounded crest and/or root
∠2 Specified root radius; optional rounded crest
∠3 * Pitch series number precedes the designation

FIGURE 18-4 Summary of thread series designations.

REPRESENTING EXTERNAL SCREW THREADS

Screw threads are further classified into two basic types: (1) *external threads* which are produced on the outside of a part, and (2) *internal threads* which are cut on the inside of the part. Letter symbols are sometimes used to designate the type: **A** for external threads and **B** for internal threads.

Internal and external threads may be represented on mechanical drawings by one of several methods: (1) a *pictorial* representation which shows the threads as they appear to the eye, (2) a *schematic* representation, or (3) a *simplified* representation. Figure 18-5 shows how each type of external thread can be represented.

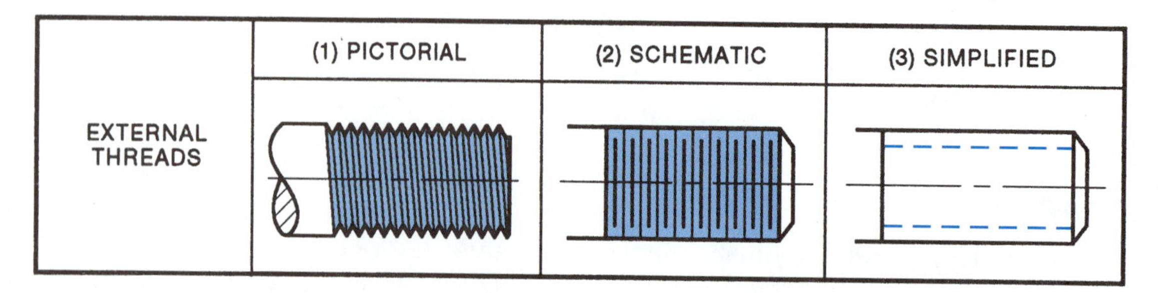

FIGURE 18-5 Representation of external screw threads (ANSI).

DIMENSIONING EXTERNAL SCREW THREADS

The representation of each thread is accompanied by a series of dimensions, letters, and numbers which, when combined, give full specifications for cutting and measuring the threads. The standard practices recommended by the American National Standards Institute (ANSI) for specifying and dimensioning external screw threads are shown in figure 18-6.

While the specifications as noted in figure 18-6 give all the information needed to describe a screw thread, the seven items are not always used in the notation on a drawing. For example, the thread class ⑤ may be covered by a general note which applies to all threaded parts. Such a note may appear elsewhere on the drawing. In other cases, the length of thread ⑦ sometimes appears as a dimension instead of appearing in the thread notation.

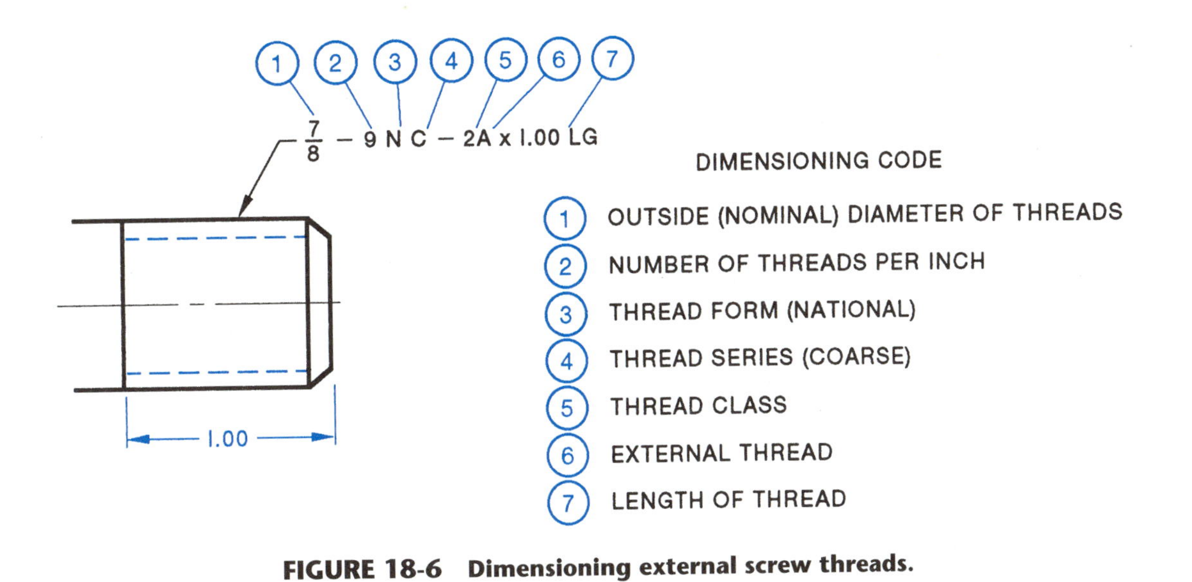

FIGURE 18-6 Dimensioning external screw threads.

PIPE THREAD REPRESENTATION

In addition to the screw threads just covered, there are three other common types of ANSI threads that are used on pipes. One type is the *Straight pipe thread*. This type of thread is noted on a drawing by the diameter of the thread followed by **NPS**. Thus, the notation, **1**" - **NPS**, indicates that (a) the part contains the standard number of threads in the pipe series for a **1**" diameter, and (b) the threads are straight.

In the second type of pipe thread, the threads are cut along a standard taper. The tapered pipe threads are used where a tight, leakproof seal is needed between the parts that are joined. Such threads are called *National Pipe Taper* threads, designated by **NPT**. They are specified on pictorial, schematic, or simplified drawings by the diameter and the type. The notation, **1**" - **NPT**, means (a) that there are the standard number of threads on the part for the **1**" diameter in the pipe thread series, and (b) both the internal and external threads are cut on the standard pipe taper.

A third type of pipe thread is used on drawings of parts for extremely high pressure applications. *Dryseal pressure-tight threads* are identified by the addition of the letter **F** after the **T** (taper) or **S** (straight) thread form designations. For example, ***NPTF*** designates threads in the *Dryseal taper series;* ***NPSF***, threads in the *Dryseal straight pipe thread series.*

While a number of different forms and thread series have been described, only the commonly used American National Standard (ANSI), Unified System, and International Standards Organization (SI) Metric Thread System are applied in this basic text.

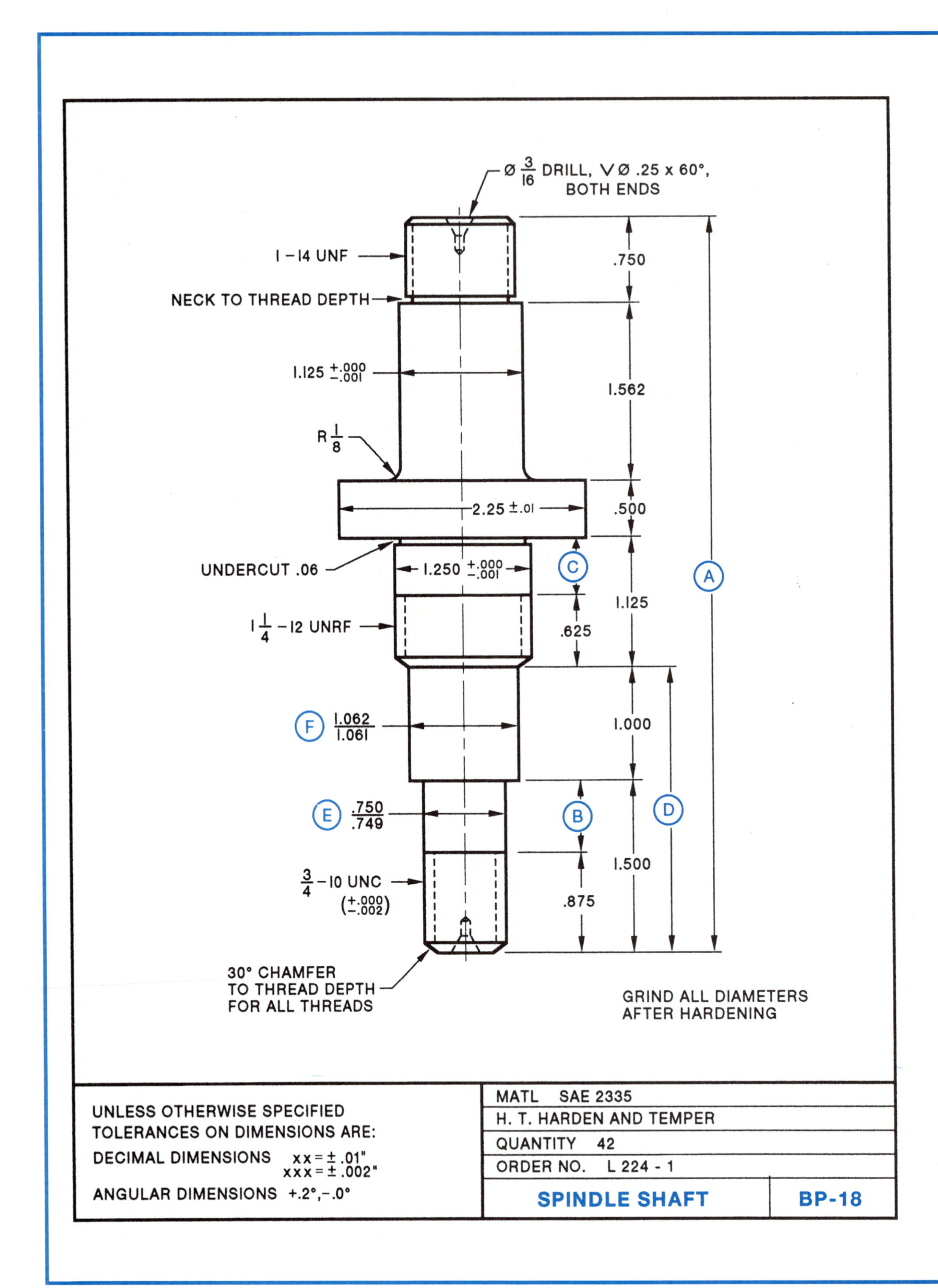

Ø 3/16 DRILL, V Ø .25 x 60°, BOTH ENDS
1 - 14 UNF
NECK TO THREAD DEPTH
1.125 +.000 -.001
R 1/8
2.25 ±.01
UNDERCUT .06
1.250 +.000 -.001
1 1/4 - 12 UNRF
F 1.062 / 1.061
E .750 / .749
3/4 - 10 UNC (+.000 -.002)
30° CHAMFER TO THREAD DEPTH FOR ALL THREADS
.750
1.562
.500
C
A
1.125
.625
1.000
B
D
1.500
.875
GRIND ALL DIAMETERS AFTER HARDENING
UNLESS OTHERWISE SPECIFIED TOLERANCES ON DIMENSIONS ARE:
DECIMAL DIMENSIONS xx = ± .01" xxx = ± .002"
ANGULAR DIMENSIONS +.2°, -.0°
MATL SAE 2335
H. T. HARDEN AND TEMPER
QUANTITY 42
ORDER NO. L 224 - 1
SPINDLE SHAFT
BP-18

SPINDLE SHAFT (BP-18)

1. What material is used for the part?
2. Give the basic overall length Ⓐ of the shaft.
3. What system of representation is used for the threaded portions?
4. At how many places are threads cut?
5. Start at the bottom of the part and give all the thread diameters.
6. Name the three thread series that the letters **UNC**, **UNRF**, and **NC** specify.
7. How many threads per inch are to be cut on the 3/4", 1 1/4" and 1" diameters?
8. Give dimensions Ⓑ Ⓒ and Ⓓ.
9. Give the upper and lower limit diameter for the 3/4" threaded portion.
10. What is the length of the 3/4" threads? The 1 1/4" threads?
11. Give the upper and lower limit dimensions of the 1 1/16" diameter portion.
12. What angle are the chamfers at the starting end of each thread cut?
13. What tolerance is specified for angular dimensions?
14. What is the upper and lower limit of size on the 1 1/8" diameter?
15. How long is that part of the Shaft which has the 1 1/4" - 12 thread?
16. Give the angle of the countersink.
17. What is the largest diameter that the 2.25" diameter can be machined to?
18. Give the diameter to which the center holes are countersunk.
19. How deep is the undercut on the 1 1/4 diameter?
20. Name the final machining operation for all diameters after the part is heat treated.
21. Classify the tolerances for ground diameters Ⓔ and Ⓕ.

ASSIGNMENT—UNIT 18

Student's Name ________________

1. ____________
2. ____________
3. ____________

   ____
4. ____________
5. ____________

   ____

   ____
6. UNC ________

   ____

   UNRF ______

   ____

   NC ________

   ____
7. 3/4" ________

   1 1/4" ______

   1" ________
8. Ⓑ ________

   Ⓒ ________

   Ⓓ ________
9. Upper ______

   Lower ______
10. 3/4" ______

    1 1/4" ______
11. Upper ______

    Lower ______
12. ____________
13. ____________
14. Upper ______

    Lower ______
15. ____________
16. ____________
17. ____________
18. ____________
19. ____________
20. ____________
21. ____________

    ____

Unit 19

REPRESENTING AND SPECIFYING INTERNAL AND LEFT-HAND THREADS

REPRESENTATION OF INTERNAL THREADS

One of the most practical and widely used methods of producing internal threads is to cut them with a round, formed thread-cutting tool called a *tap*. The tap is inserted in a hole which is the same diameter as the root diameter of the thread to be produced. As the tap with its multiple cutting edges is turned, it advances into the hole to cut a thread. The thread is a specified size and shape to correspond with a mating threaded part. This process is called *tapping* and the hole which is threaded is called a *tapped hole*. When the hole is a drilled hole, the drill is called a *tap drill*.

A part may be threaded internally throughout its entire length or to a specified depth, either by tapping or another machining process. Some threads are bottomed at the same depth as the tap drill. In other cases, the tap drill hole may be deeper. Pictorial, schematic, and simplified forms of representing such threads are used in figure 19-1. Section (A) illustrates through threads; Section (B) illustrates threads in a blind hole.

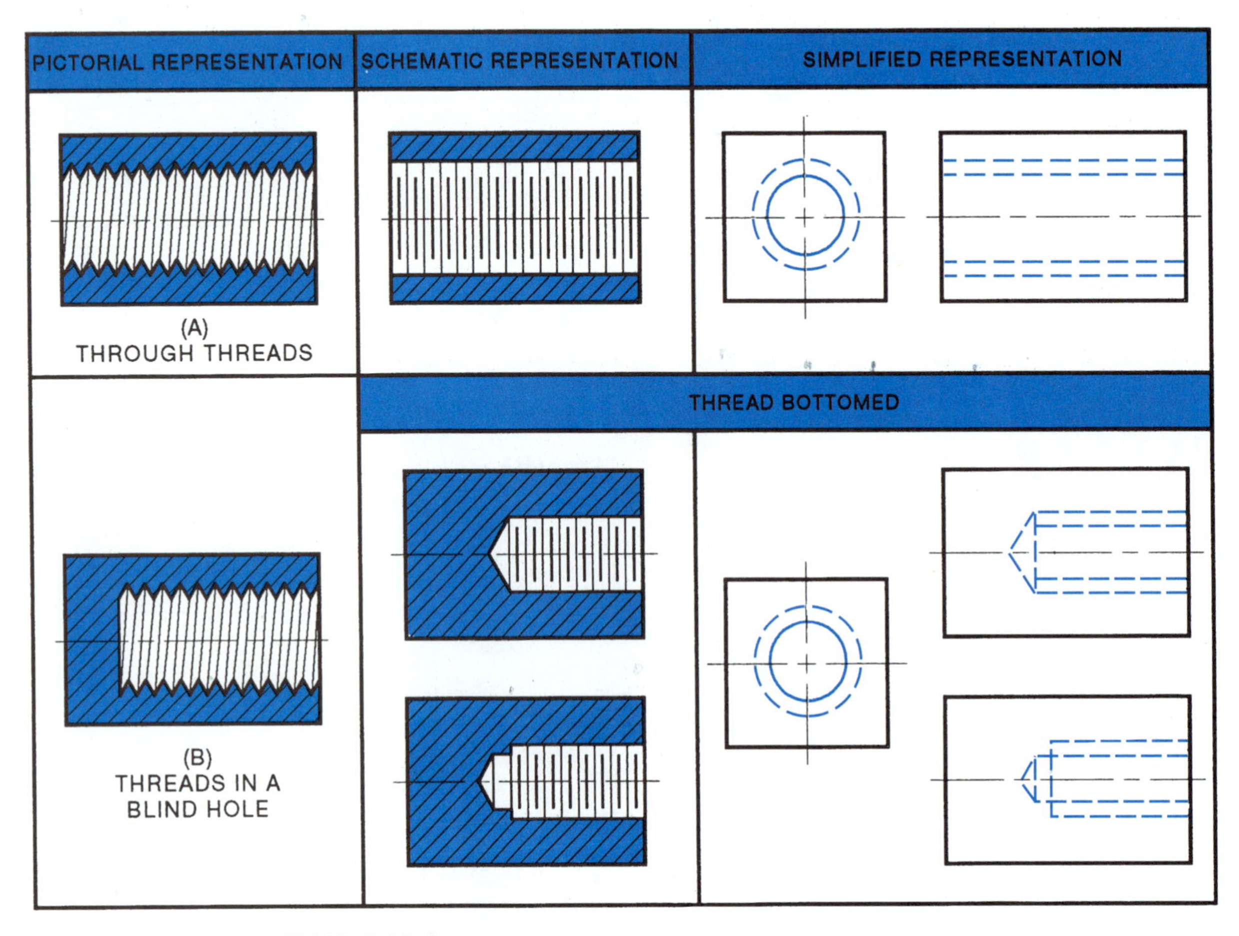

FIGURE 19-1 Representation of internal threaded holes.

DIMENSIONING A THREADED HOLE

The same system of dimensioning that is used to specify external threads applies to internal threads. However, in the case of internal threads, the thread length is specified as the *depth of thread*. The recommended dimensioning of a tapped hole is illustrated in figure 19-2.

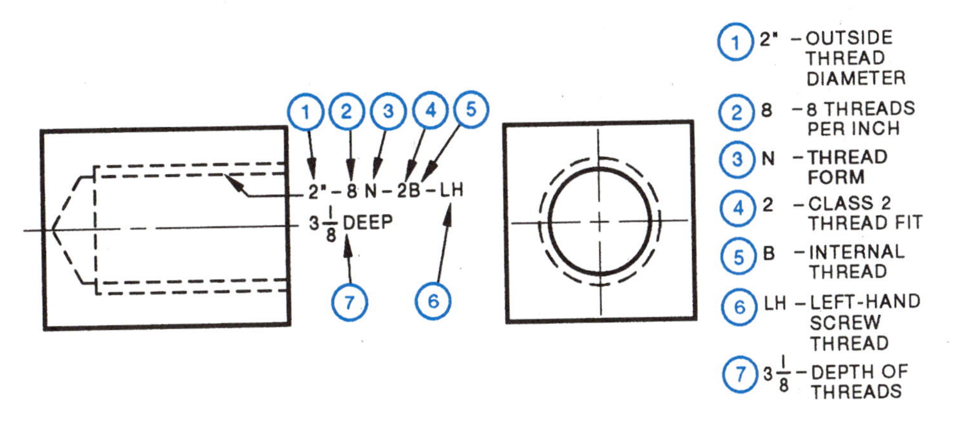

FIGURE 19-2 Dimensioning a tapped hole.

THREAD CLASS: UNIFIED AND AMERICAN NATIONAL FORM THREADS

The accuracy between an external and an internal thread is specified in trade handbooks and manufacturer's tables as a *thread class*. In the old classes of thread fits, there were equal allowances (tolerances) on the pitch diameters of both the external and internal threads. Today, the new standards require a larger pitch diameter tolerance on the internal thread.

There are three regular classes of fits for external and internal threads. Each of the three classes is distinguished from one another by the amount of clearance between the sides, crests, and roots of the internal and external threads.

The variation is from the loose *Class 1* fit to *Class 3* which requires a high degree of accuracy. Fits that are Class 1A (where A designates external threads) and Class 1B (B for internal threads) provide the greatest allowance and tolerances between the fitted parts. The Class 2A and Class 2B fit is the most widely used as it provides an acceptable minimum of clearance between the mating parts. The Class 3A and Class 3B fit is used for great accuracy. When the class of fit is given, it is usually included on the drawing as part of the screw thread notation, figure 19-2. The current thread classes 1, 2, and 3 replace the former American National Standard 1, 2, and 3 classes of fit.

SPECIFYING LEFT-HAND THREADS

Screw threads are cut either right-hand or left-hand depending on the application, figure 19-3. As the terms imply, a right-hand thread is advanced by turning clockwise ↻, or to the right; the left-hand thread is advanced by turning counterclockwise ↺, or to the left.

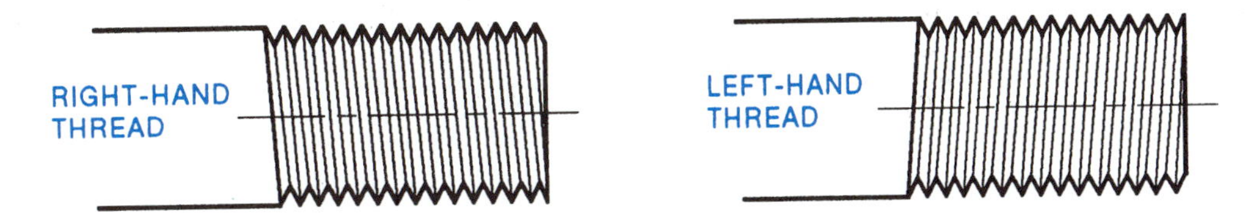

FIGURE 19-3 Examples of right- and left-hand threads.

No notation is made on a drawing for right-hand threads, as this direction of thread is assumed unless otherwise noted. However, the letters **LH** are included on the drawing specifications to indicate when a left-hand thread is required as shown in figure 19-2 by the **LH** location at ⑥.

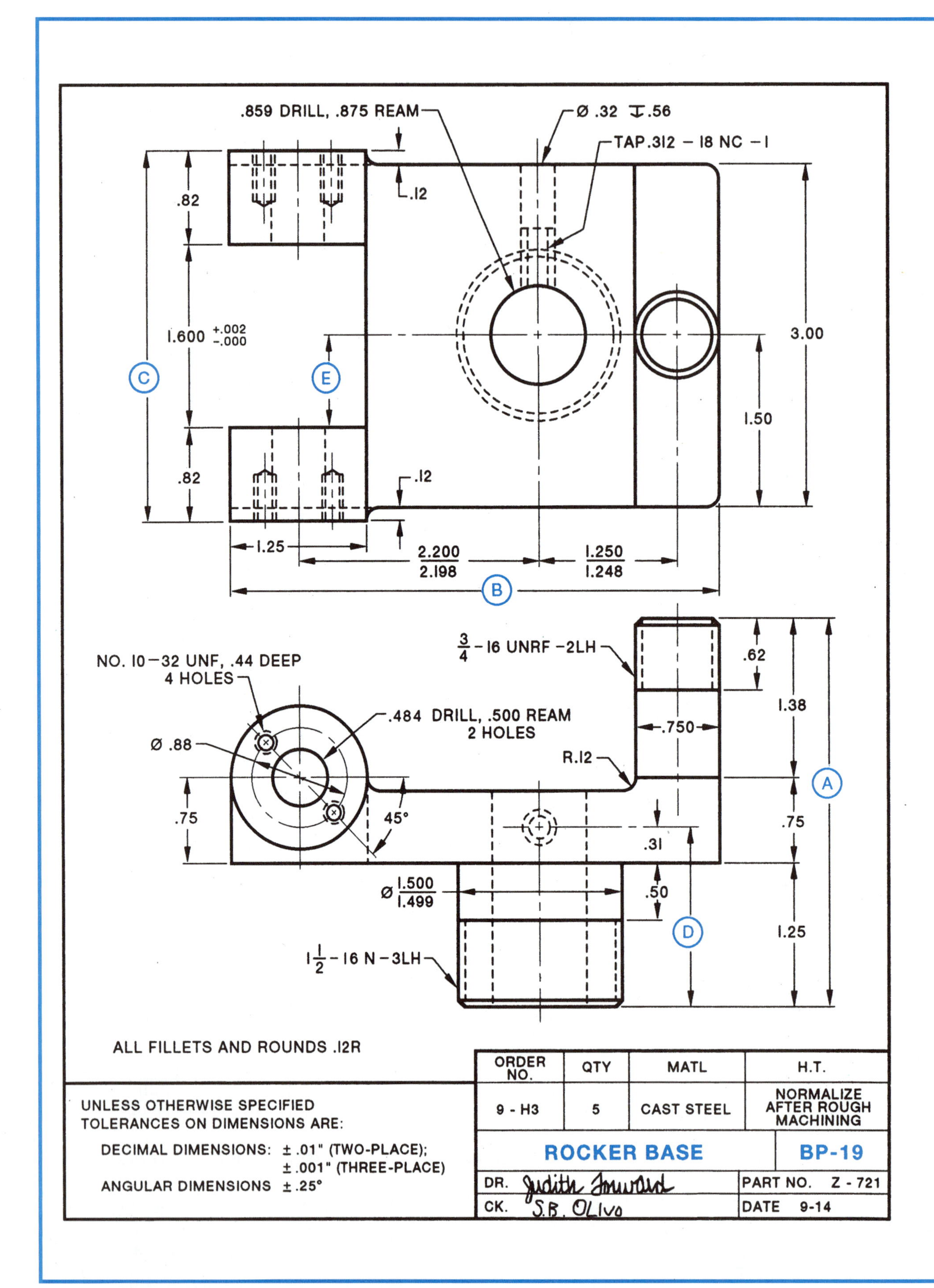
.859 DRILL, .875 REAM
Ø .32 ⊥.56
TAP .312 – 18 NC – 1
.12
.82
1.600 +.002 -.000
C
E
3.00
1.50
.82
.12
1.25
2.200
2.198
1.250
1.248
B
3/4 – 16 UNRF – 2LH
.62
NO. 10 – 32 UNF, .44 DEEP
4 HOLES
1.38
.484 DRILL, .500 REAM
2 HOLES
.750
Ø .88
R.12
A
.75
45°
.75
.31
Ø 1.500
1.499
.50
D
1.25
1 1/2 – 16 N – 3LH
ALL FILLETS AND ROUNDS .12R
ORDER NO.
QTY
MATL
H.T.
9 - H3
5
CAST STEEL
NORMALIZE AFTER ROUGH MACHINING
UNLESS OTHERWISE SPECIFIED TOLERANCES ON DIMENSIONS ARE:
DECIMAL DIMENSIONS: ± .01" (TWO-PLACE); ± .001" (THREE-PLACE)
ANGULAR DIMENSIONS ± .25°
ROCKER BASE
BP-19
DR. Judith Forward
CK. S.B. Olivo
PART NO. Z - 721
DATE 9-14

ASSIGNMENT—UNIT 19

ROCKER BASE (BP-19)

1. What material is specified?
2. What tolerances are allowed on:
   (a) Decimal dimensions?
   (b) Fractional dimensions?
   (c) Angular dimensions?
3. What heat-treating process is required after rough machining?
4. Determine overall nominal dimensions Ⓐ Ⓑ Ⓒ.
5. What is the radius of all fillets and rounds?
6. Compute dimensions Ⓓ and Ⓔ.
7. How many holes are to be reamed .500"?
8. How many external threads are to be cut?
9. What thread series is used for the internal and external threads?
10. What does **3/4 - I6UNRF - 2LH** mean?
11. What are the upper and lower limits of the .750″ diameter portion?
12. What does **11/2 - 16N - 3LH** mean?
13. What does **.312 - 18NC - I** mean?
14. What is the lower limit diameter of the unthreaded 1 1/2″ diameter portion?
15. Determine the length of the **1 1/2"** - **16** threaded portion.
16. What diameter drill is used for the **.875**" reamed hole?
17. How many holes are to be threaded **10 - 32UNF**?
18. How deep are the holes to be threaded?
19. Give the angle to the horizontal at which the **10 - 32UNF** holes are to be drilled.
20. Give the diameter of the circle on which the **10 - 32UNF** tapped holes are located.
21. Identify the system of dimensioning used on the drawing of the Rocker Base.
22. Change the decimal and angular tolerances so they apply in one direction (+) only. Then, determine the upper and lower limit dimensions for Ⓐ Ⓑ Ⓒ Ⓓ Ⓔ and the 45° angle.

Student's Name ____________

1. ________
2. (a) ________
   (b) ________
   (c) ________
3. ________
4. Ⓐ ________
   Ⓑ ________
   Ⓒ ________
5. ________
6. Ⓓ ________
   Ⓔ ________
7. ________
8. ________
9. ________
10. 3/4 ________
    16 ________
    UNRF ________
    2 ________
    LH ________
    ____
11. Upper ________
    Lower ________
12. 1 1/2 ________
    16 ________
    N ________
    3 ________
    LH ________
13. .312 ________
    18 ________
    NC ________
    1 ________
14. ________
15. ________
16. ________
17. ________
18. ________
19. ________
20. ________
21. ________
22. Ⓐ ________ Ⓓ ________
    Ⓑ ________ Ⓔ ________
    Ⓒ ________ 45°∠ ________

Unit 20

DIMENSIONING TAPERS AND MACHINED SURFACES

Blueprints of parts that uniformly change in size along their length show that the part is *tapered*. On a round piece of work, the taper is the difference between the diameter at one point and a diameter farther along the length. The drawing may give the large and small diameters at the beginning and end of the taper or one diameter and the length of the taper, figure 20-1. The taper is usually specified on drawings where dimensions are given in inches by a note which gives the **TAPER PER FOOT** or the **TAPER PER INCH**. For example, a taper of one-half inch per foot is expressed:

.50 TAPER PER FOOT

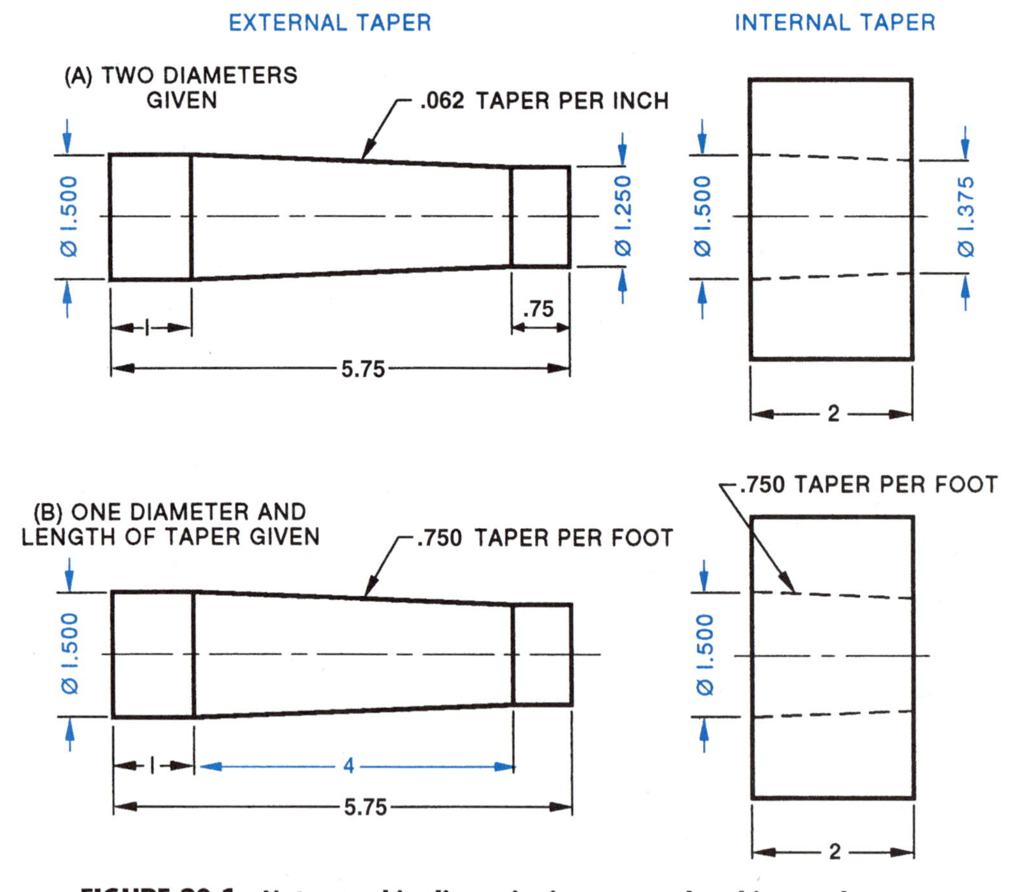

FIGURE 20-1 Notes used in dimensioning external and internal tapers.

The newest ANSI standard for dimensioning *conical* (*cone-shaped*) tapered surfaces recommends the use of a ratio followed by the taper symbol (-⊳). The ratio represents the difference in diameter in relation to the length of the taper. For example, a part that is dimensioned in customary units (**CIU**) of **.25**" in **5.00**" has a **CIU** taper ratio per unit (inch) length of **.25:5.00** and is shown on a drawing as -⊳ **0.2:1**. Similarly, a conical part with a metric diameter difference of **20 mm** and length of **100 mm** has a *taper ratio* (**TR**) of **20.0:100.0** or **1:5** -⊳ (METRIC).

CALCULATING OUTSIDE AND INSIDE TAPER MEASUREMENTS

To find the total taper (T_t) *when two diameters of a taper are given,* subtract the diameter of the small end (d) from the diameter of the large end (D) of the taper. $T_t = D - d$. Applied to the external taper in figure 20-1(A), T_t = 1.500" − 1.250" = .250".

To find the total taper (T_t) *when one diameter, the taper per foot* (T_{pf}), *and the length of taper* (ℓ) *are given,* follow two steps.

STEP 1 Divide the taper per foot by twelve to get the taper per inch.

STEP 2 Multiply the taper per inch by the length of the taper in inches.

In the case of the internal taper, figure 20-1(B), the taper per inch (T_{pi}) = .750 ÷ 12 = .0625". The total taper (T_t) = .0625" × 2" = .125".

FINISHED SURFACES

Surfaces that are machined to a smooth finish and to accurate dimensions are called *finished surfaces.* These surfaces were identified by the former 60° V ANSI symbol, the √ symbol which was accepted by the Canadian Standards Association (CSA) and the ANSI. Presently the symbol √ is used to indicate that material is to be removed by machining the designated surface.

The bottom of the finish symbol touches the surface to be machined. Figure 20-2 shows a steel forging. This same forging as it looks when machined is illustrated in figure 20-3. The working drawing, figure 20-4, shows how the finish marks are placed. Note that these marks may appear on the hidden edge lines as well as on the object lines,

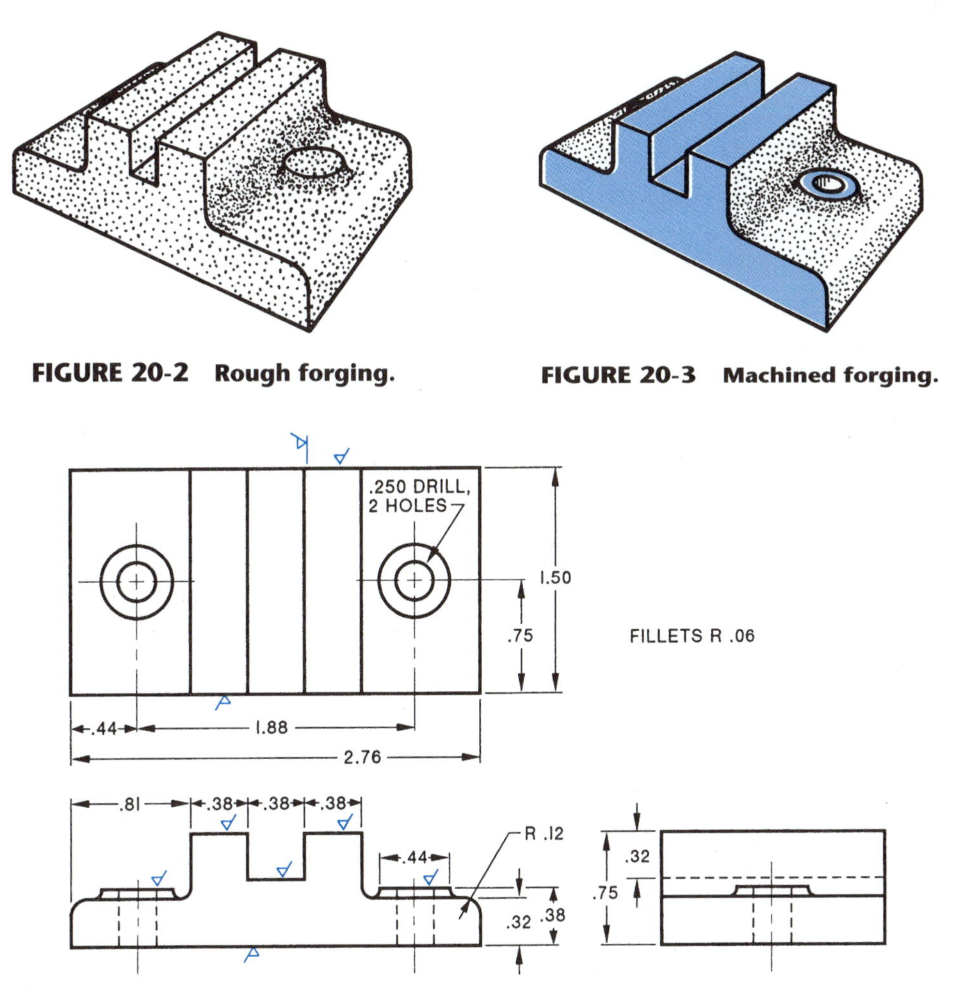

FIGURE 20-2 Rough forging.

FIGURE 20-3 Machined forging.

FIGURE 20-4 Working drawing of forging showing application of finish marks.

with the exception of the small drilled holes. On large holes that require accurate machining, the ✓ is used once again. Extension lines and/or leaders are also used with the finish symbol to identify surfaces which are to be machined.

FINISH ALL OVER

When a casting, forging, or welded part is to be finished all over, the drawing is simplified by omitting the finish symbols and adding the note: **FINISH ALL OVER.** The abbreviation **FAO** may be used in place of the phrase, finish all over.

ANSI SYMBOLS FOR SURFACE FINISH PROCESSES

Four basic ANSI symbols for denoting the process by which a surface may be finished are shown in figure 20-5, column (a). A brief description of the process which each symbol identifies on a drawing is given in column (b).

ANSI Surface Finish Symbol (a)	General Processes Identified on Drawings (b)
√	The surface may be produced by any manufacturing process.
√ (with circle)	The surface is to be produced by a nonmachining process such as casting, forging, welding, rolling, powder metallurgy, etc. There is to be no subsequent machining of the surface.
√ (with bar)	Material removal from the surface by machining is required.
.05″ √ or 1.5mm √	The addition of a decimal or metric value indicates the amount of stock to be removed by machining.

FIGURE 20-5 Basic ANSI surface finish symbols.

SURFACE TEXTURE MEASUREMENT TERMS AND SYMBOLS

Surface texture, surface finish, surface roughness, and *surface characteristics* are generally used interchangeably in the shop and laboratory. However, application of these terms to drawings must follow precise ANSI standards. Surface texture symbols and values provide specific standards against which finished parts may be accurately produced, uniformly inspected, and measured. Common surface texture characteristics are identified in figure 20-6, and a description of basic terms follows.

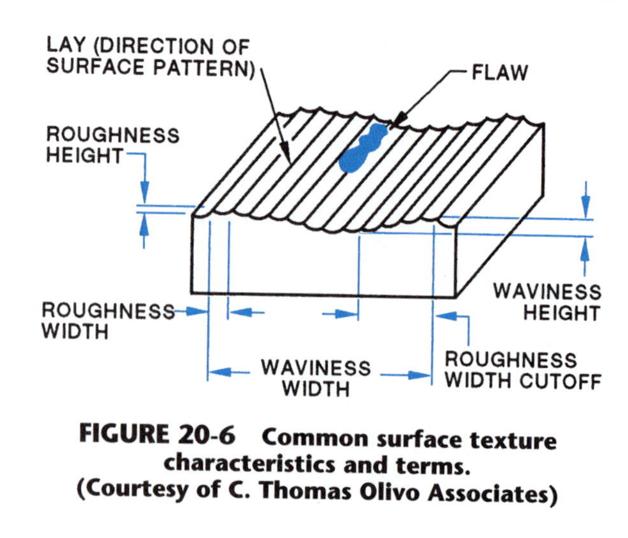

FIGURE 20-6 Common surface texture characteristics and terms. (Courtesy of C. Thomas Olivo Associates)

Surface texture. As applied to drawings, surface texture denotes repetitive or random deviations from a nominal (specified) surface. These variations form the *pattern* of the surface.

Surface finish. The symbol √‾‾‾ and a surface roughness measurement value are used to indicate the roughness height in microinches (μin) or micrometers (μm). For instance, if the roughness height limit for a machined surface is 125 microinches (125 μin), this measurement is shown on a drawing as ${}^{125}\surd$‾‾‾ .

Microinch or micrometer value. The microinch (μin) and micrometer (μm) are the base units of surface roughness measurement. The prefix micro indicates a millionth part of an inch or meter, respectively.

Lay. Every machining and superfinishing process produces lines that form six basic patterns. The patterns are identified on drawings using *lay symbols.* Symbols that give the lay direction are: (⊥) perpendicular, (II) parallel, (X) angular, (M) multidirectional, (C) circular, and (R) radial.

Waviness. A series of waves in a surface may be produced by machining, stresses or defects in a material, heat treating, handling, or other processing. *Waviness* is measured by height (in thousandths of an inch) and width (in other decimal parts of an inch). *Waviness width* refers to the measurement between successive peaks or valleys.

Flaws. Defects in a surface such as scratches, burrs, casting, forging, machining, or other imperfections are referred to as flaws.

REPRESENTATION AND INTERPRETATION OF SURFACE TEXTURE MEASUREMENTS

The ANSI surface texture system requires a series of measurement values to be used with the symbol. The values specify the degree of accuracy to which surface irregularities (*roughness*) are to be measured. The pattern of marks produced by machining (*lay*), the spacing of the irregularities (*waviness*), and scratches and other surface imperfections (*flaws*) are represented by still other symbols and measurement values. The placement of the surface texture roughness, and waviness symbols and values, is shown graphically in figure 20-7.

Figure 20-8(A) illustrates the way lay symbols and surface texture specifications should appear on a drawing. Interpretation of the lay direction and each surface texture measurement is given at (B).

A. PLACEMENT OF MAXIMUM ROUGHNESS WIDTH RATING TO RIGHT OF LAY SYMBOL — LAY SYMBOL — ⊥ 0.020

B. PLACEMENT OF ROUGHNESS WIDTH CUTOFF — 0.100

C. PLACEMENT OF MAXIMUM VALUE ROUGHNESS HEIGHT — 63

D. PLACEMENT OF PERMISSIBLE RANGE OF ROUGHNESS HEIGHT — 63 50

E. PLACEMENT OF MAXIMUM WAVINESS HEIGHT AND WIDTH VALUE — HEIGHT — 0.0002-1.0 — WIDTH

F. PLACEMENT OF MINIMUM VALUE FOR CONTACT OR BEARING AREA — 95%

FIGURE 20-7 Placement of measurements and specifications in relation to a surface texture symbol. (Courtesy of C. Thomas Olivo Associates)

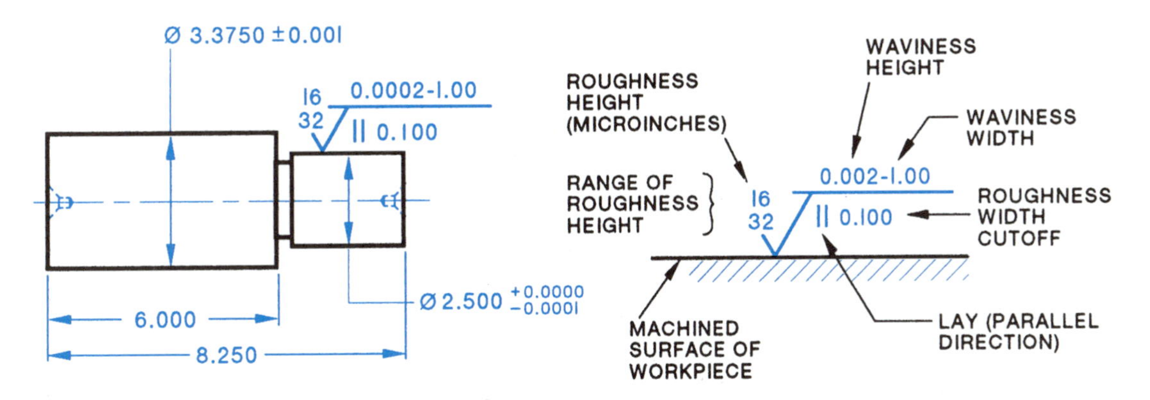

FIGURE 20-8 Representation and interpretation of surface texture symbols and measurement values. (Courtesy of C. Thomas Olivo Associates)

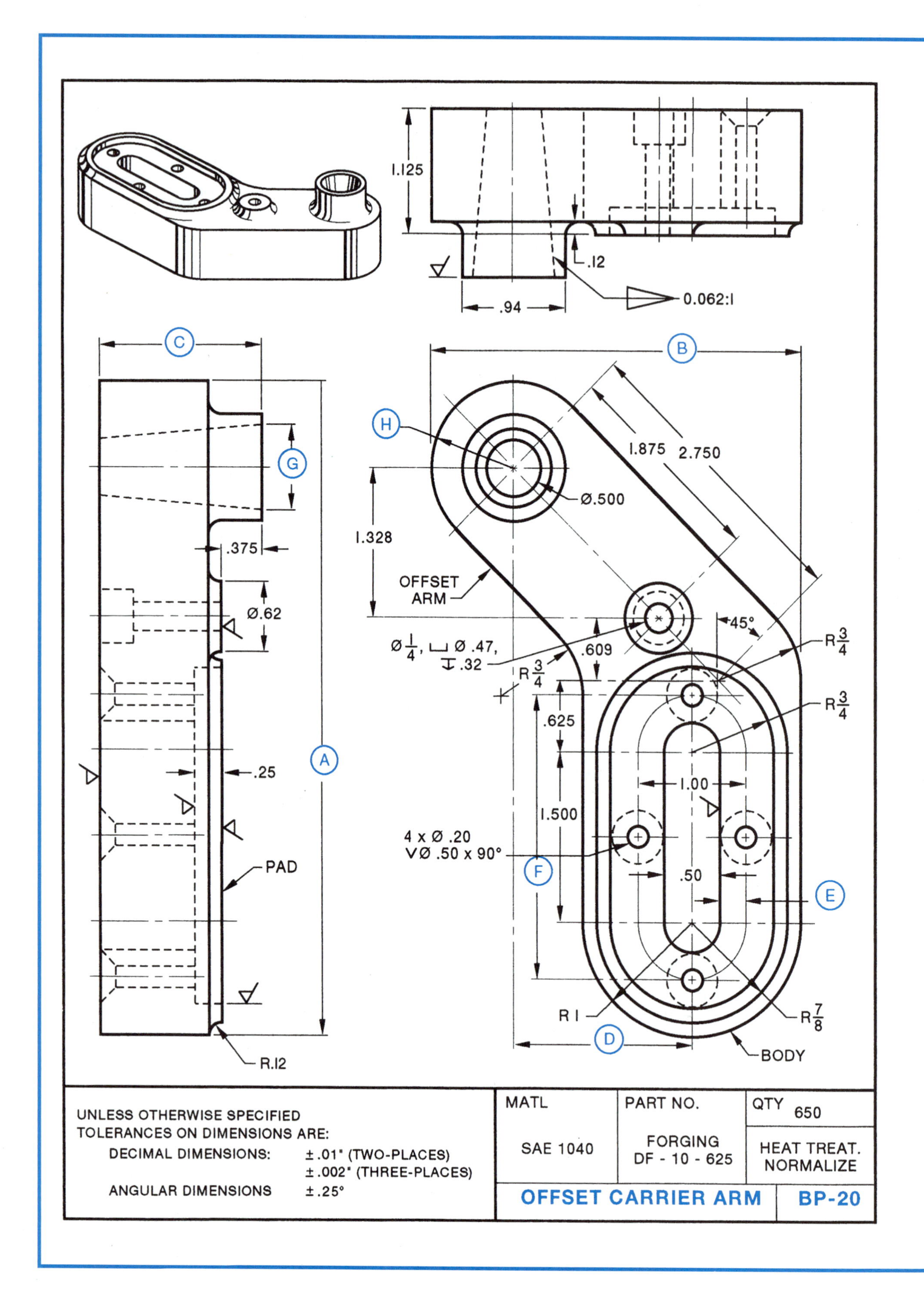

1.125
.12
.94
0.062:1
C
B
G
H
1.875
2.750
Ø.500
.375
1.328
OFFSET ARM
Ø.62
45°
Ø 1/4, ⌴ Ø .47, ↧ .32
.609
R 3/4
R 3/4
R 3/4
.625
.25
A
1.00
1.500
4 x Ø .20
VØ .50 x 90°
F
.50
E
PAD
R 1
R 7/8
D
BODY
R.12
UNLESS OTHERWISE SPECIFIED
TOLERANCES ON DIMENSIONS ARE:
DECIMAL DIMENSIONS: ±.01" (TWO-PLACES)
±.002" (THREE-PLACES)
ANGULAR DIMENSIONS ±.25°
MATL
SAE 1040
PART NO.
FORGING
DF - 10 - 625
QTY 650
HEAT TREAT.
NORMALIZE
OFFSET CARRIER ARM
BP-20

ASSIGNMENT—UNIT 20

OFFSET CARRIER ARM (BP-20)

1. Name the three views.
2. What is the part number?
3. How many surfaces are to be machined?
4. Give the angle at which the arm is offset from the body.
5. What is the center-to-center distance between the holes in the offset arm?
6. Find the length of the elongated slot.
7. What is the overall length of the pad?
8. How is the counterbored hole specified?
9. What diameter drill is used for the countersunk holes?
10. Give the angle of the countersunk holes.
11. Compute the nominal diameter at the large end of the tapered hole (G).
12. What tolerance is allowed on:
    (a) Decimal dimensions,
    (b) Angular dimensions,
13. What is the basic overall length (A)?
14. Find the basic overall width (B).
15. Find the maximum overall thickness (C).
16. What are dimensions (D) and (H)?
17. If the elongated slot is machined .48 would it be over, under, or within the specified limits?
18. What is the distance (E)?
19. Give center-to-center distance (F).
20. What heat treatment is required?
21. What classification of tolerances applies to all dimensions?
22. Indicate the meaning of the √ symbol on BP-20.
23. Give (a) the dimensions and (b) state surface texture characteristics for (I), (J), (K), and (L).

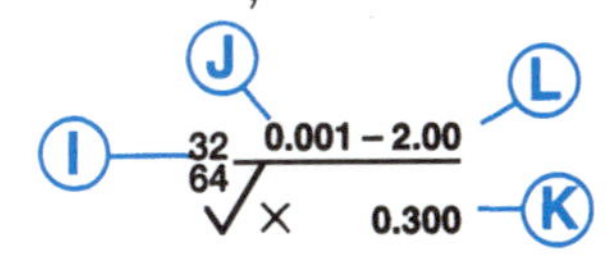

24. Tell what the lay symbol × means.

Student's Name ____________

1. ____________
____
____
2. ____________
3. ____________
4. ____________
5. ____________
6. ____________
7. ____________
8. ____________
____
9. ____________
10. ____________
11. (G) = ____________
12. (a) ____________
(b) ____________
13. (A) = ____________
14. (B) = ____________
15. (C) = ____________
16. (D) = ____________
(H) = ____________
17. ____________
18. (E) = ____________
19. (F) = ____________
20. ____________
21. ____________
22. ____________
____
____
23. Dimension (a) Characteristic (b)
(I) ____________
(J) ____________
(K) ____________
(L) ____________
24. ____________
____
____

Unit 21

DIMENSIONING WITH SHOP NOTES

The draftsperson often resorts to the use of notes on a drawing to convey to the mechanic all the information needed to make a part. Notes such as those used for drilling, reaming, counterboring, or countersinking holes are added to ordinary dimensions.

A note may consist of a very brief statement at the end of a leader, or it may be a complete sentence which gives an adequate picture of machining processes and all necessary dimensions. A note is found on a drawing near the part to which it refers. This unit includes the types of machining notes that are found on drawings of knurled surfaces, chamfers, grooves, and keyways. A sample of a typical change note is also given.

DIMENSIONING KNURLED SURFACES

The term *knurl* refers to a raised diamond-shaped surface or straight line impression in the surface of a part, figure 21-1A. The dimensions and notes which furnish sufficient information for the technician to produce the knurled part are also given, figure 21-1B and figure 21-1C. The pitch of the knurl, which gives the number of teeth per linear inch, is the size. The standard pitches are: coarse (14P), medium (21P), and fine (33P).

Newer ANSI standards for knurls and knurling use the letter P to precede the pitch. Where parts are to be press fit, the minimum acceptable (toleranced) diameter is indicated on the drawing. A knurled surface is represented by diamond or straight line pattern symbols or by using the designation of the knurl as straight or diamond. Common methods of representing knurls are shown in figure 21-1B and figure 21-1C. A partial pattern is shown where the symbol clarifies a drawing.

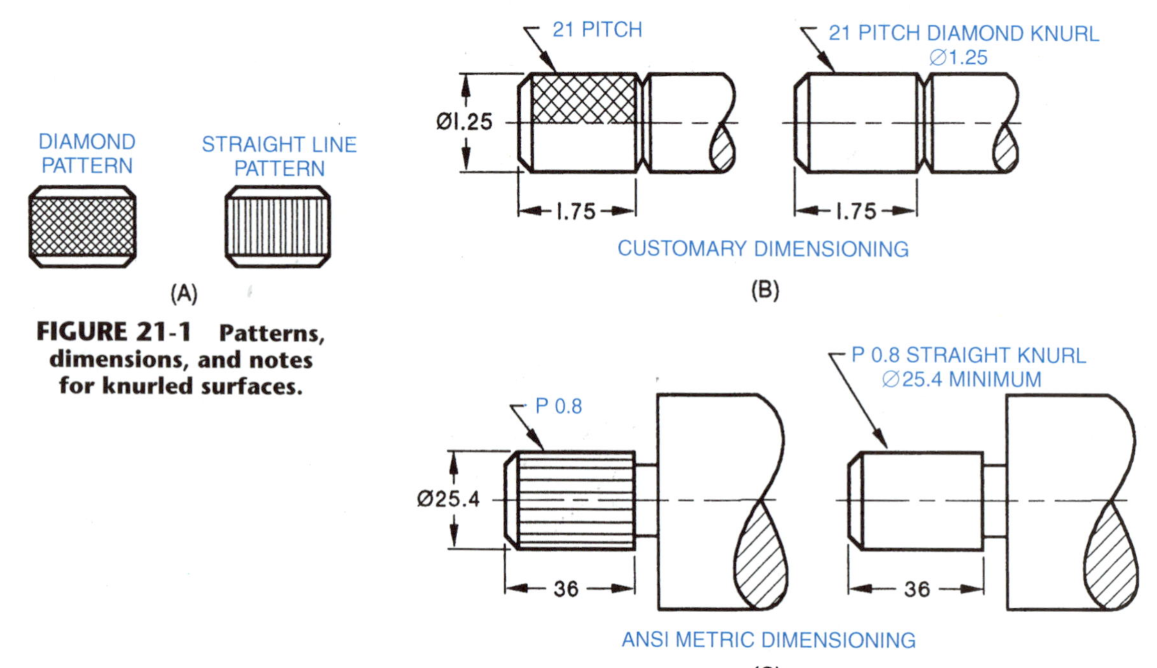

FIGURE 21-1 Patterns, dimensions, and notes for knurled surfaces.

The pitch of a knurl is usually stated on drawings in terms of the number of teeth as shown in figure 21-1B or fractional part per millimeter, figure 21-1C. The pitch is followed by the type of knurl.

DIMENSIONING CHAMFERS AND GROOVES

When a surface must be cut away at a slight bevel, or have a groove cut into it, the drawing or blueprint gives complete machining information in the form illustrated in figure 21-2 and figure 21-3. The angle is dimensioned as an included angle when it relates to a design requirement.

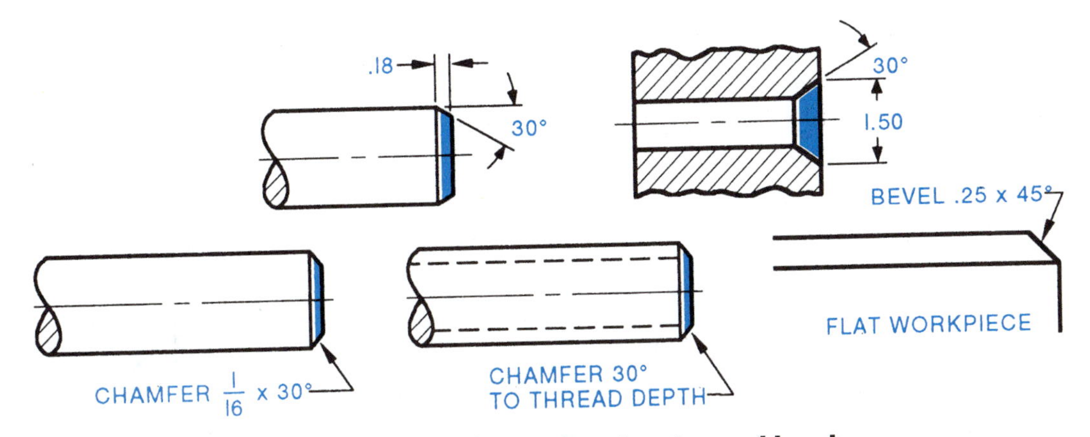

FIGURE 21-2 Dimensioning chamfers and bevels.

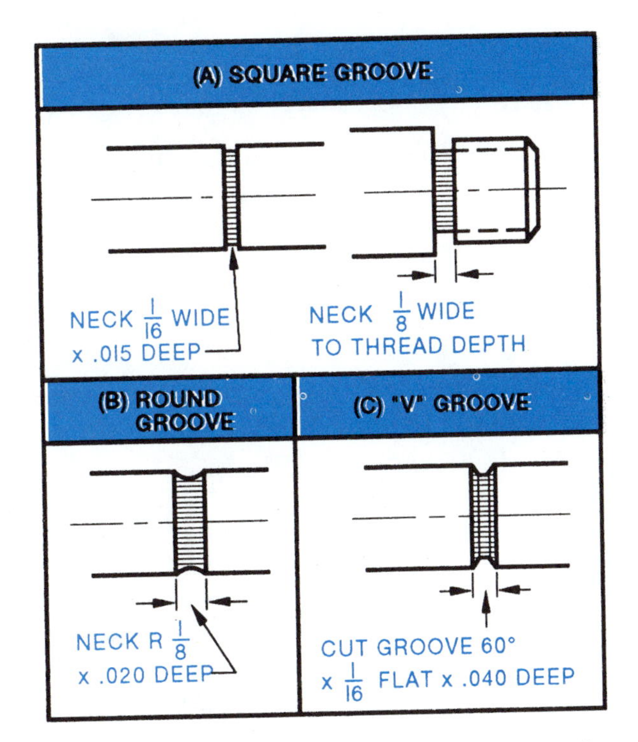

FIGURE 21-3 Dimensioning three common types of grooves.

The process of producing a groove is often referred to as *undercutting* or *necking*. The simplified ANSI method of dimensioning a round or square (plain) undercut is to give the width followed by the diameter. For example, a note such as .12 × ∅ 1.25 indicates a groove width of .12" and a 1.25" diameter at the bottom of the undercut. If a round groove is required, the radius is one-half the width dimension.

KEYS, KEYWAYS, AND KEYSEATS

Parts that are assembled and disassembled in a particular position to transmit motion from one member to another are *keyed* together. This means a square or flat piece of steel is partly seated in a shaft or groove called a *keyseat*. The key extends an equal distance into a corresponding groove (*keyway*) in a second member (*hub*). Another general type of key is semicircular in shape. It fits into a round machined groove in the shaft. This type is known as a *Woodruff key*.

Dimensions and specifications as they appear on drawings for square and flat keys, keyseats, and keyways are shown in figure 21-4. Drawings often include a height dimension which gives the measurement from the circumference to the depth to which the keyway and keyseat are machined.

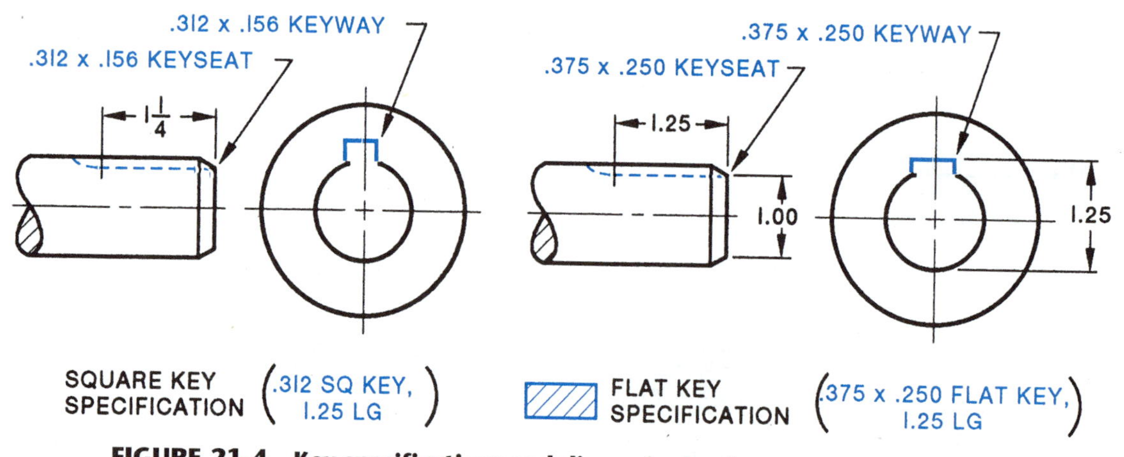

FIGURE 21-4 Key specifications and dimensioning keyseats and keyways.

CHANGE NOTES

The specifications and dimensions of parts are frequently changed on working drawings. An accurate record is usually made on the tracing and blueprint to indicate the nature of the change, the date, and who made the changes.

Changes that are minor in nature may be made without altering the original lines on the drawing. One of the easiest ways of making such changes is shown in figure 21-5.

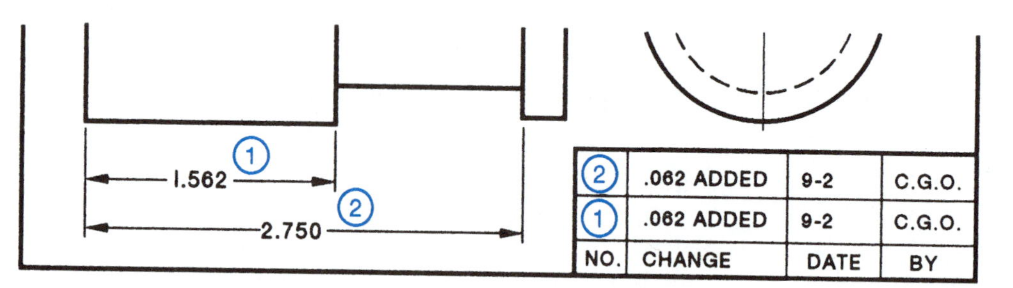

FIGURE 21-5 Application of change notes.

DESIGN SPECIFICATIONS AND CLASSES OF FITS

Drawings of interchangeable parts contain design data that deal with *fits*. The term *fit* denotes a relationship between two mating parts.* A *fit* is produced by an allowance for clearance, interference, and either clearance or interference in what is known as a *transition fit*. A drawing dimensioned for a fit provides information about the range of tightness between two mating parts. When the parts move in relation to each other, the dimensions must provide for a *positive clearance*. The amount of positive clearance depends on the kind of material in the parts, the nature of the motion, lubrication, temperature, and other forces and factors. Machined parts with a positive clearance are dimensioned for a *clearance fit*. A *negative clearance* is required with parts that are to be forced together as a single unit. Parts requiring a negative clearance are dimensioned for an *interference fit*.

TERMS USED WITH ALLOWANCES AND TOLERANCES FOR DIFFERENT CLASSES OF FITS

ANSI and other standards for size limits and dimensioning requirements for Classes of Fits are provided in Handbook Tables. General purpose fits include: *forced and driving fits, running and sliding fits,* and *locational fits*. The following is a summary of common terms used to classify and dimension fits:

Basic Size. This dimension provides a theoretically exact size, form, or position of a surface, point, or feature. A basic dimension is often identified on a drawing as a dimension within a rectangular box as, for example, **1.750** or **45.25°** .

Nominal Size. This term designates a particular dimension, unit, or detail without reference to specific limits of accuracy.

Actual Size. The true measurement of a part after it has been produced is its actual size.

Limit Dimensions. Limit Dimensions relate to acceptable maximum and minimum sizes (dimensions).

Tolerance. The total amount a part can vary from the basic size and still be useable. (An interpretation of how these dimensions are applied to drawings follows.)

NOMINAL SIZE **2.500**

BASIC SIZE **2.500**

BASIC SIZE WITH TOLERANCE ADDED **2.500 ± 0.001**
LIMIT DIMENSION (Acceptable Maximum and Minimum Sizes)
LARGEST SIZE **2.501"** = UPPER LIMIT
SMALLEST SIZE **2.499"** = LOWER LIMIT

TOLERANCE = UPPER LIMIT − LOWER LIMIT

UPPER LIMIT	**2.501**
LOWER LIMIT	**2.499**
TOLERANCE =	**+ .002**

*An *allowance* is a prescribed, intentional difference in the dimensions of mating parts.

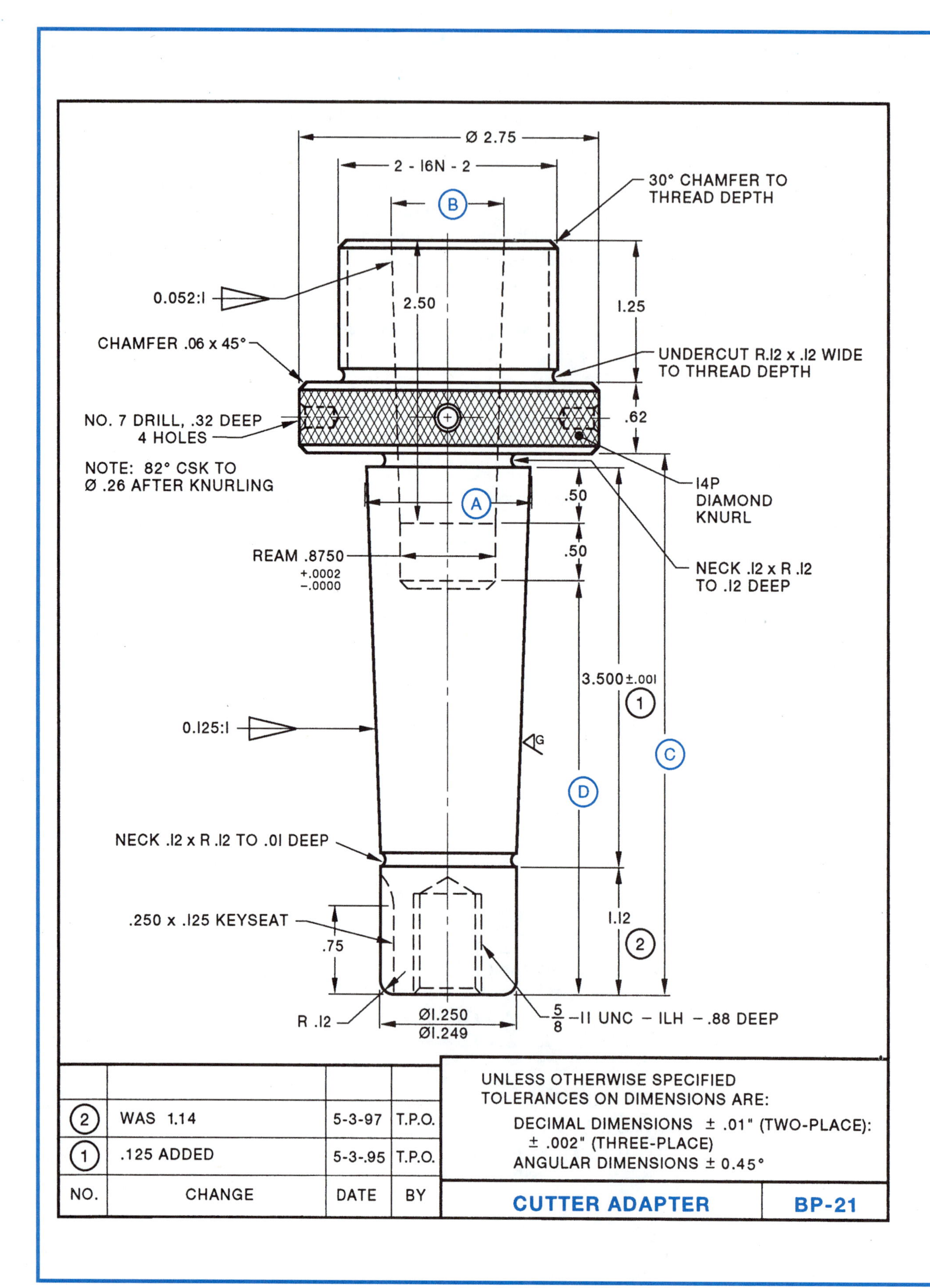

NO.	CHANGE	DATE	BY
2	WAS 1.14	5-3-97	T.P.O.
1	.125 ADDED	5-3-95	T.P.O.

UNLESS OTHERWISE SPECIFIED TOLERANCES ON DIMENSIONS ARE:
DECIMAL DIMENSIONS ± .01" (TWO-PLACE): ± .002" (THREE-PLACE)
ANGULAR DIMENSIONS ± 0.45°

CUTTER ADAPTER | **BP-21**

ASSIGNMENT—UNIT 21

CUTTER ADAPTER (BP-21)

1. What tolerances are given for:
   (a) Angular dimensions?
   (b) Decimal dimensions?
2. Give the body taper specifications.
3. What is the taper per inch for the body?
4. What is the diameter at the small end of the outside taper?
5. Determine dimension Ⓐ.
6. Give the taper for the hole.
7. What is the taper per inch for the hole?
8. Give the diameter at the small end of the inside taper.
9. Compute dimension Ⓑ.
10. Give the outside thread specifications.
11. What does the LH in the thread note for the shank end specify?
12. What thread class is the 2" threaded nose?
13. What operation is performed on the 2.75" diameter?
14. How is this operation specified?
15. Determine the maximum diameter for the reamed hole.
16. Give the note for drilling the four holes in the largest diameter.
17. Specify the chamfer for the knurled area.
18. Specify the angle of chamfer for the 2" threaded nose.
19. How deep is the 2" threaded chamfer?
20. Give the original length of the tapered body before any change was made.
21. Compute maximum length of dimension Ⓒ.
22. What matching operation is indicated by the finish symbol G√ ?
23. Determine dimension Ⓓ.
24. Give the dimensions of the keyseat.
25. Determine the overall length of the adapter.
26. Name three different types of fits.
27. Describe briefly the purpose of (a) positive clearance and (b) negative clearance.
28. Give the limit dimensions for the reamed hole.

Student's Name ______________

1. (a) ______________
   (b) ______________
2. ______________
3. ______________
4. ______________
5. Ⓐ ______________
6. ______________
7. ______________
8. ______________
9. Ⓑ ______________
10. ______________
11. ______________
12. ______________
13. ______________
14. ______________
    ____
15. ______________
16. ______________
    ____
17. ______________
18. ______________
19. Deep ______________
20. ______________
21. Ⓒ ______________
22. ______________
23. Ⓓ ______________
24. Width ______________
    Depth ______________
25. ______________
26. (a) ______________
    (b) ______________
    (c) ______________
27. (a) ______________
    ____
    ____
    (b) ______________
    ____
    ____
28. ______________

SECTION 4 The SI Metric System

Unit 22

METRIC SYSTEM DIMENSIONING, DIAMETRAL DIMENSIONS, AND ISO SYMBOLS

The International System of Units (SI) was established by agreement among many nations to provide a logical interconnected framework for all measurements used in industry, science, and commerce. SI is a modernized accepted adaptation of many European metric systems. SI metrics relates to six basic units of measurement: (1) length, (2) time, (3) mass, (4) temperature, (5) electric current, and (6) luminous intensity. Multiples and submultiples of these basic units are expressed in decimals. All additional SI units are derived from the six basic units.

THE SI METRIC SYSTEM OF MEASUREMENT

Prefixes are used in the metric system to show how the dimension relates to a basic unit. For example, *deci-* means one-tenth of the basic unit of measure. Thus, 1 decimeter = 0.1 of a meter; 1 deciliter = 0.1 liter; and 1 decigram = 0.1 gram. Similarly, *centi-* = one hundredth; and *milli-* = one thousandth.

Dimensions larger than the basic unit of measure are expressed with the following prefixes: *deka-* = ten times greater; *hecto-* = one hundred times; and *kilo-* = 1000 times greater. To repeat, the most common unit of metric measure used on engineering drawings is the millimeter (mm). The next frequently used unit is the meter (m). Surface finishes are given in *micrometers* (millionths of a meter), using the symbol μm.

CONVERTING METRIC AND CUSTOMARY INCH SYSTEM DIMENSIONS

Trade and engineering handbooks contain conversion charts which simplify the process of determining in one system the equivalent value of a dimension given in the other system. Sections of a conversion table are illustrated in figure 22-1 to show customary inch units and equivalent metric units. Figure 22-1 also includes a drill series ranging from #80 to #1, letter size drills from (A) to (Z), and fractional dimensions from .001" to 1.000" . Note that millimeter equivalents, correct to four decimal places, are shown for both the drill sizes and the decimals which represent fractional parts of an inch.

The millimeter range of the table is from 0.0254 mm to 25.4000 mm, corresponding to 0.001" to 1.000", respectively. The millimeter equivalent of a measurement in the inch system may be found by multiplying the decimal value by 25.40.

ROUNDING OFF LINEAR DIMENSIONS

Many dimensions that are converted from inches to millimeters or from millimeters to inches are rounded off. With 1 mm = 0.03937" , conversions can be made to very high limits of accuracy. This is done by calculating the decimal value to a greater number of digits than is required within the range of tolerances.

Drill No. or Letter	Inch Standard		SI Metric Standard
	Fractional or Decimal Size		mm Size
		.001	0.0254
		.002	0.0508
		.003	0.0762
		.004	0.1016
		.005	0.1270
		.006	0.1524
		.007	0.1778
		.008	0.2032
		.009	0.2286
		.010	0.2540
		.011	0.2794
		.012	0.3048
80	.0135	.013	0.3302
79	.0145	.014	0.3556
		.015	0.3810
	1/64	.0156	0.3969
78		.016	0.4064
		.017	0.4318

Drill No. or Letter	Inch Standard		SI Metric Standard
	Fractional or Decimal Size		mm Size
		.401	10.1854
		.402	10.2108
		.403	10.2362
Y		.404	10.2616
		.405	10.2870
		.406	10.3124
	13/32	.4062	10.3187
		.407	10.3378
		.408	10.3632
		.409	10.3886
		.996	25.2984
		.997	25.3238
		.998	25.3492
		.999	25.3746
		1.000	25.4000

FIGURE 22-1 Portions of a conversion table showing drill sizes and millimeter equivalents to inch dimensions.

Production costs are related directly to the degree of accuracy required to produce a part. Therefore, the limits of dimensions should be rounded off to the least number of digits in the decimal dimension which will provide the greatest tolerance and ensure interchangeability.

A decimal dimension may be rounded off by increasing the last required digit by (1) if the digit which follows on the right is (5) or greater or by leaving the last digit unchanged if the digit to the right is less than (5).

For example,	1.5875"	rounded off to three decimal places	= 1.588"
	1.5874 mm	rounded off to three decimal places	= 1.587 mm
	1.646 mm	rounded off to two decimal places	= 1.65 mm
	1.644"	rounded off to two decimal places	= 1.64"

When dimensions on a drawing need to be divided, the part designer may round off the dimensions to the nearest even decimal. This practice permits parts to be economically machined to tolerances within a required number of decimal places.

DIAMETRAL DIMENSIONS

The symbol ∅ on a drawing indicates that the part has cylindrical sections. For example, the part shown in figure 22-2 may be fully described in one view by using the diametral dimension symbol ∅ for the cylindrical surfaces. The one-view drawing gives the dimensions of each of the three cylindrical surfaces and the bore diameter. The ∅ symbol may precede or follow the dimensional value.

PROJECTION SYMBOLS

The International Organization for Standardization (ISO) recommends the use of a projection symbol on drawings that are produced in one country for use among many countries, figure 22-3. The projection symbols are intended to promote the accurate exchange of technical information through drawings.

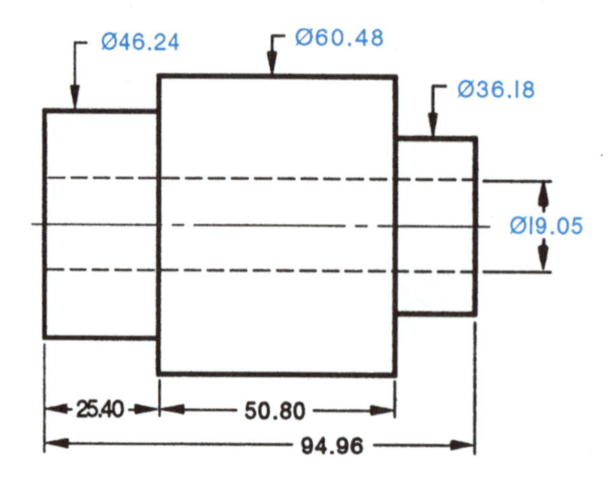

FIGURE 22-2 Use of diametral dimension symbol.

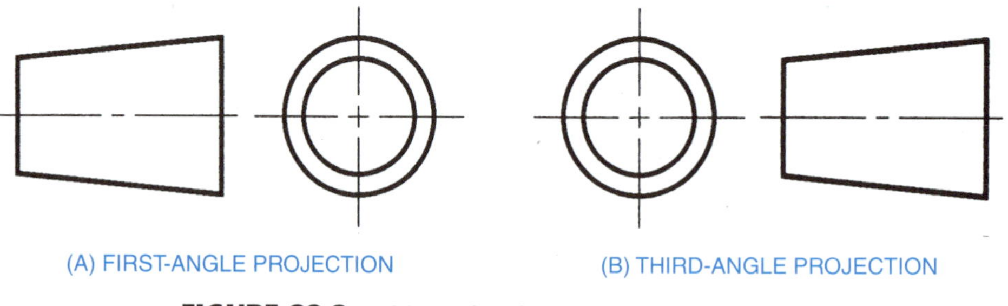

FIGURE 22-3 ISO projection symbols (enlarged).

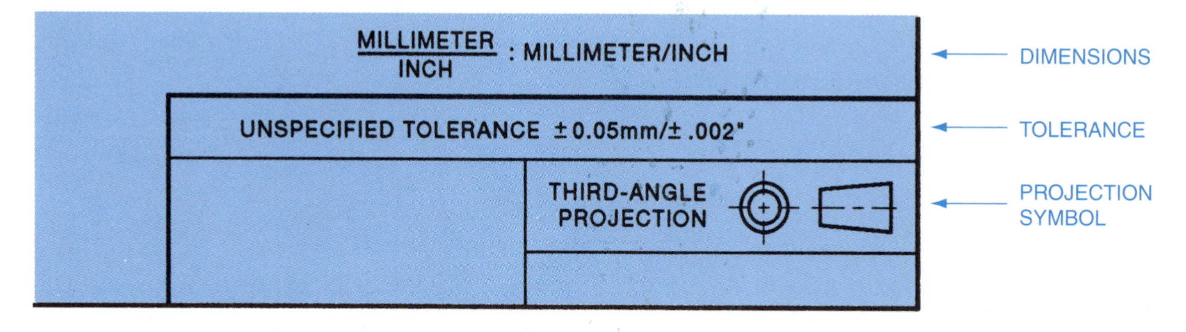

FIGURE 22-4 Dimension, tolerance, and symbol notes in a title block.

The United States and Canada use the third-angle system of projection for drawings. Other countries, however, use a different system which is known as the first-angle projection system. The purpose of introducing the ISO projection symbols is to indicate that there is a continuously increasing international exchange of drawings for the production of interchangeable parts. Thus, the symbol tells whether the drawing follows the third-angle or the first-angle projection system.

The ISO projection symbol, the notation on tolerances, and information on whether metric and/or inch dimensions are used on the drawing should appear as notes either within the title block or adjacent to it, figure 22-4. The designation **THIRD-ANGLE PROJECTION** is not always included with the symbol on a drawing.

LINEAR UNIT SYMBOLS ON DRAWINGS

When a limited number of customary inch dimensions appear on a millimeter-dimensioned drawing, the **IN** symbol is recommended, following the dimension. Similarly, when a few SI metric dimensions appear on an inch-dimensioned drawing, the **mm** symbol is placed after each millimeter value.

UNILATERAL AND BILATERAL MILLIMETER TOLERANCING

The practice for *unilateral millimeter tolerancing* on SI metric drawings (where either the + or − tolerance is zero) is to use a single zero without a plus (+) or minus (−) sign. For example if a positive (+) millimeter tolerance of 0.25 and a zero negative tolerance applies to a 50 mm dimension, the toleranced dimension on a SI metric drawing is $50^{+0.25}_{0}$. If a − 0.25 mm and zero positive tolerance is required, the toleranced dimension is given as $50^{0}_{-0.25}$.

A bilateral tolerance such as + 0.75 mm and − 0.1 mm, applied to a 125 mm dimension, appears on a drawing as $125^{+0.75}_{-0.10}$. Note that one or more zeros are added so both the plus (+) and minus (−) values have the same number of decimal places.

DIMENSIONING RADII FROM UNLOCATED CENTERS

In standard practice center, extension, and dimension lines are usually given on a drawing for locating and dimensioning a radius. Many times the center is not important since the location of a radius (arc) is controlled by other features or dimensions. Dimensioning in such cases is simplified by conveniently placing the radius dimensions as shown in figure 22-5.

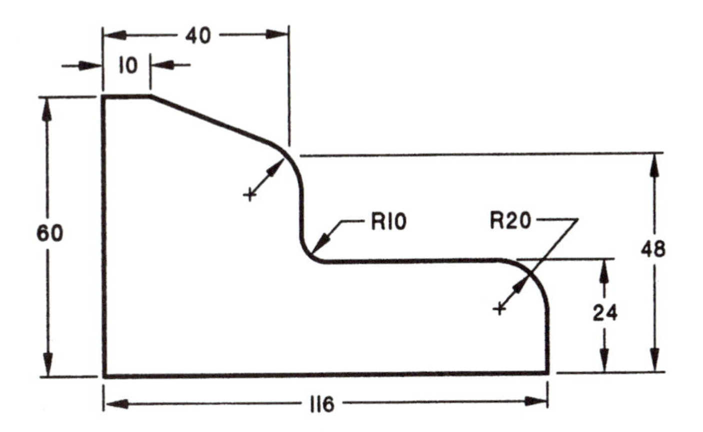

FIGURE 22-5 Placement of leaders and radii dimensions with unlocated centers.

DIMENSIONING CHORDS AND ARCS

Dimensioning of chords and arcs according to ANSI standards is illustrated in figure 22-6. The symbol ⌒ is used to indicate an arc.

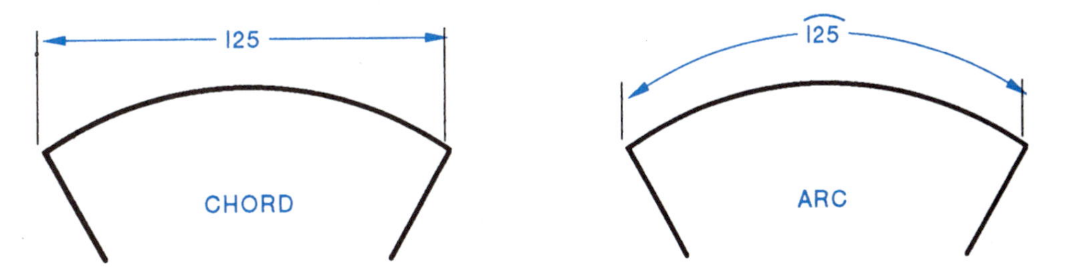

FIGURE 22-6 Representation of chord and arc dimensions.

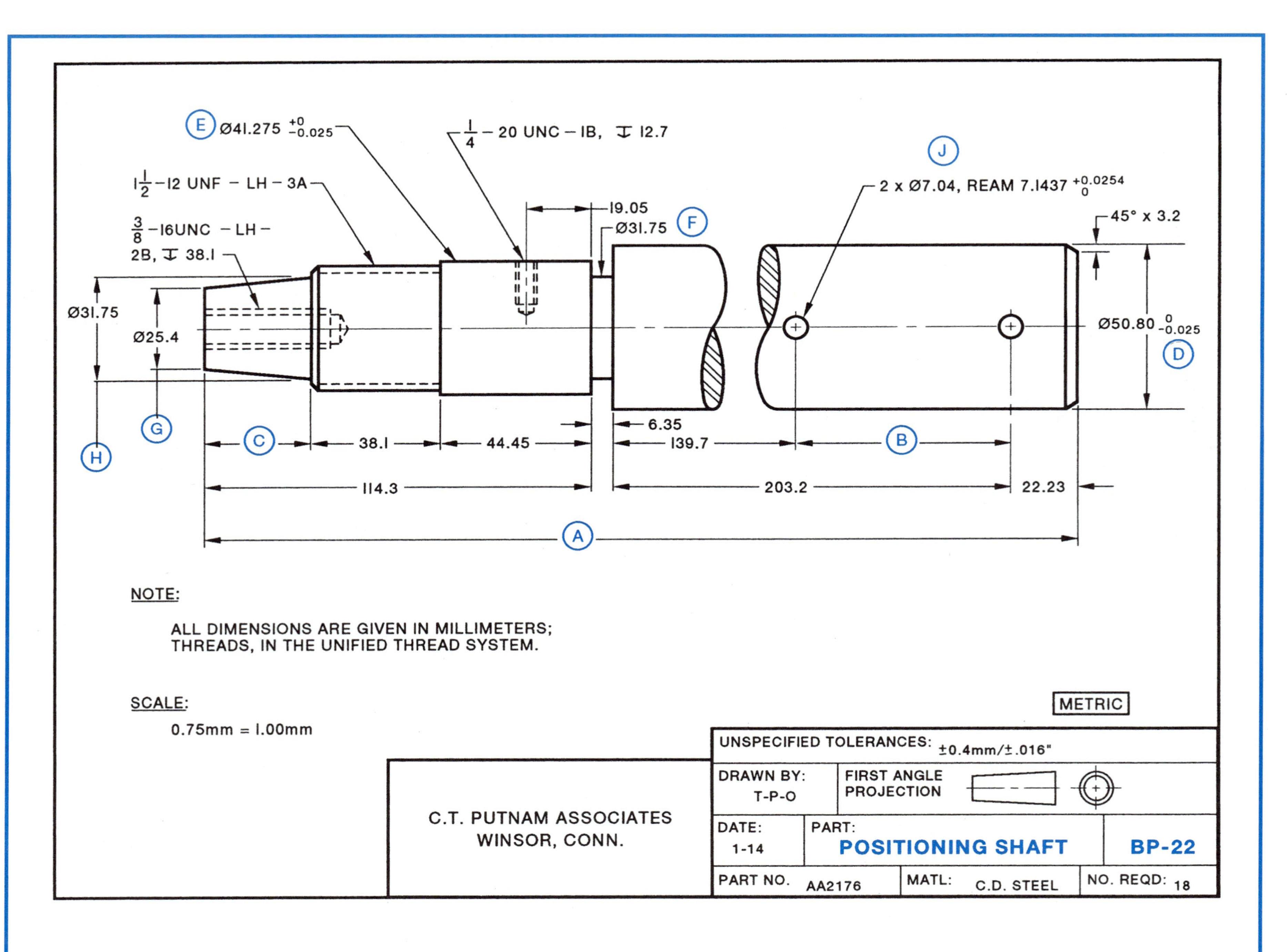

Ø41.275 +0 -0.025
1/4 – 20 UNC – IB, ⟂ 12.7
1 1/2 – 12 UNF – LH – 3A
3/8 – 16UNC – LH – 2B, ⟂ 38.1
19.05
Ø31.75
2 x Ø7.04, REAM 7.1437 +0.0254 0
45° x 3.2
Ø31.75
Ø25.4
Ø50.80 0 -0.025
6.35
38.1
44.45
139.7
114.3
203.2
22.23
A
B
C
D
E
F
G
H
J
NOTE:
ALL DIMENSIONS ARE GIVEN IN MILLIMETERS; THREADS, IN THE UNIFIED THREAD SYSTEM.
SCALE:
0.75mm = 1.00mm
METRIC
UNSPECIFIED TOLERANCES: ±0.4mm/±.016"
DRAWN BY: T-P-O
FIRST ANGLE PROJECTION
DATE: 1-14
PART: POSITIONING SHAFT
BP-22
PART NO. AA2176
MATL: C.D. STEEL
NO. REQD: 18
C.T. PUTNAM ASSOCIATES
WINSOR, CONN.

POSITIONING SHAFT (BP-22)

1. State what system of projection is used.
2. Indicate what unit of measurement is used for dimensioning.
3. Give the tolerance that applies on (a) metric dimensions and (b) dimensions converted to the inch system, when no tolerance is specified.
4. Compute the basic metric dimension for overall length Ⓐ rounded off to two decimal places.
5. Determine the basic center-to-center distance between the two reamed holes Ⓑ, in millimeters.
6. Find the length of the tapered portion Ⓒ.
7. Determine the equivalent customary (inch) units for each linear dimension, rounded off to two decimal places. Use a four-place millimeter/inch conversion chart, if available.
8. Convert each metric diametral dimension to its equivalent inch dimension, correct to three decimal places.
9. Give the letter size drill which corresponds to the drill size of the two reamed holes.
10. State what system of representation is used for the threaded sections.
11. Convert the depth of each tapped hole to its equivalent in the inch system.
12. Tell what each part of the three thread dimensions designates.
13. Give the minimum and maximum limits of diametral dimensions Ⓓ and Ⓔ in metric and inch dimensions, correct to three decimal places.
14. Determine the minimum and maximum limits of diametral dimensions Ⓕ, Ⓖ, and Ⓗ in metric and inch dimensions, rounded to two decimal places.
15. Determine the (a) taper per foot of the tapered portion, correct to two decimal places, and (b) convert the taper per foot to its millimeter equivalent.
16. Compute the lower and upper dimensional limits of each linear dimension in both the metric and inch systems. Round off the dimensions to two decimal places.

ASSIGNMENT—UNIT 22

Student's Name ____________________

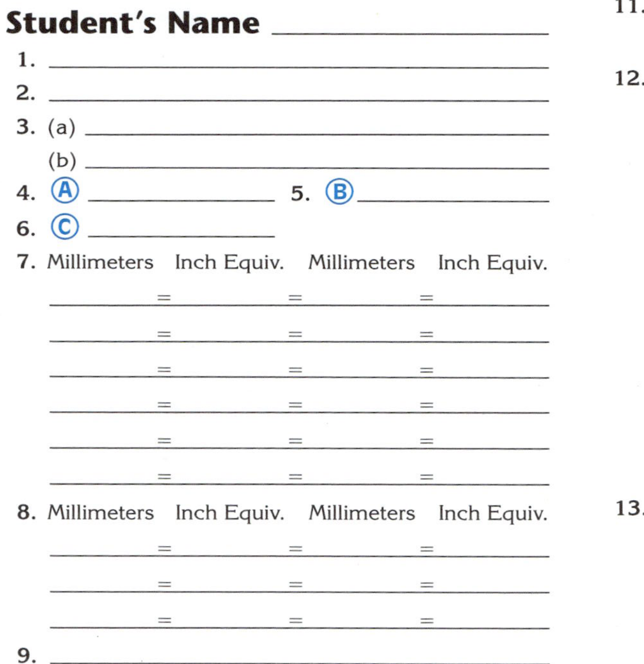

1. ____________________
2. ____________________
3. (a) ____________________
   (b) ____________________
4. Ⓐ __________ 5. Ⓑ __________
6. Ⓒ __________
7. Millimeters Inch Equiv. Millimeters Inch Equiv.
   ______ = ______ ______ = ______
   ______ = ______ ______ = ______
   ______ = ______ ______ = ______
   ______ = ______ ______ = ______
   ______ = ______ ______ = ______
   ______ = ______ ______ = ______
8. Millimeters Inch Equiv. Millimeters Inch Equiv.
   ______ = ______ ______ = ______
   ______ = ______ ______ = ______
   ______ = ______ ______ = ______
9. ____________________

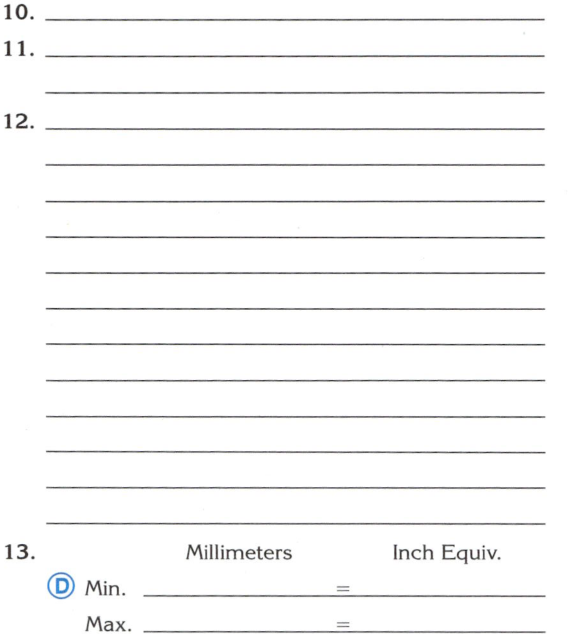

10. ____________________
11. ____________________
12. ____________________

13. Millimeters Inch Equiv.
   Ⓓ Min. ______ = ______
   Max. ______ = ______
   Ⓔ Min. ______ = ______
   Max. ______ = ______

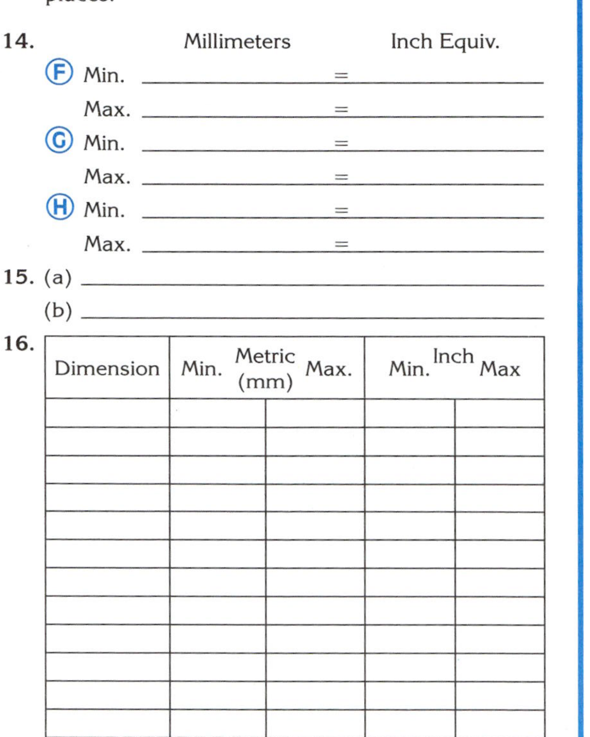

14. Millimeters Inch Equiv.
   Ⓕ Min. ______ = ______
   Max. ______ = ______
   Ⓖ Min. ______ = ______
   Max. ______ = ______
   Ⓗ Min. ______ = ______
   Max. ______ = ______
15. (a) ____________________
   (b) ____________________
16.

Dimension	Metric (mm) Min.	Metric (mm) Max.	Inch Min.	Inch Max

Unit 23

FIRST-ANGLE PROJECTION AND DIMENSIONING

PROJECTION BASED ON SYSTEM OF QUADRANTS

The position each view of an object occupies, as treated thus far in this text, is based upon the United States and Canadian standard of third-angle projection. Third-angle projection is derived from a theoretical division of all space into four quadrants. The horizontal plane in figure 23-1 represents an **X**-axis. The vertical plane is the **Y**-axis. The four quadrants produced by the two planes are shown as **I**, **II**, **III**, and **IV**.

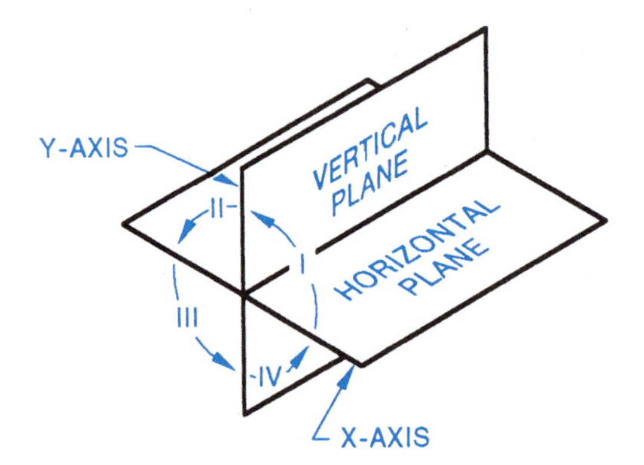

FIGURE 23-1 Four quadrants.

It is possible to place an object in any one of the four quadrants. Views of the object may then be projected. The third quadrant was adopted in the United States and Canada because the projected views of an object occupy a natural position. Drawings produced by such standards of projection are comparatively easy to interpret. Each view is projected so the object is represented as it is seen. The principles of third-angle projection are reviewed graphically in figure 23-2.

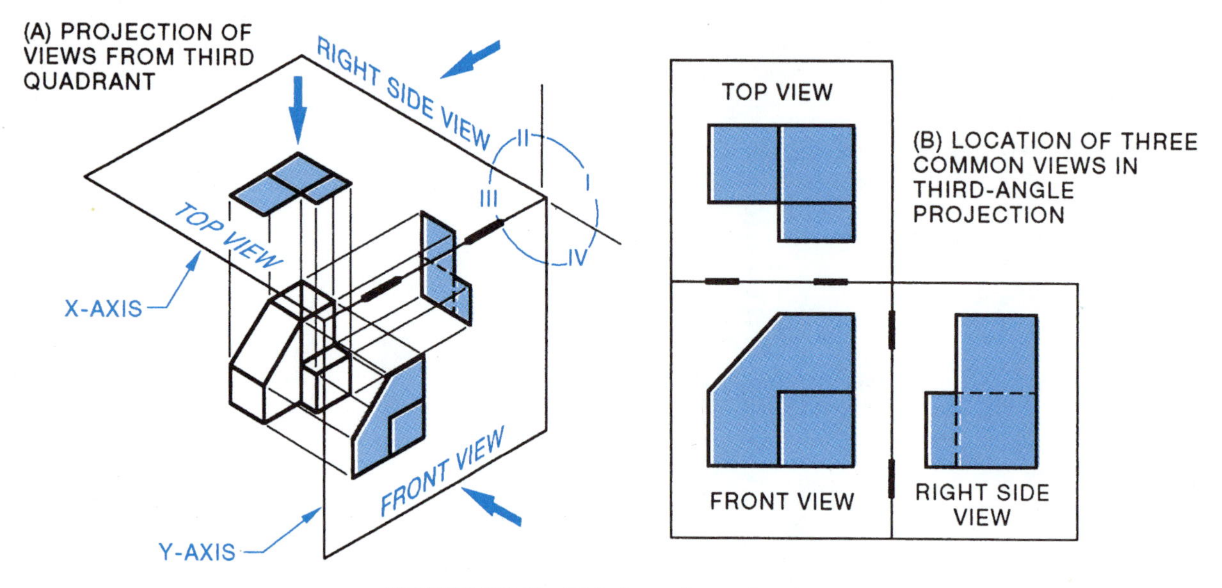

FIGURE 23-2 Third-angle projection.

METRIC: FIRST-ANGLE PROJECTION

Increasing worldwide commerce and the interchange of materials, instruments, machine tools, and other precision-made parts and mechanisms require the interpretation of drawings which have been prepared according to different systems of projection. The accepted standard of projection of the Common Market and other countries of the world relates to the first quadrant (**I**). The system, therefore, is identified as *first-angle projection.*

Dimensions given in the metric system are used with first-angle drawings. As world leaders in business and industry, the United States and Canada are also using a great number of first-angle projection drawings.

The quadrants and **X** and **Y** axes in first-angle projection are shown in figure 23-3. Arrows are used in figure 23-3A to illustrate positions from which the object in quadrant **I** may be viewed. The three common views are named Front View, Top View, and Left-Side View. These views are positioned in figure 23-3B as they would appear (without the imaginary projection box outline) on a drawing.

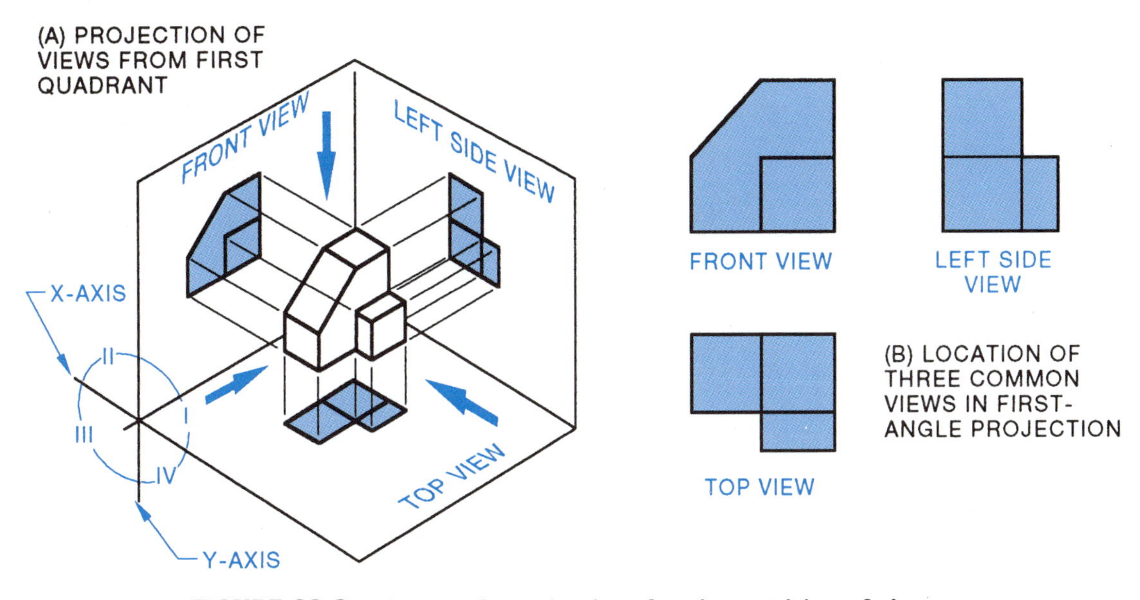

FIGURE 23-3 First-angle projection showing position of views.

In contrast with third-angle projection, the top view in first-angle projection appears under the front view. Similarly, the left-side view is drawn in the position occupied by the right-side view in third-angle projection. This positioning, which is not natural, makes it more difficult to visualize and interpret first-angle drawings. Each view in first-angle projection is projected through the object from a surface of the object to the corresponding projection plane. However, the views in both first-angle and third-angle projection provide essential information for the craftsperson to produce or assemble a single part or a complete mechanism.

DIMENSIONING FIRST-ANGLE DRAWINGS

Dimensions on first-angle drawings are in metric units of measure. When a numerical value is less than 1, the first numeral is preceded by a zero and a decimal point. For example, the decimal .733 is dimensioned to read 0.733. Certain countries of Europe use a comma in the place of a decimal point. In this case the decimal 0.733 is dimensioned to read 0,733. Unless all measurements on the drawing are in millimeters, this value is followed by the unit of measurement, like mm, to read 0,733. Whole numbers do not require a zero to the right of the decimal point. Twenty four (24 mm) would not be written as 24.0; it would be 24 mm.

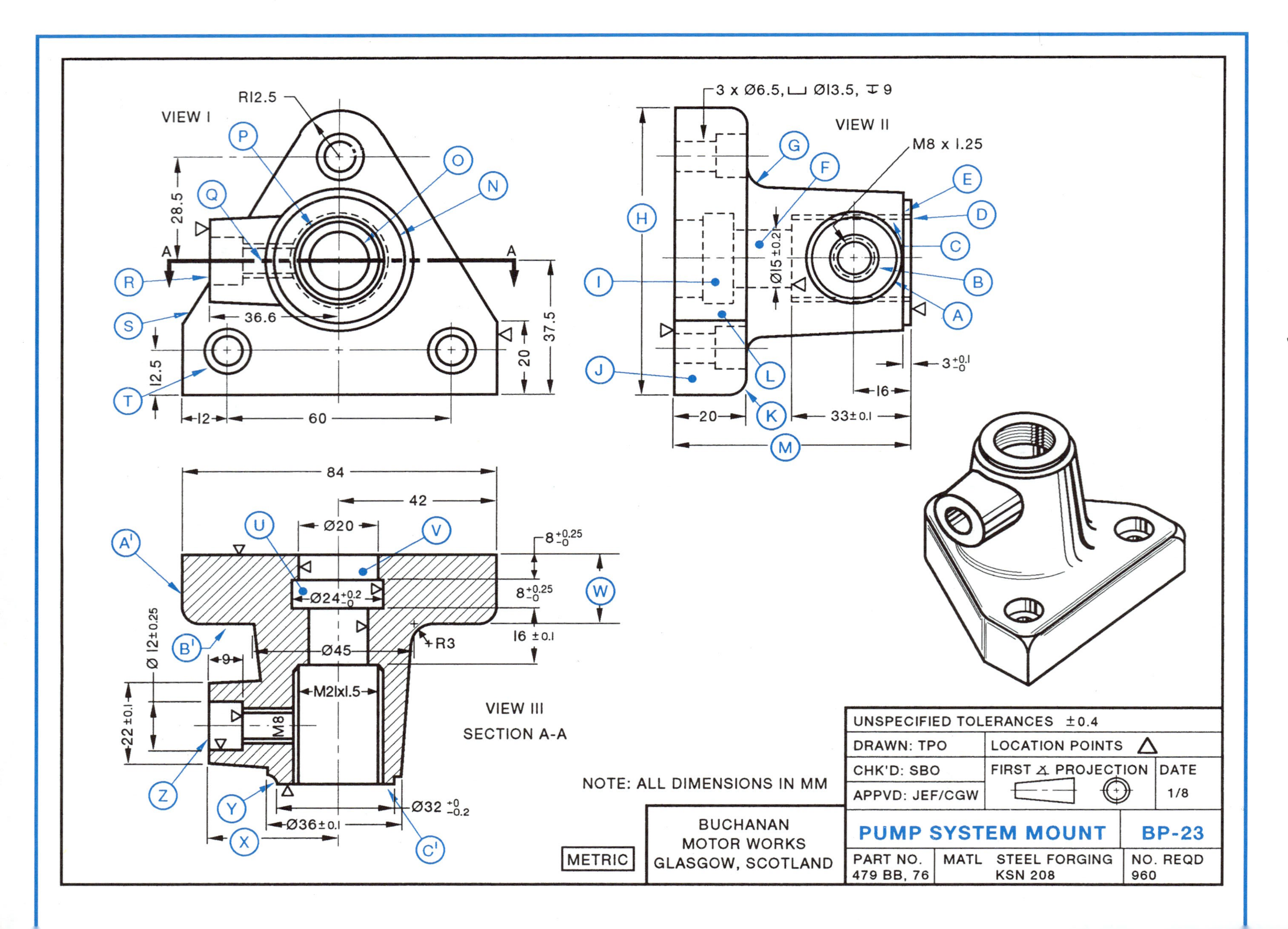

VIEW I
R12.5
28.5
36.6
37.5
20
12.5
12
60
A
VIEW II
3 x Ø6.5, ⌴ Ø13.5, ↧ 9
M8 x 1.25
Ø15 ±0.2
3 +0.1 −0
16
20
33±0.1
VIEW III
SECTION A-A
84
42
Ø20
Ø24 +0.2 −0
8 +0.25 −0
8 +0.25 −0
16 ±0.1
R3
Ø45
M21x1.5
M8
Ø 12±0.25
9
22±0.1
Ø36±0.1
Ø32 +0 −0.2
NOTE: ALL DIMENSIONS IN MM
METRIC
BUCHANAN
MOTOR WORKS
GLASGOW, SCOTLAND
UNSPECIFIED TOLERANCES ±0.4
DRAWN: TPO
LOCATION POINTS
CHK'D: SBO
FIRST ∡ PROJECTION
DATE
1/8
APPVD: JEF/CGW
PUMP SYSTEM MOUNT
BP-23
PART NO.
479 BB, 76
MATL
STEEL FORGING
KSN 208
NO. REQD
960

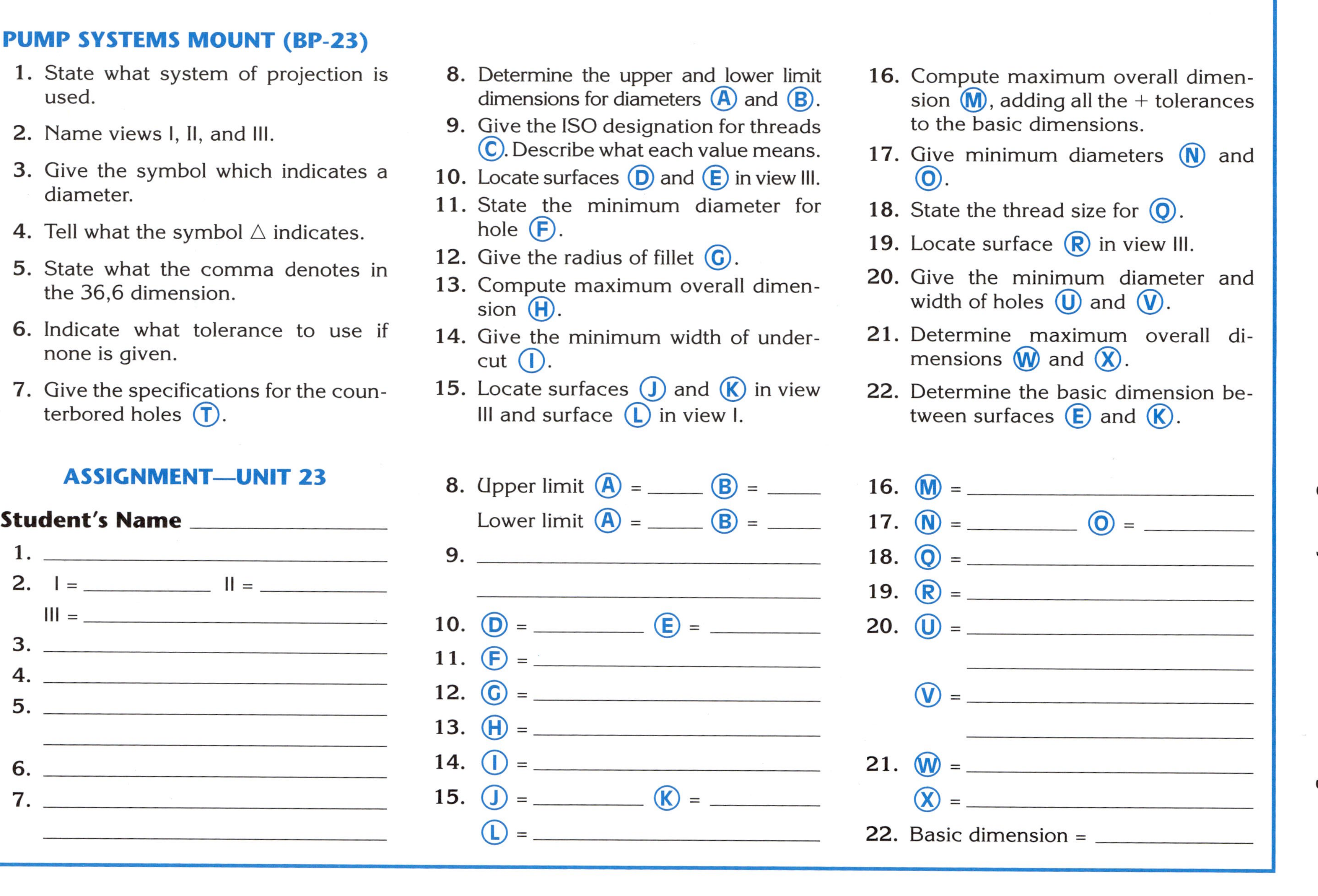

PUMP SYSTEMS MOUNT (BP-23)

1. State what system of projection is used.
2. Name views I, II, and III.
3. Give the symbol which indicates a diameter.
4. Tell what the symbol △ indicates.
5. State what the comma denotes in the 36,6 dimension.
6. Indicate what tolerance to use if none is given.
7. Give the specifications for the counterbored holes Ⓣ.
8. Determine the upper and lower limit dimensions for diameters Ⓐ and Ⓑ.
9. Give the ISO designation for threads Ⓒ. Describe what each value means.
10. Locate surfaces Ⓓ and Ⓔ in view III.
11. State the minimum diameter for hole Ⓕ.
12. Give the radius of fillet Ⓖ.
13. Compute maximum overall dimension Ⓗ.
14. Give the minimum width of undercut Ⓘ.
15. Locate surfaces Ⓙ and Ⓚ in view III and surface Ⓛ in view I.
16. Compute maximum overall dimension Ⓜ, adding all the + tolerances to the basic dimensions.
17. Give minimum diameters Ⓝ and Ⓞ.
18. State the thread size for Ⓠ.
19. Locate surface Ⓡ in view III.
20. Give the minimum diameter and width of holes Ⓤ and Ⓥ.
21. Determine maximum overall dimensions Ⓦ and Ⓧ.
22. Determine the basic dimension between surfaces Ⓔ and Ⓚ.

ASSIGNMENT—UNIT 23

Student's Name ____________

1. ____________
2. I = ________ II = ________
   III = ____________
3. ____________
4. ____________
5. ____________
   ____
6. ____________
7. ____________
   ____
8. Upper limit Ⓐ = _____ Ⓑ = _____
   Lower limit Ⓐ = _____ Ⓑ = _____
9. ____________
   ____
10. Ⓓ = ________ Ⓔ = ________
11. Ⓕ = ____________
12. Ⓖ = ____________
13. Ⓗ = ____________
14. Ⓘ = ____________
15. Ⓙ = ________ Ⓚ = ________
    Ⓛ = ____________
16. Ⓜ = ____________
17. Ⓝ = ________ Ⓞ = ________
18. Ⓠ = ____________
19. Ⓡ = ____________
20. Ⓤ = ____________
    ____
    Ⓥ = ____________
    ____
21. Ⓦ = ____________
    Ⓧ = ____________
22. Basic dimension = ____________

Unit 24

DUAL SYSTEM OF SCREW THREADS, DIMENSIONING, AND TOLERANCES

THREAD NOTES: BASIC THREAD AND TOLERANCE CALLOUTS

Thread notes provide complementary information to that given by graphically representing and dimensioning screw threads on a drawing. Dimensions and other data contained in thread notes for SI metric screw threads follow ISO standards. The necessary information for producing general-purpose metric screw threads is arranged in a standardized sequence in what is called a *basic metric* and/or *tolerancing thread callout*. The callout is referenced to the thread by using a dimensioning leader. Figure 24-1 identifies the sequencing of data for SI metric screw threads.

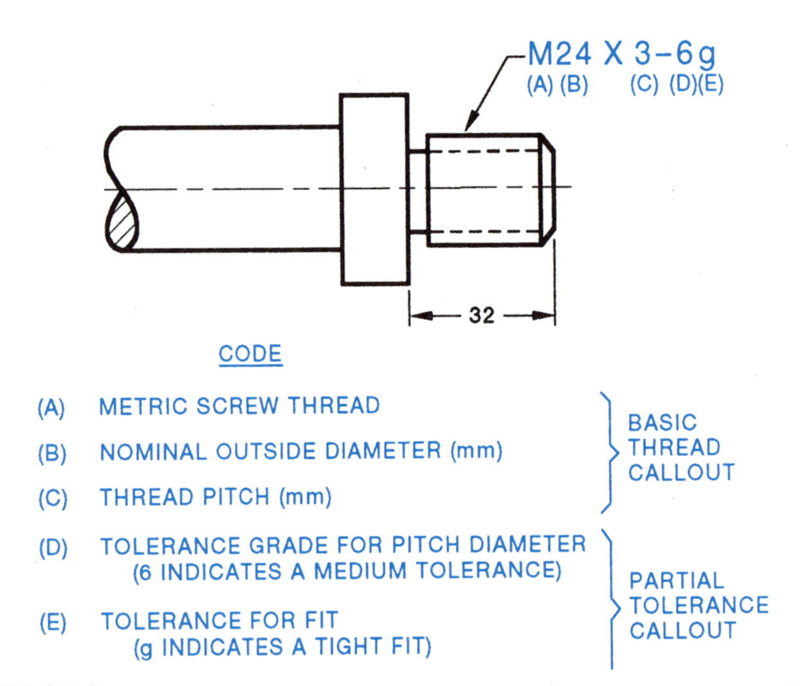

FIGURE 24-1 Basic and partial tolerance callout on a SI metric screw thread.

DEPTH (DEEP) SYMBOL

The symbol (⌴) followed by a dimension provides another standard method of dimensioning the depth of a screw thread, slot, step, or other part feature. An application of the symbol in relation to the threading of a part is illustrated in figure 24-2.

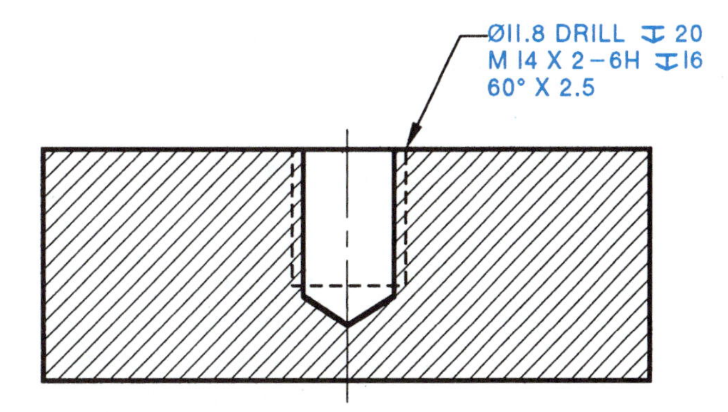

FIGURE 24-2 Application of depth (deep) symbol and a metric screw thread note.

The preceding symbol and thread note provide the following information. A **14** mm (nominal outside diameter) metric screw thread with a pitch of **2** mm is required. A medium tolerance (grade 6) with no specified thread fit allowance is to be used. The hole to be tapped is drilled with an **11.8** mm diameter drill to a depth of **20** mm (⌴**20**). The thread depth is **16** mm (⌴**16**) and the drilled hole is chamfered at a **60°** angle × **2.5** mm.

REPRESENTATION AND CONVERSION TABLES OF SI METRIC SCREW THREADS

Screw threads may be represented and dimensioned on drawings according to either American Unified Screw Thread or SI metric specifications. The general range of SI metric threads is from **0.30 UNM** (Unified Metric) to **1.40 UNM**, and from **M 1.6** × **0.35** to **M200** × **3**.

Screw thread conversion tables contain engineering data on major (nominal) outside diameters and pitch for SI metric screw threads, American Unified thread series sizes, and the best SI metric equivalents for selected American screw thread sizes.

For example, a **1/2 – 20 UNF (½ (.500) – 20 UNF)** designation on a drawing indicates an outside thread diameter of **.50**" or **.500**", a pitch (distance between two identical points on successive threads) of **20** threads per inch, and a Unified Fine Thread Series. Although currently there is no precise metric equivalent, the outside diameter and pitch (expressed in metric units) are **12.7** mm and **1.27** mm, respectively. This example may be dual dimensioned as shown in figure 24-3.

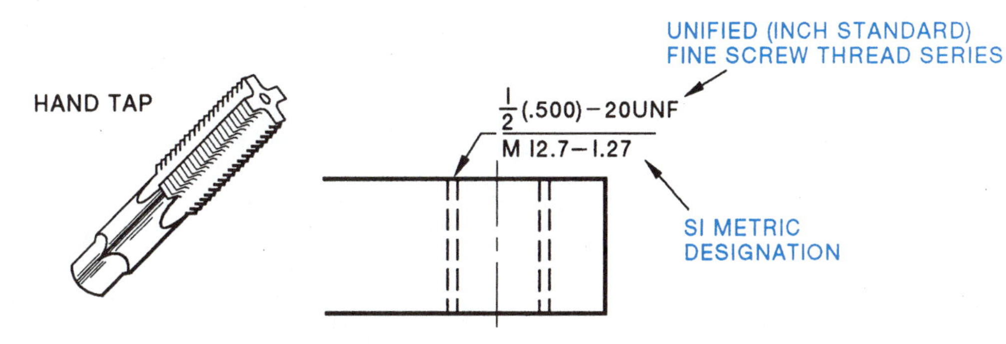

FIGURE 24-3 Dual dimensioning a tapped hole.

CONTROLLING DIMENSIONS

When a drawing contains dimensions in both the inch and the metric systems, the dimension in which the product was designed (*controlling dimension*) and the country of origin (overseas or North America) appears above the dimension line. The converted value appears below the line, figure 24-4.

Three common techniques of representing a *product design controlling dimension,* using a customary inch dimension, are illustrated in figure 24-4 A, B, and C. The customary inch dimension is given above the dimension line; the metric equivalent millimeter dimension, below the dimension line. Note that this technique of dimensioning applies to nominal dimensions (figure 24-4 A and D) as well as the toleranced dimensions given for dimensions B, C, E, and F). Similarly, where the controlling dimension is given in metric, the millimeter dimension appears above the dimension line; the customary inch equivalent dimension, below the line.

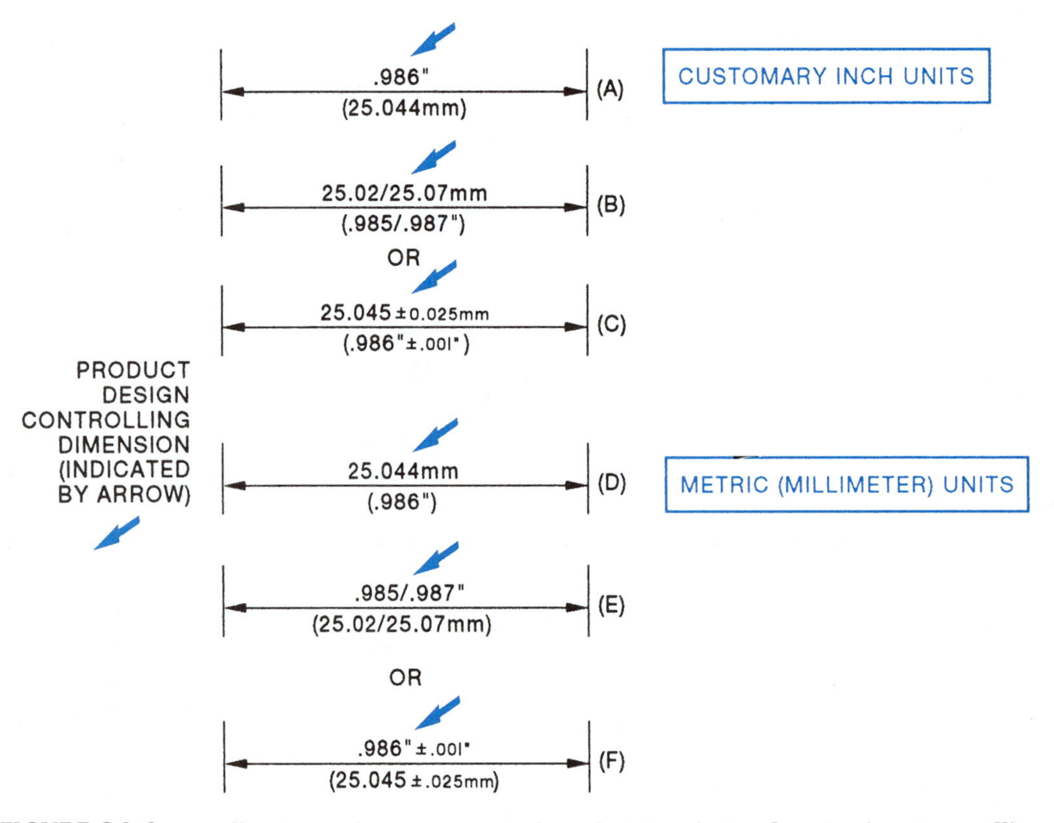

FIGURE 24-4 Applications of Customary Inch and SI Metric Product Design Controlling Dimensions.

DUAL DIMENSIONING: CUSTOMARY INCH AND SI METRIC

Dual-dimensioned parts may be produced in any country regardless of whether Customary inch system dimensions or SI metric dimensions, or both, are used. Dual dimensioning implies that each dimension is given both in metric and English units. The metric dimension (including tolerances) is given on one side of the dimension line; the inch dimension is given on the opposite side of the line. For example, if dual dimensioning is used on a dimension of 83 millimeters (3.268"), the dimension is shown as:

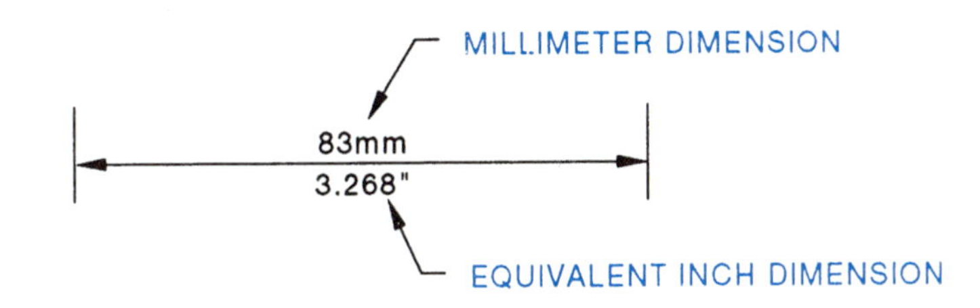

If a tolerance of ±0.2 mm applies to the base dimension of 83 mm, the dimension and tolerance in the customary inch system equals 3.268" ±.008". The dual dimension is represented as:

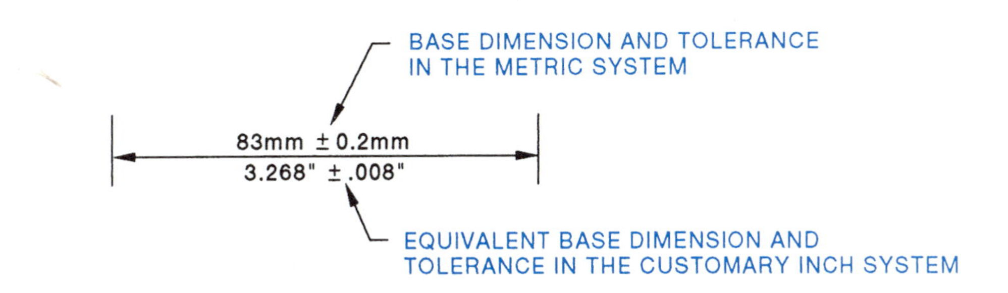

Instead of a dimension line between the dimensions in the two systems, some industries use the slash (/) symbol between the metric and customary inch units for dual dimensioning.

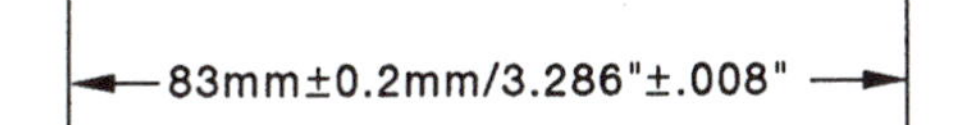

UNILATERAL AND EQUAL/UNEQUAL BILATERAL TOLERANCES

The previous examples showed equal tolerances in both directions (±). However, some tolerances may be unequal or unilateral. If unequal bilateral tolerances of $^{+0.2}_{-0.1}$ mm are applied to the 83 mm dimension, the drawing (with the scale and dimensional systems indicated) is dimensioned as,

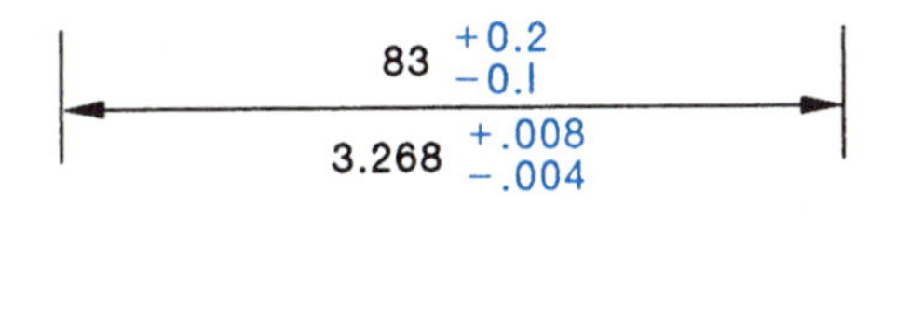

OR

83 $^{+0.2}_{-0.1}$/3.268 $^{+.008}_{-.004}$

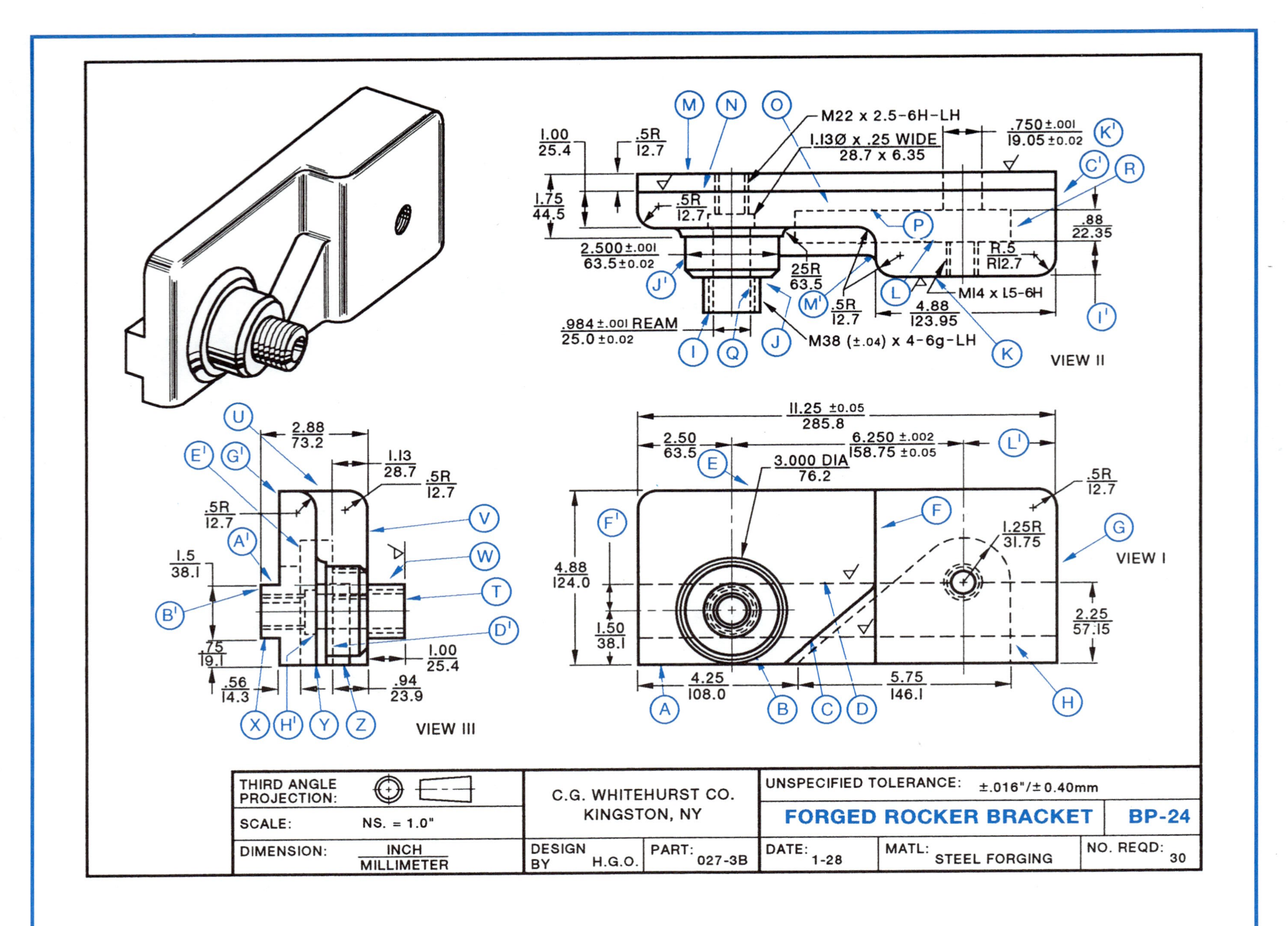
M22 x 2.5-6H-LH
1.13Ø x .25 WIDE
28.7 x 6.35
.750 ±.001
19.05 ±0.02
1.00
25.4
.5R
12.7
1.75
44.5
.88
22.35
R.5
R12.7
2.500 ±.001
63.5 ±0.02
25R
63.5
M14 x 1.5-6H
4.88
123.95
.984 ±.001 REAM
25.0 ±0.02
M38 (±.04) x 4-6g-LH
VIEW II
2.88
73.2
1.13
28.7
1.5
38.1
.75
19.1
.56
14.3
.94
23.9
VIEW III
11.25 ±0.05
285.8
2.50
63.5
6.250 ±.002
158.75 ±0.05
3.000 DIA
76.2
1.25R
31.75
4.88
124.0
1.50
38.1
2.25
57.15
4.25
108.0
5.75
146.1
VIEW I
THIRD ANGLE PROJECTION:
SCALE: NS. = 1.0"
DIMENSION: INCH / MILLIMETER
C.G. WHITEHURST CO.
KINGSTON, NY
DESIGN BY H.G.O.
PART: 027-3B
UNSPECIFIED TOLERANCE: ±.016"/±0.40mm
FORGED ROCKER BRACKET
BP-24
DATE: 1-28
MATL: STEEL FORGING
NO. REQD: 30

FORGED ROCKER BRACKET (BP-24)

1. Assume View I is the front view. Name Views II and III.
2. Name the dimensions above the line if these represent the product design dimensions.
3. State what dual dimensioning implies.
4. Determine what system of tolerancing is used.
5. Locate surfaces (A), (C), (D) and (E) in View III. Give the corresponding letter.
6. Give the letters in View II that identify surfaces (F), (G) and (H).
7. Locate surfaces (I), (J), (K), (L), (M), (N), (O) and (P) in View III.
8. Determine the distance between the center of the .750" hole and surface (G) correct to two decimal places. Give the dimension in metric and customary inch units.
9. Give the center distance (F') in millimeters and inches, correct to two decimal places.
10. Locate thickness (I') in both dimensioning systems, correct to two decimal places.
11. Indicate the tolerance to use if none is specified.
12. Interpret the meaning of the screw thread designation **M38×4=6g=LH.**
13. Give the outside diameter and pitch (in millimeters) for each metric screw thread.
14. Convert each metric screw thread to its equivalent American screw thread.
15. Give the nominal diameter of (J) and the upper and lower dimensional limits of (J') in inch units, to three decimal places.
16. Determine the minimum and maximum metric dimensions for (K') and (L'), correct to two decimal places.

ASSIGNMENT—UNIT 24

Student's Name ____________

1. ____________
2. ____________
3. ____________
____
4. ____________
5. (A) = ______ (D) = ______
(C) = ______ (E) = ______
6. (F) = ____ (G) = ____ (H) = ____
7. (I) = ____ (L) = ____ (O) = ____
(J) = ____ (M) = ____ (P) = ____
(K) = ____ (N) = ____ ____
8. ____________
9. ____________
10. ____________
11. ____________
12. ____________
____
____
____
____
13. ____________
____
____
14. ____________
____
____
15. (J) = ____________
(J') = ____________
16. (K') = ____________
(L') = ____________

SECTION 5 Working Drawings

Unit 25

DETAIL DRAWINGS AND ASSEMBLY DRAWINGS

A *working drawing* is one which includes all the information necessary to successfully and accurately complete a job. A *detail drawing* is a working drawing which includes a great deal of data including the size and shape of the project, what kinds of materials should be used, how the finishing should be done, and what degree of accuracy is needed. Every detail must be given. An *assembly drawing,* on the other hand, may have very little detail. The purpose of this type of technical drawing is to depict how the machine is to be put together (assembled). See Figure 25-1.

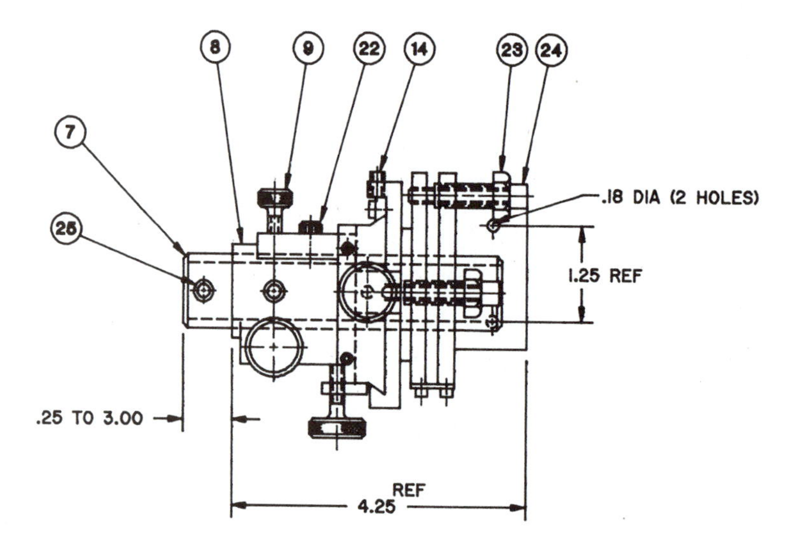

FIGURE 25-1 Partial Assembly Drawing (copyright Thomas P. Olivo)

A well-prepared working drawing will follow the style and practices of the office or industry where it is to be used, but there are certain rules which must be followed. For instance, to ensure sharp contrast, proper line technique must be used; dimensions and notes must be accurate and easy to read; and standard terms and abbreviations must be used. When the drawing is complete, it should be very carefully checked by a second person as a precaution against error.

DETAIL DRAWINGS

Detail drawings provide all information necessary to manufacture a part. Typically only one part is detailed in each drawing; however, there may be multiple views of that part. In some cases, though, several parts of a machine may be detailed on one sheet. Sometimes a separate drawing is made for each of several departments such as pattern-making, welding, or machining. A drawing of this type will contain only dimensions and information needed by the department for which it is made.

In the language of technical drawing, an object is drawn giving its shape and its size. This is true whether it is a simple machine, a steam or gas engine, a building, an automobile, or a satellite. The only part which may not need to be drawn is a *standard part,* one which can probably be bought from an outside supplier more economically than it can be manufactured. These parts include screws, keys, pins, and bolts. They do not need to be drawn, but they do need to be included in the general information which is part of each sheet.

All or some of the following items should be included in detail drawings.

1. Multiple view, as necessary
2. Material used to make the part
3. Dimensions
4. General notes and specific manufacturing information
5. Identification of the project name, the part, and the part number
6. Name or initials of who worked on or with the drawing
7. Any engineering changes and related information.

Usually, the detail drawings have information that is classified into three groups:

1. Shape description which describes or shows the shape of the part
2. Size description which shows the size and location of features of the part
3. Specifications regarding items such as material, finish, and heat treatment.

ASSEMBLY DRAWINGS

A very large number of products contain more than one part to them. An assembly drawing identifies each part and shows how they all fit together. The value of assembly drawings is that they show how the parts fit together, the overall look of the construction, and dimensions needed for installation. Hidden lines and unnecessary details are often omitted. In some situations, sectional or auxiliary views are given.

When using CAD, individual parts (details) can be merged together to create an assembly or a working drawing before the parts are actually made. Three-dimensional (3-D) models can be created in such a manner that it is possible to superimpose images and to graphically measure clearances. If the parts have been designed or drawn incorrectly, the errors will frequently be obvious and appropriate correction can be made. All these details make the final print accurate and the resultant parts function correctly.

There are actually a variety of assembly drawings. Some are:

1. *Layout assembly drawings* used as a first step in developing a new product.
2. *Installation assembly drawings* are necessary to show how to install large pieces of equipment.
3. *Diagram assembly drawings* use conventional symbols to show the approximate location and/or sequence of parts to be assembled.
4. *Working assembly drawings* are fully dimensioned and noted. For very simple products they can be used as detail drawings.
5. *Exploded assembly drawings* pictorially show parts laid out in their correct order of assembly as if they had been exploded (figure 25-2).

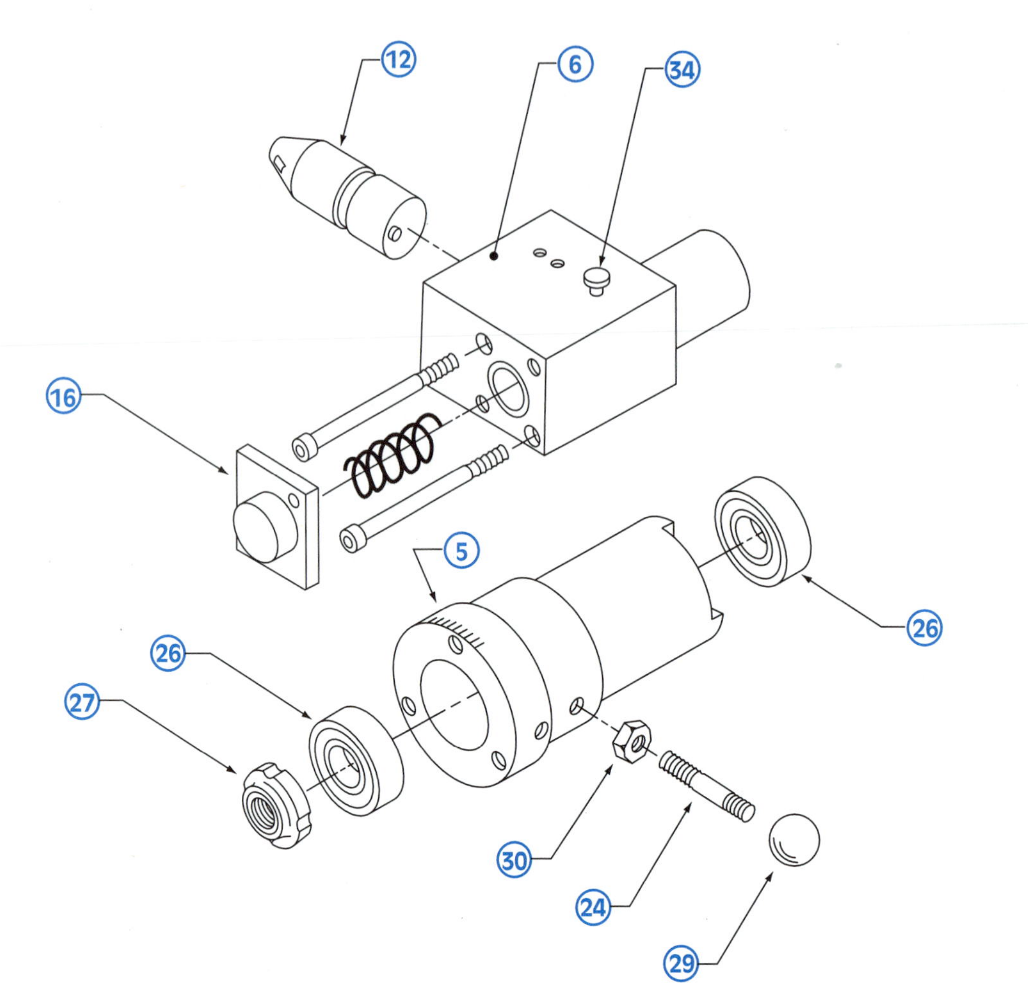

FIGURE 25-2 Exploded Assembly Drawing (copyright Thomas P. Olivo)

TITLE BLOCK

This part of a drawing is an outlined rectangular space in the lower right-hand corner of the drawing or across the bottom (figure 25-3A). The latter is referred to as a record strip. Within this box is contained a variety of information including:

1. The title of the drawing
2. The name of the company and its location
3. The name of the machine or unit of which the drawing is a subassembly
4. The number of the drawing
5. The part number
6. The number of parts to be made
7. The material from which the part is to be made
8. Drafting form record (who drew, checked, and approved the drawing, including the date of each).
9. Tolerances
10. Finishes
11. Heat treatment
12. Scale

13. Angle of projection (1st or 3rd angle)
14. Geometric drawing symbols used on the drawing
15. The sheet number (1 of 2, 2 of 2, etc.)
16. The date on which it was drawn
17. The weight of the part

PARTS LIST (BILL OF MATERIALS)

The graphically depicted parts of the product are keyed to a parts list with reference numbers. This parts list usually appears right above the title block. It will include the different parts that go into the assembly shown on the drawing, the description and name of each part, the quantity needed of each part, and any materials specifications. The parts should be listed in general order of size or importance. If parts are to be purchased from a vendor, the name or an identification code should be listed.

If a drawing is complicated, containing many parts, the parts list may be on a separate piece of paper which is attached to the drawing. (See figure 25-3B.)

Revisions Block

Sometimes, after a drawing has been released to a production department, it is necessary to make design revisions. This may be for design improvement, errors found in the original drawing, production problems, or other reasons. These changes must be approved and shown on the drawing in what is known as a *revisions* or *change block*. It is usually located to the left or above the title block. Typically the entry includes a brief description of the change and an identifying letter referencing it to the specific location on the drawing and the date. The initials of the draftsperson making the change and those approving it are required. (See figure 25-3C.)

ITEM	QTY	PART NUMBER	DESCRIPTION
25	2	DDS-A2-24	SCREW,SET,NYLON PT NORDEX
24	2		SOC. HD. CAPSCR,#10-32x3/8 LG ST.ST.
23	2	77A025	JACK SCREW
22	1	77A024	DOG PT SET SCREW
21	1	77B011	SLIDE
20	1	DFS-A1-6	SOC SET SCREW NORDEX
19	4	DKS-A3-5	SOC HD. SCREW (4-40x3/8) NORDEX
18	3	EPS-E1-9	PIN NORDEX
17	20	DKS-A3-3	SOC HD. SCREW (4-40x1/2) NORDEX
16	2	BAS-A1-4	COLLAR
15	4		WASHER, FLAT #4 ST.ST.
14	4	DFS-A1-8	SOC SET SCREW (4-40x1/2) NORDEX
13	1	77A018	GEAR
12	2	AGB-A2-4	BEARING NORDEX
11	1	77A023	THUMB SCREW
10	2	77A022	THUMB SCREW
9	1	77A021	THUMB SCREW
8	1	77B009	HOUSING
7	1	77B012	RACK SLEEVE
6	2	77A019	SPRING
5	2	77A020	GIB
4	1	77B008	PLATE BRACKET
3	1	77B010	SLIDE
2	1	77A017	PLATE
1	REF	AS SPEC	SHAFT TYPE SI SPERRY PROD.

PARTS LIST

AUTOMATION INDUSTRIES, INC.
SPERRY DIVISION

REVISIONS		
NO.	DATE	BY
A		
B		
C	8/26/98	
4		
5		

MANIPULATOR ASSEMBLY BP-42

DRAWN BY | SCALE 1/1 | MATERIAL
CHKD | DATE 1/15/98 | DRAWING NO. 77D025
TRACED | APPD

B, C, A

FIGURE 25-3 **(A) Title block, (B) Parts list, (C) Revision block. Copyright Thomas P. Olivo.**

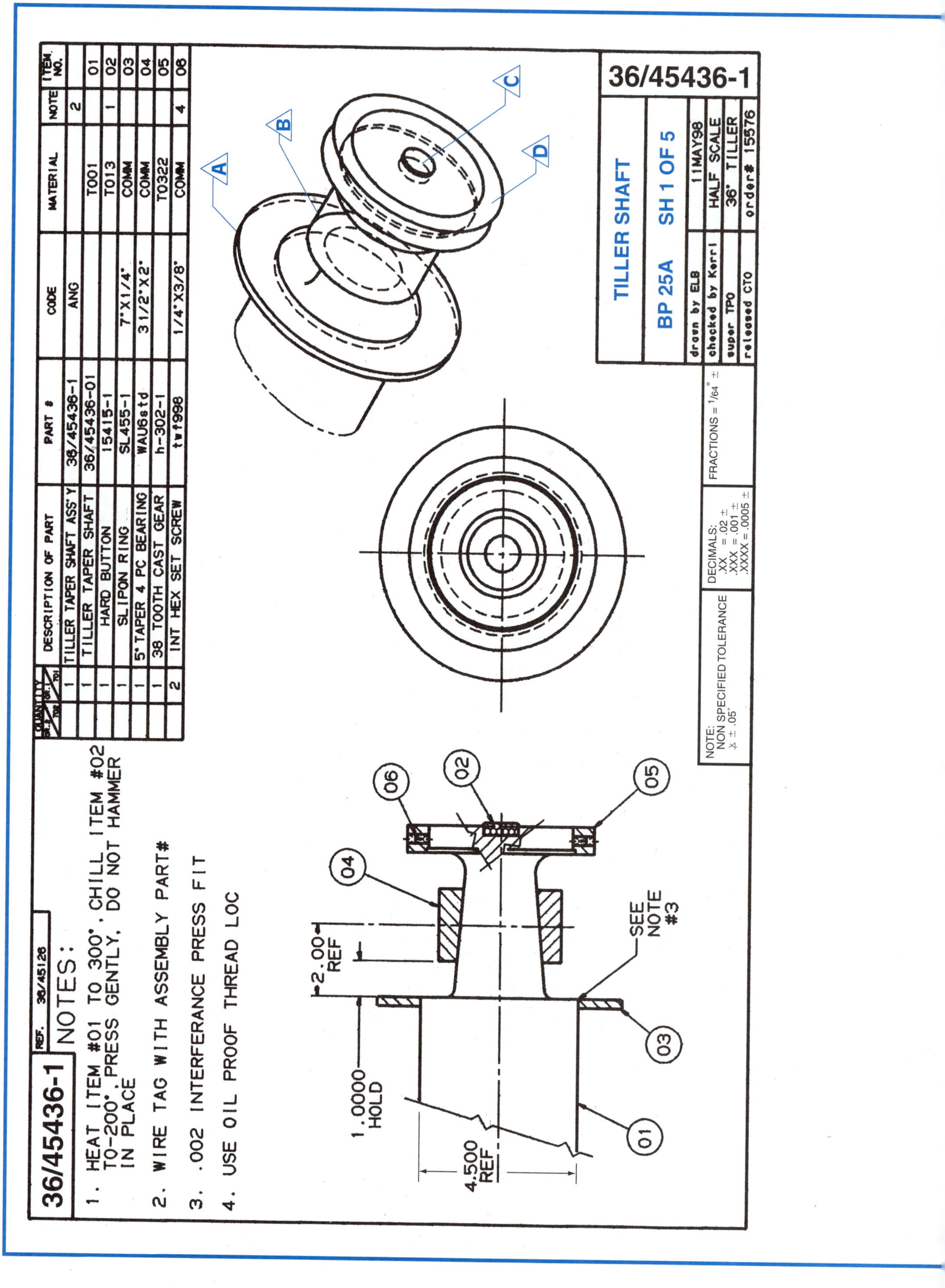

36/45436-1

REF. 36/45126

NOTES:

1. HEAT ITEM #01 TO 300°, CHILL ITEM #02 TO-200°, PRESS GENTLY, DO NOT HAMMER IN PLACE
2. WIRE TAG WITH ASSEMBLY PART#
3. .002 INTERFERANCE PRESS FIT
4. USE OIL PROOF THREAD LOC

QUANTITY GR.2 T02	QUANTITY GR.1 T01	DESCRIPTION OF PART	PART #	CODE	MATERIAL	NOTE	ITEM NO.
	1	TILLER TAPER SHAFT ASS'Y	36/45436-1	ANG		2	
	1	TILLER TAPER SHAFT	36/45436-01		T001		01
	1	HARD BUTTON	15415-1		T013	1	02
	1	SLIPON RING	SL455-1	7"X1/4"	COMM		03
	1	5° TAPER 4 PC BEARING	WAU6std	31/2"X2"	COMM		04
	1	38 TOOTH CAST GEAR	h-302-1		T0322		05
	2	INT HEX SET SCREW	twf998	1/4"X3/8"	COMM	4	06

NOTE:
NON SPECIFIED TOLERANCE
∡ ± .05°

DECIMALS:
.XX = .02 ±
.XXX = .001 ±
.XXXX = .0005 ±

FRACTIONS = 1/64" ±

TILLER SHAFT

BP 25A SH 1 OF 5

drawn by ELB	11MAY98
checked by Kerrl	HALF SCALE
super TPO	36" TILLER
released CTO	order# 15576

36/45436-1

36/45436

REF. 36/45126

NOTES:

1. HEAT ITEM #01 TO 300°, CHILL ITEM #02 TO-200°, PRESS GENTLY, TO NOT HAMMER IN PLACE
2. WIRE TAG WITH ASSEMBLY PART#
3. .002 INTERFERANCE PRESS FIT
4. USE OIL PROOF THREAD LOC

QUANTITY SH.2 702	QUANTITY SH.1 701	DESCRIPTION OF PART	PART #	CODE	MATERIAL	NOTE	ITEM NO.
	1	TILLER TAPER SHAFT ASS'Y	36/45436-1	ANG		2	
	1	TILLER TAPER SHAFT	36/45436-01	4.500 OD	T001		01
	1	HARD BUTTON	15415-1		T013	1	02
	1	SLIPON RING	SL455-1	7.00 ODX.25	COMM		03
	1	5° TAPER 4 PC BEARING	WAU6std	.350 ODX2.000	COMM		04
	1	38 TOOTH CAST GEAR	h-302-1		T0322		05
	2	INT HEX SET SCREW	twf998	1/4" X3/8"	COMM	4	06

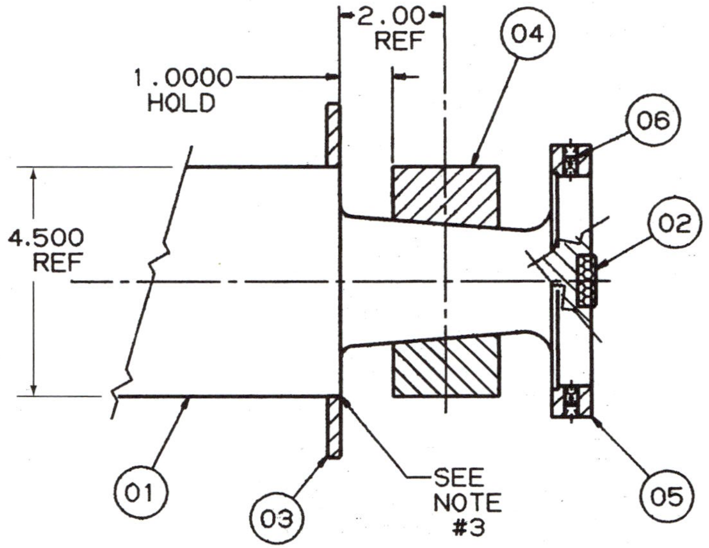

TILLER SHAFT

BP 25A SH 2 OF 5

drawn by ELB	11MAY98
checked by Kerri	HALF SCALE
super TPO	36" TILLER
released CTO	order# 15576

36/45436

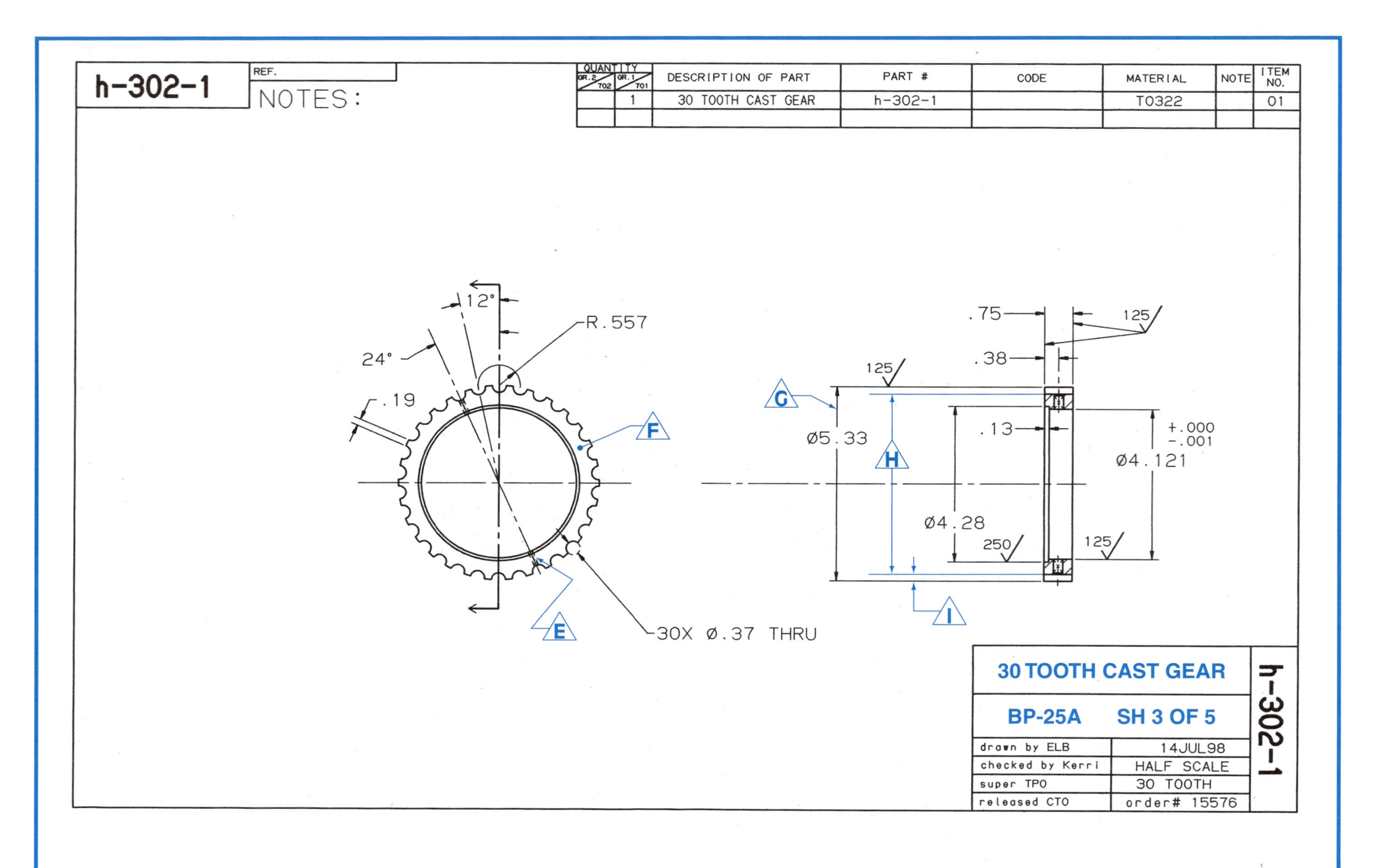
h-302-1
REF.
NOTES:
QUANTITY
GR.2
GR.1
T02
T01
1
DESCRIPTION OF PART
30 TOOTH CAST GEAR
PART #
h-302-1
CODE
MATERIAL
T0322
NOTE
ITEM NO.
01
12°
R.557
24°
.19
F
E
30X Ø.37 THRU
.75
125
.38
125
G
Ø5.33
H
.13
+.000
-.001
Ø4.121
Ø4.28
250
125
I
30 TOOTH CAST GEAR
BP-25A
SH 3 OF 5
drawn by ELB
14JUL98
checked by Kerri
HALF SCALE
super TPO
30 TOOTH
released CTO
order# 15576
h-302-1

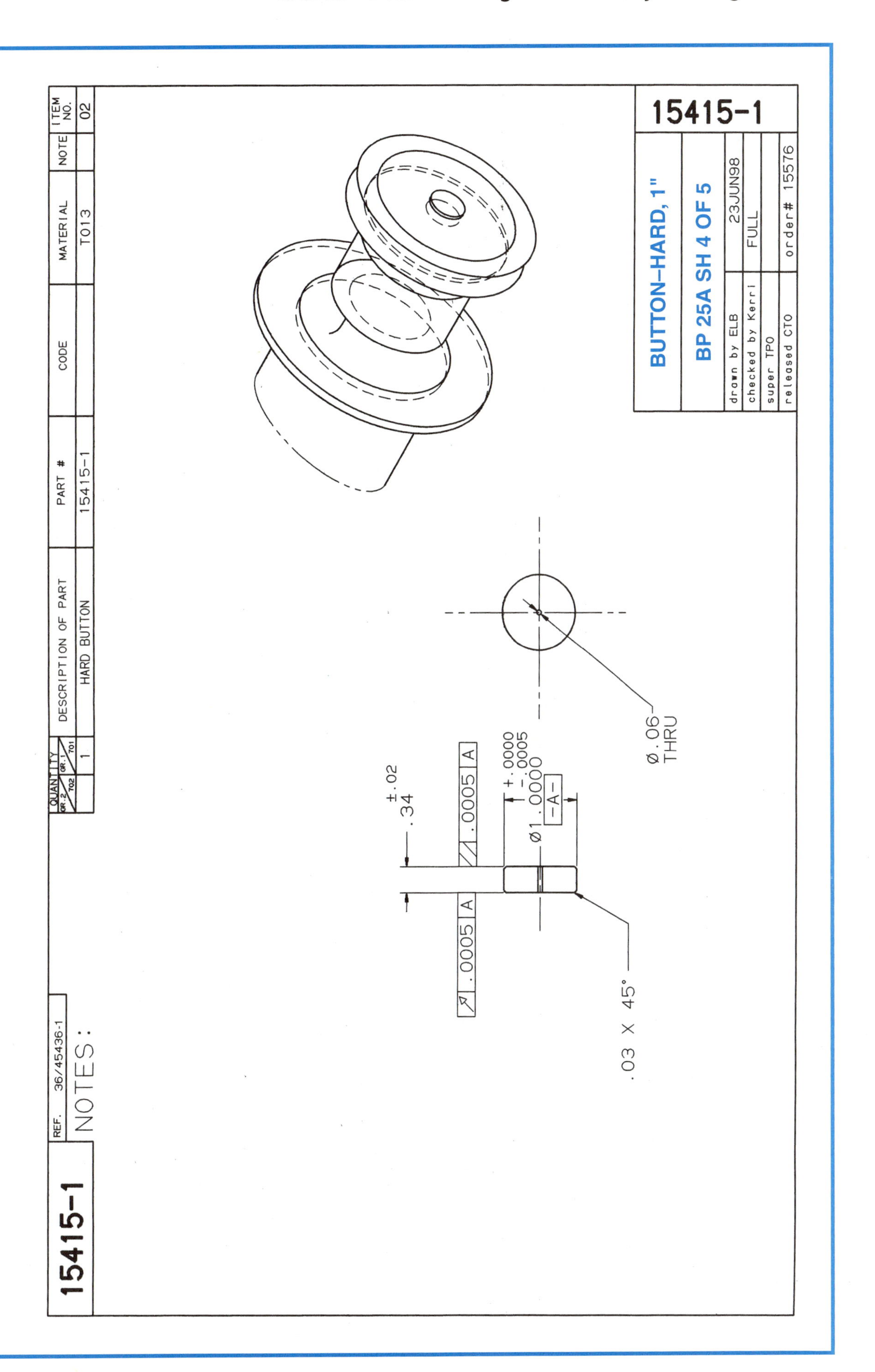

15415-1
BUTTON-HARD, 1"
BP 25A SH 4 OF 5
drawn by ELB
23JUN98
checked by Kerri
FULL
super TPO
released CTO
order# 15576
ITEM NO.
02
NOTE
MATERIAL
T013
CODE
PART #
15415-1
DESCRIPTION OF PART
HARD BUTTON
QUANTITY
GR.2
GR.1
702
701
1
REF. 36/45436-1
NOTES:
15415-1
Ø.06
THRU
±.02
.34
.0005 A
.0005 A
+.0000
-.0005
Ø1.0000
-A-
.03 X 45°

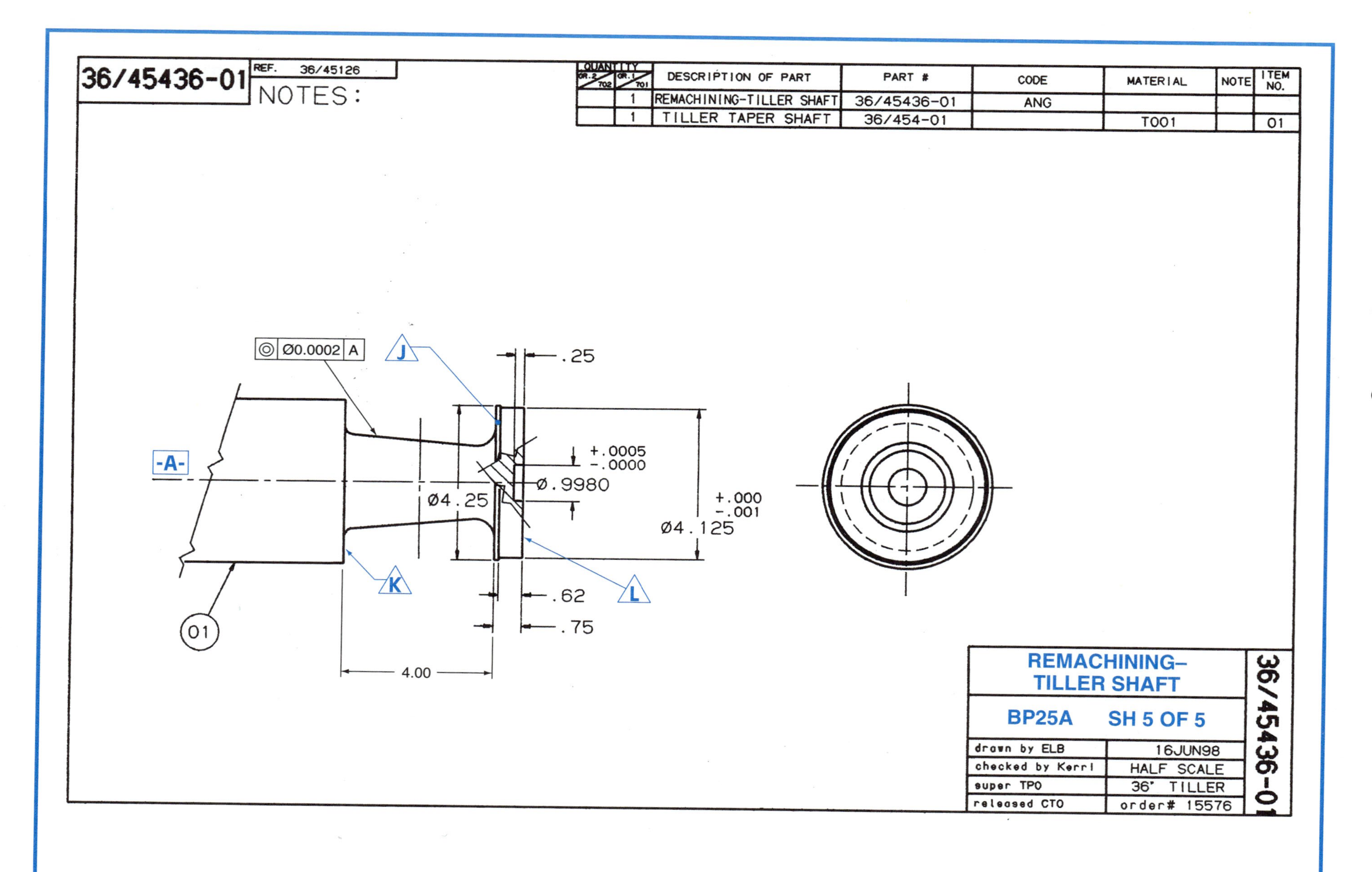
36/45436-01
REF. 36/45126
NOTES:
QUANTITY
DESCRIPTION OF PART
PART #
CODE
MATERIAL
NOTE
ITEM NO.
1
REMACHINING-TILLER SHAFT
36/45436-01
ANG
1
TILLER TAPER SHAFT
36/454-01
T001
01
Ø0.0002 A
J
K
L
-A-
01
.25
+.0005
-.0000
Ø.9980
Ø4.25
+.000
-.001
Ø4.125
.62
.75
4.00
REMACHINING–
TILLER SHAFT
BP25A SH 5 OF 5
drawn by ELB
16JUN98
checked by Kerri
HALF SCALE
super TPO
36" TILLER
released CTO
order# 15576
36/45436-01

ASSIGNMENT A - UNIT 25

Tiller Taper Shaft Assembly (BP-25A)

Student's Name ____________

The following questions relate to the assembly drawing

1. Print the name of the following parts: A, B, C, D
2. Give the drawing number for the tiller taper shaft assembly.
3. How many detail and assembly drawings make up the tiller taper shaft working drawing?
4. How many parts are needed to make a tiller tapper shaft assembly?
5. How many of these parts are commercial or purchased parts?

The following questions relate to the bearing.

1. Tell why the section lines of (04) go in 2 different directions.
2. Give the minimum taper angle of the bearing.
3. What is the part number of the bearing?
4. How many pieces make up the bearing?
5. What is the basic height and thickness of the bearing?

The following questions relate to the ring.

1. What is the thickness of the ring?
2. What is the basic OD of the ring?
3. What is the minimum hole size of the ring when using the upper limit of the shaft & lower limit of the ring?
4. Explain the meaning of "1.0000 Hold".
5. Explain why the 2.00 REF is given.

The following questions relate to the gear.

1. What surface finish sizes are given on the cast gear?
2. What is the smoothest surface of the gear?
3. What do the hidden lines at E indicate?
4. Give the part number of the gear.
5. What manufacturing process is used to make the gear?
6. Using a reference handbook give the following gear terms: G, H, I
7. How many teeth does the gear have?
8. Determine the basic clearance between the tiller shaft and the .13 gear slot diameter.

The following questions relate to the button.

1. How is the button attached to the shaft?
2. Research, then explain why part 01 will contract allowing the large part to fit into the smaller part.
3. Calculate the smallest allowable difference between the tiller taper shaft and the button.
4. Determine the maximum distance the button will stick out beyond the shaft hole.

The following questions relate to the shaft.

1. Determine the lower limit of the thickness of J.
2. Explain what each item in the geometric tolerance frame of the tiller shaft means. (See unit 30.)
3. Determine the upper limit distance between surfaces K & E

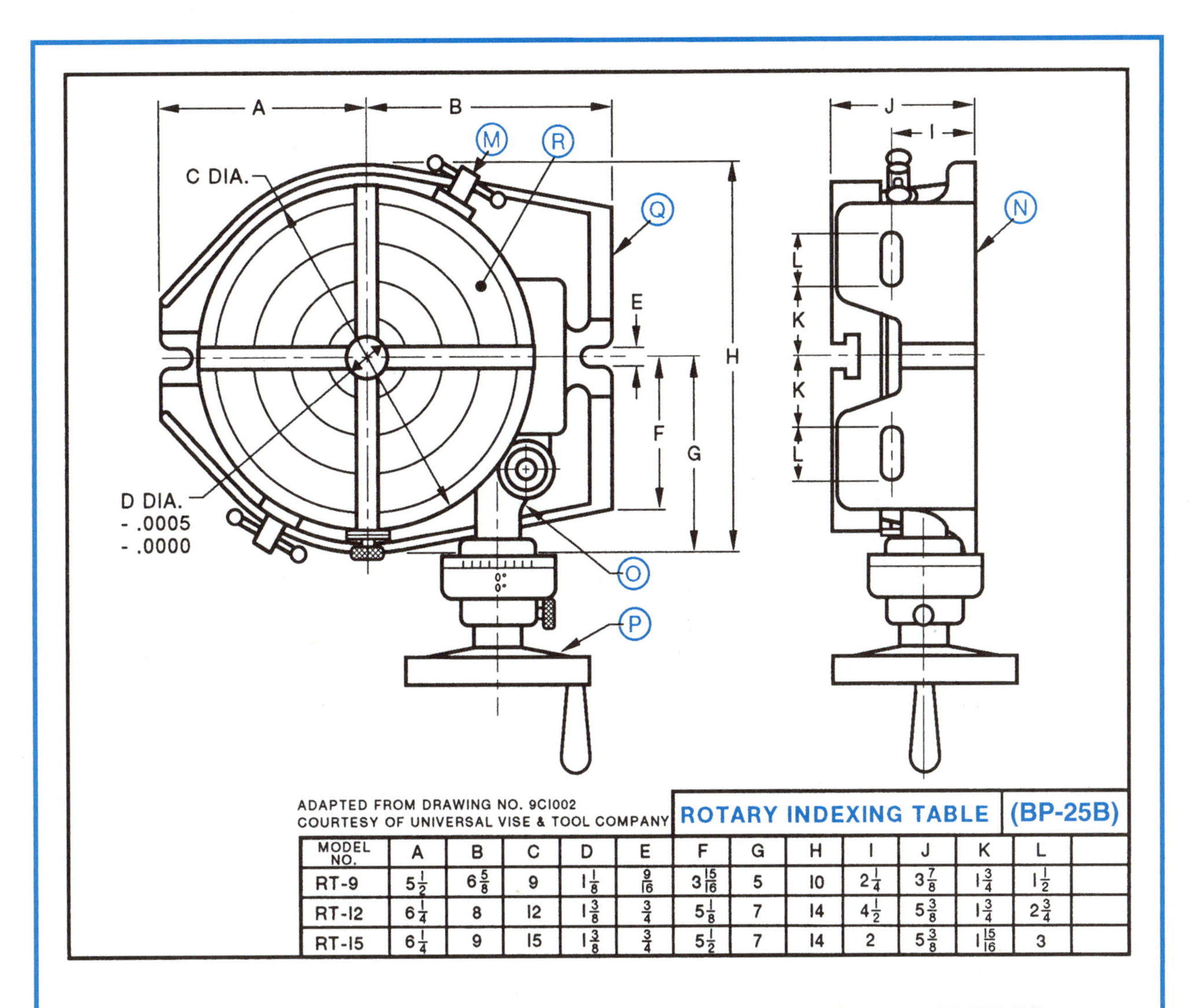

MODEL NO.	A	B	C	D	E	F	G	H	I	J	K	L	
RT-9	$5\frac{1}{2}$	$6\frac{5}{8}$	9	$1\frac{1}{8}$	$\frac{9}{16}$	$3\frac{15}{16}$	5	10	$2\frac{1}{4}$	$3\frac{7}{8}$	$1\frac{3}{4}$	$1\frac{1}{2}$	
RT-12	$6\frac{1}{4}$	8	12	$1\frac{3}{8}$	$\frac{3}{4}$	$5\frac{1}{8}$	7	14	$4\frac{1}{2}$	$5\frac{3}{8}$	$1\frac{3}{4}$	$2\frac{3}{4}$	
RT-15	$6\frac{1}{4}$	9	15	$1\frac{3}{8}$	$\frac{3}{4}$	$5\frac{1}{2}$	7	14	2	$5\frac{3}{8}$	$1\frac{15}{16}$	3	

ROTARY INDEXING TABLE (BP-25B)

1. a. Identify the type of drawing by which the Rotary Indexing Table is represented.
   b. Name the two views.
2. State two important functions that are served by this drawing.
3. Provide model numbers for which specifications are given.
4. Identify the letter on the drawing that represents each of the following parts or features:
   a. Rotary Table Plate
   b. Base
   c. Right-Angle Plate
   d. Table Locking Screws
   e. Indexing Crank
   f. Gear Reduction Unit
5. Use the dimensional information given in the drawing specifications. Record the sizes as indicated in the table for each specified item as follows.
   a. RT-9 Ⓐ Ⓑ Ⓒ and Ⓓ.
   b. RT-12 Ⓔ Ⓕ Ⓖ and Ⓗ.
   c. RT-15 Ⓘ Ⓙ Ⓚ and Ⓛ.

ASSIGNMENT B—UNIT 25

Student's Name ________________

1. a. ________________
   b. ________________
2. (1) ________________
   ____
   (2) ________________
   ____
3. Models ________________
4. a. ◯ c. ◯ e. ◯
   b. ◯ d. ◯ f. ◯
5. a. A = ____ b. E = ____ c. I = ____
   B = ____ F = ____ J = ____
   C = ____ G = ____ K = ____
   D = ____ H = ____ L = ____

Unit 26

CUTTING PLANES, FULL SECTIONS, AND SECTION LINING

An exterior view shows the object as it looks when seen from the outside. The inside details of such an object are shown on the drawing by hidden lines.

As the details inside the part become more complex, additional invisible lines are needed to show the hidden details accurately. This tends to make the drawing increasingly more difficult to interpret. One technique the draftsperson uses on such drawings to simplify them is to cut away a portion of the object. This exposes the inside surfaces. On cutaway sections, all of the edges that are visible are represented by visible edge or object lines.

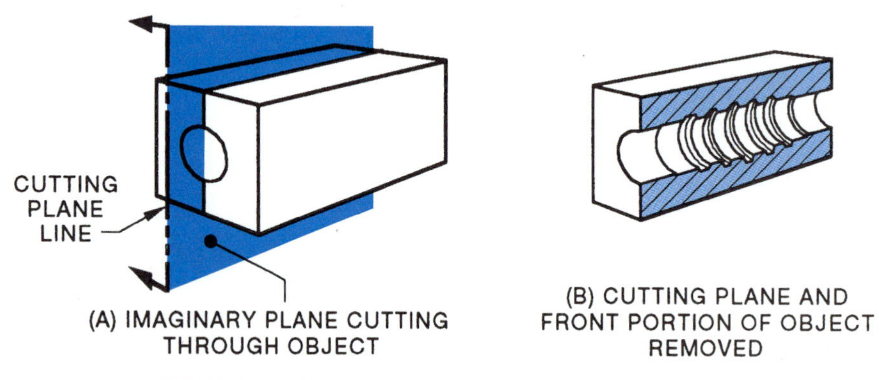

FIGURE 26-1 Cutting plane and its application.

To obtain a sectional view, an imaginary cutting plane is passed through the object as shown in figure 26-1A. Figure 26-1B shows the front portion of the object removed. The direction and surface through which the cutting plane passes is represented on the drawing by a cutting plane line. The exposed surfaces, which have been theoretically cut through, are further identified by a number of fine slant lines called *section* or *cross-hatch* lines. If the cutting plane passes completely through the object, the sectional view is called a *full section*. A section view is often used in place of a regular view.

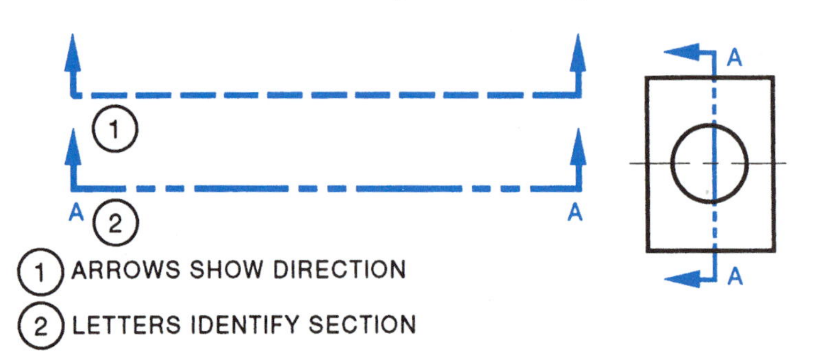

FIGURE 26-2 The cutting plane line.

CUTTING PLANE LINES

The cutting plane line is either a thick (heavy) line with one long and two short dashes or a series of thick (heavy), equally spaced long dashes. Examples are shown in figure 26-2. The line represents the edge of the cutting plane. The arrowheads on the ends of the cutting plane line show the direction in which the section is viewed. A capital letter is usually placed at the end of the cutting plane line to identify the section.

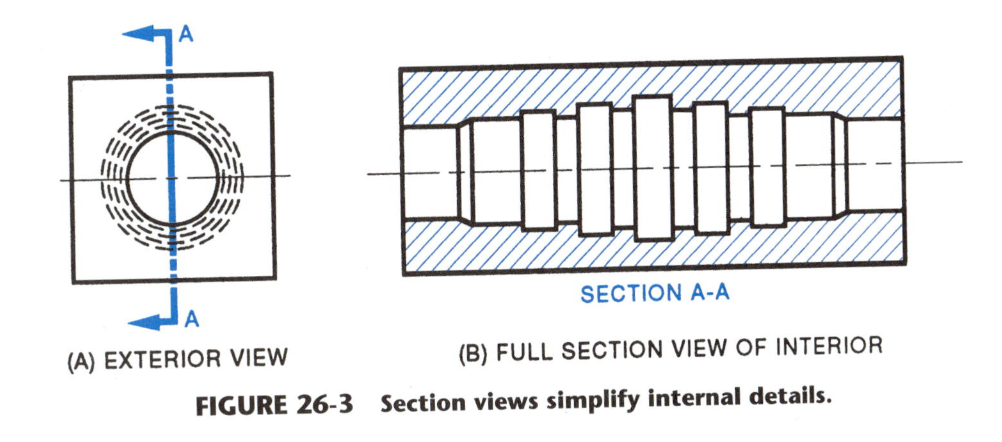

FIGURE 26-3 Section views simplify internal details.

CROSS HATCHING OR SECTION LINING

The interpretation of a sectional drawing is simplified further by *cross hatch* or *section lines*. These uniformly spaced thin lines are drawn at a 45° angle as shown in figure 26-3. It is easy to distinguish one part from another on a cutaway section when section lines are used.

Section lines also serve the purpose of identifying the kind of material. Combinations of different lines and symbols are used for such metals as cast iron and steel and other groups of related materials. Section lining symbols of common materials are illustrated in figure 26-4.

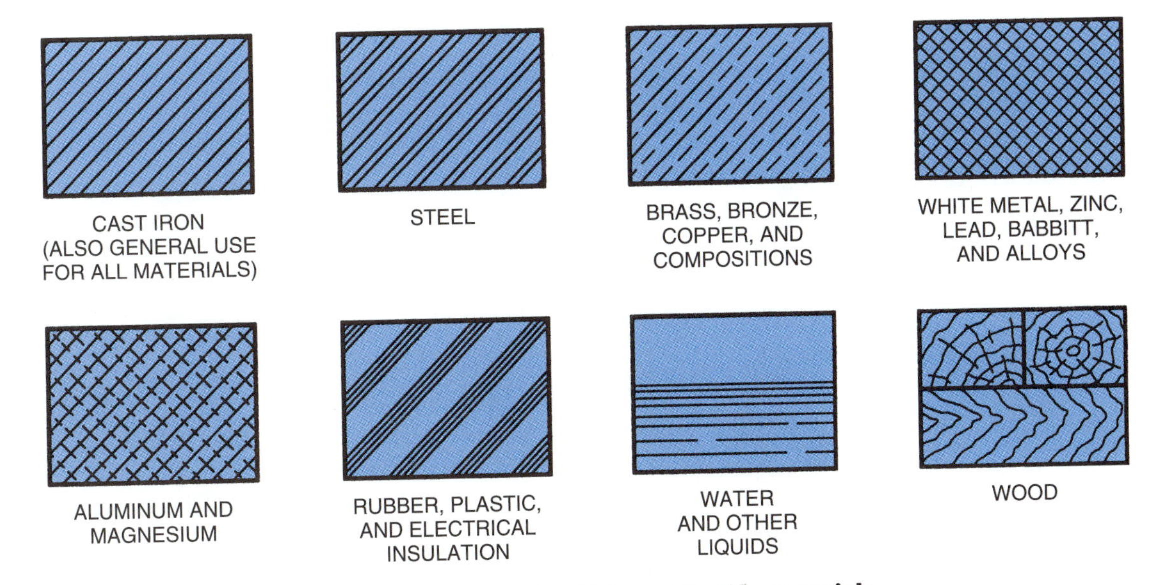

FIGURE 26-4 Section linings identify materials.

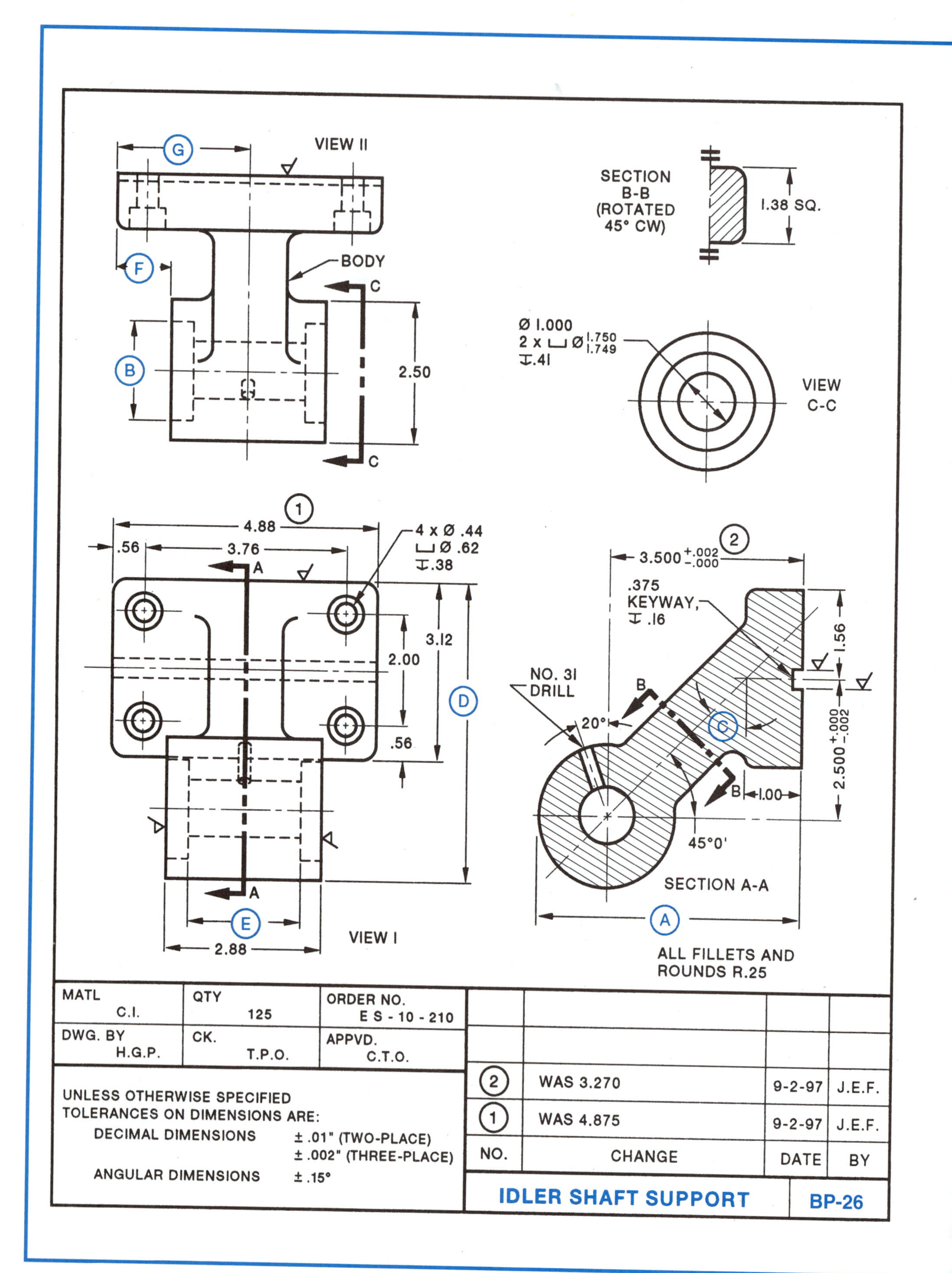

VIEW II
G
F
BODY
B
C
2.50
C
SECTION
B-B
(ROTATED
45° CW)
1.38 SQ.
Ø 1.000
2 x ⌴ Ø 1.750 1.749
↧.41
VIEW
C-C
1
4.88
.56
3.76
4 x Ø .44
⌴ Ø .62
↧.38
A
3.12
2.00
D
.56
A
E
2.88
VIEW I
2
3.500 +.002 -.000
.375
KEYWAY,
↧.16
NO. 31
DRILL
20°
B
C
1.56
2.500 +.000 -.002
B
1.00
45°0'
SECTION A-A
A
ALL FILLETS AND
ROUNDS R.25
MATL C.I.
QTY 125
ORDER NO. E S - 10 - 210
DWG. BY H.G.P.
CK. T.P.O.
APPVD. C.T.O.
UNLESS OTHERWISE SPECIFIED
TOLERANCES ON DIMENSIONS ARE:
DECIMAL DIMENSIONS ± .01" (TWO-PLACE)
± .002" (THREE-PLACE)
ANGULAR DIMENSIONS ± .15°
2 WAS 3.270 9-2-97 J.E.F.
1 WAS 4.875 9-2-97 J.E.F.
NO. CHANGE DATE BY
IDLER SHAFT SUPPORT
BP-26

ASSIGNMENT—UNIT 26

IDLER SHAFT SUPPORT (BP-26)

1. Give the specifications for the counterbored holes.
2. How deep is the counterbored portion of the holes?
3. Give the specifications for the flat keyway.
4. What size are the fillets and rounds?
5. What size drill is used for the hole drilled at an angle?
6. Give the tolerance on the 20° dimension.
7. Compute angle Ⓒ from dimensions given on the drawing.
8. Give the maximum diameter for the 1" hole.
9. What is the largest size to which diameter Ⓑ can be bored?
10. How are the machined surfaces indicated?
11. What type line shows where Section **A-A** is taken? In what view?
12. Determine dimensions Ⓐ and Ⓓ.
13. Determine from Section **B-B** what material is required.
14. How wide and thick is the body?
15. Indicate what two changes were made from the original drawing.
16. Determine nominal dimension Ⓔ.
17. Compute maximum dimension Ⓕ.
18. What is upper limit of dimension Ⓖ?
19. Why is section **A-A** a full section?
20. Show the section linings for aluminum and for steel.
21. Identify (a) the system of dimensioning and (b) the classification of the unspecified tolerances.

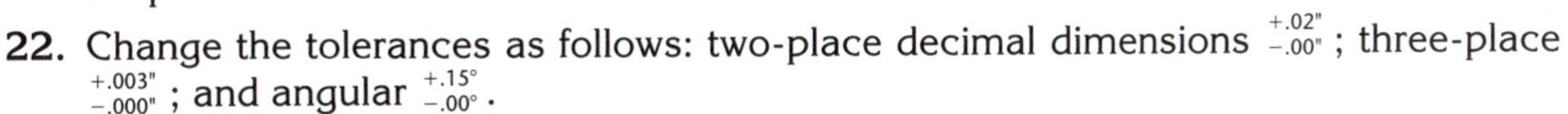

22. Change the tolerances as follows: two-place decimal dimensions $^{+.02"}_{-.00"}$; three-place $^{+.003"}_{-.000"}$; and angular $^{+.15°}_{-.00°}$.

Then compute the upper and lower limit for angle Ⓒ and dimensions Ⓐ, Ⓓ, Ⓔ, Ⓕ, and Ⓖ.

Unit 27

HALF SECTIONS, PARTIAL SECTIONS, AND FULL-SECTION ASSEMBLY DRAWINGS

HALF SECTIONS

The internal and external details of a part may be represented clearly by a sectional view called a *half section*. In a half-section view, one-half of the object is drawn in section and the other half is drawn as an exterior view, figure 27-1. The half section is used principally where both the inside and outside details are symmetrical and where a full section would omit some important detail in an exterior view.

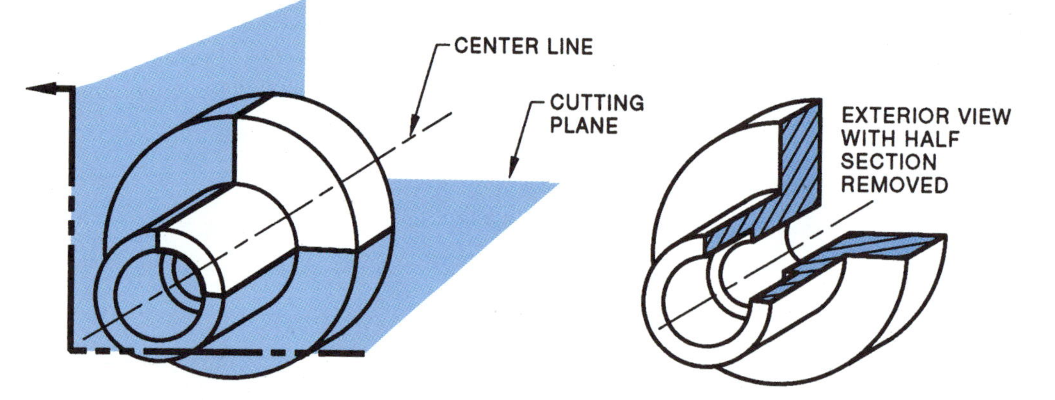

FIGURE 27-1 Half-section view.

Theoretically, the cutting plane for a half-section view extends halfway through the object, stopping at the axis or center line, figure 27-2. Drawings of simple symmetrical parts may not always include the cutting plane line or the arrows and letters showing the direction in which the section is taken. Also, hidden lines are not shown in the sectional view unless they are needed to give details of construction or for dimensioning.

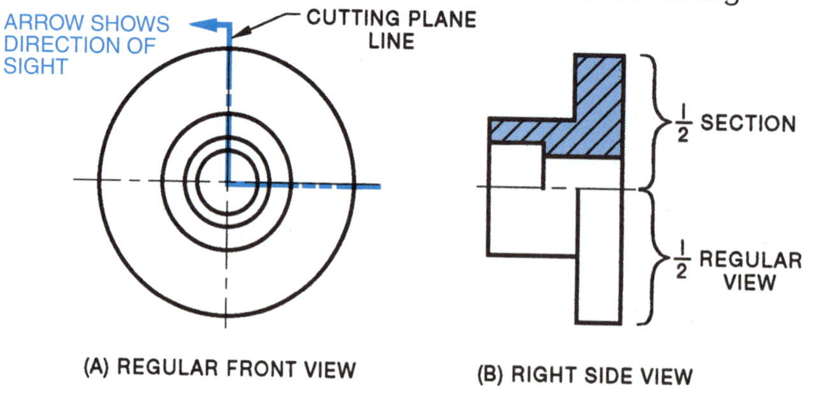

FIGURE 27-2 Cutting plane line and half section view.

BROKEN OR PARTIAL SECTIONS

On some parts, it is not necessary to use either a full or a half section to expose the interior details. In such cases, a *broken-out* or *partial section* may be used, figure 27-3. The cutting plane is imagined as being passed through a portion of the object and the part in front of the plane is then broken away. The break line is an irregular freehand line which separates the internal sectioned view and the external view.

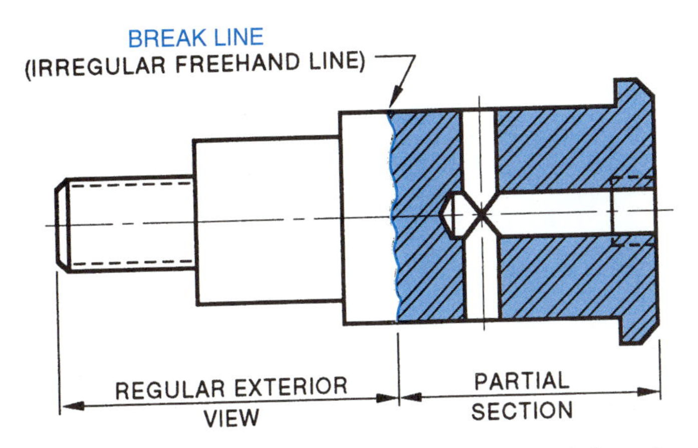

FIGURE 27-3 Example of a broken-out or partial section.

CONVENTIONAL BREAKS

A long part with a uniform cross section may be drawn to fit on a standard size drawing sheet by cutting out a portion of the length. In this manner, a part may be drawn larger to bring out some complicated details.

The cutaway portion may be represented by a conventional symbol which does two things: (1) it indicates that a portion of uniform cross section is removed, and (2) it shows the internal shape. A few conventional symbols which are accepted as standard are illustrated in figure 27-4.

ASSEMBLY DRAWINGS: TYPES AND FUNCTIONS

Full, half, partial, and cutaway sections are often included in assembly drawings to display the external shape of a portion of a complete mechanism, as well as internal features and movements of parts which make up the mechanism. Assembly drawings are valuable in design, construction, assembly, and preparation of specifications.

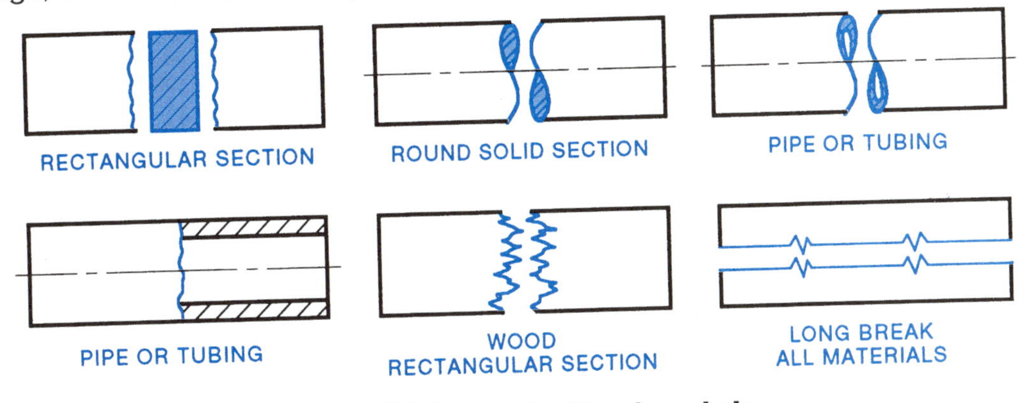

FIGURE 27-4 Standard break symbols.

An assembly drawing may include one or more standard views, with or without sections. Dimensions and details are generally omitted. Assembly drawings are identified by use. Figure 27-5 illustrates a *full-section assembly drawing* of a rotary actuator. The section clearly shows each part and movements within this pneumatic device.

FIGURE 27-5 Full-section assembly drawing showing design features and movement of parts. Courtesy of C. Thomas Olivo Associates.

Other examples include: *design working assembly drawings* (that combine the functions of a detail drawing and an assembly drawing); *pictorial assembly drawings* (produced mechanically or freehand); *installation and maintenance assembly*, and *subassembly drawings;* and *technical catalog exploded assembly drawings* (in which parts are aligned and represented in position for assembly). See Unit 25 for more information on this topic.

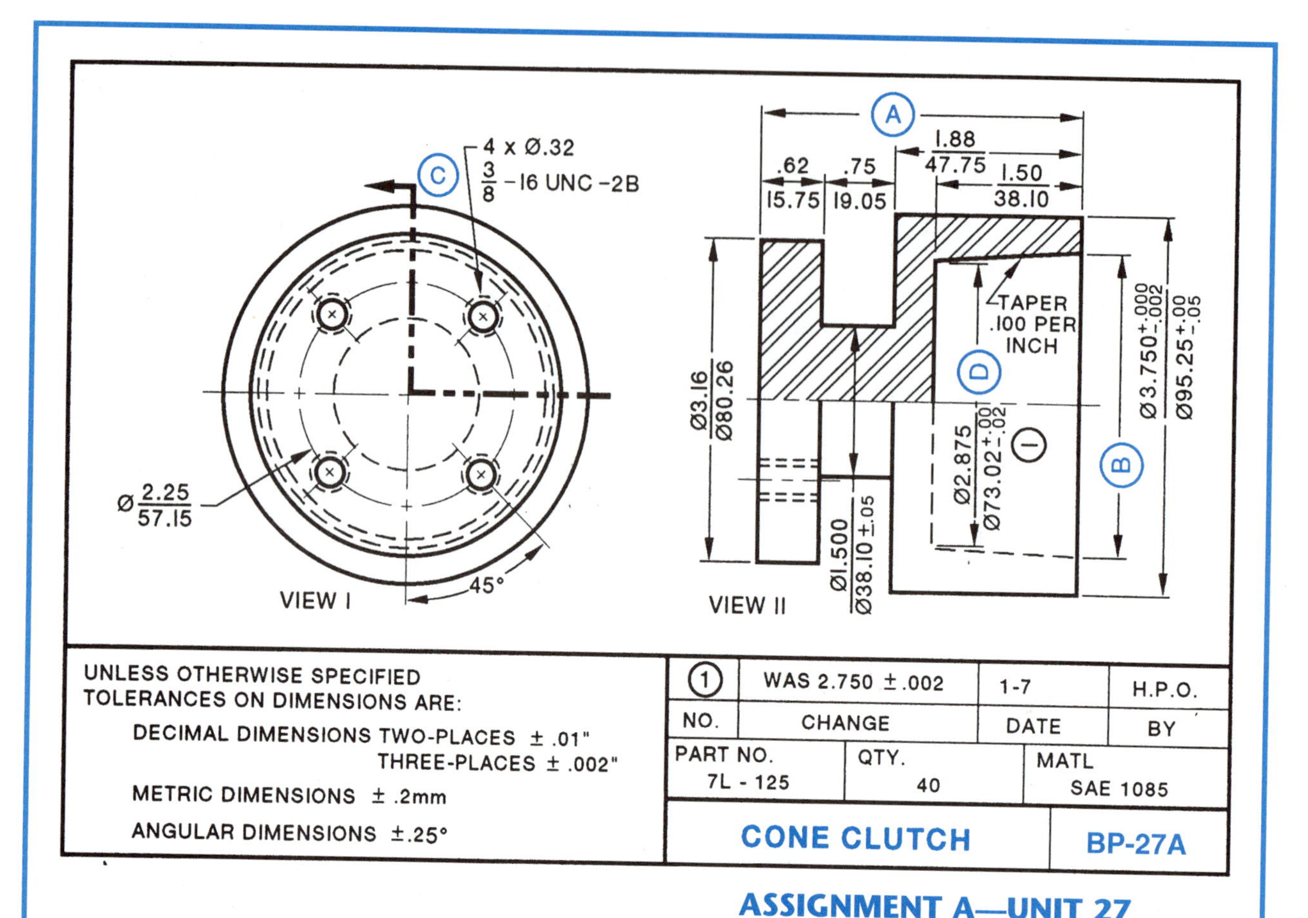

ASSIGNMENT A—UNIT 27

CONE CLUTCH (BP-27A)

Note: Compute all dimensions where applicable, in customary inch and SI metric mm values.

1. Name View I and View II in third-angle projection.
2. What type line is (C)?
3. Give the tapped holes specification.
4. What type of screw thread representation is used?
5. Give the diameter on which the threaded holes are located.
6. Give the upper limit dimension for the 45° angle.
7. Determine overall length (A).
8. Give the lower limit for the 1.500" DIA.
9. Compute nominal dimension (B).
10. Give the original diameter of (D).

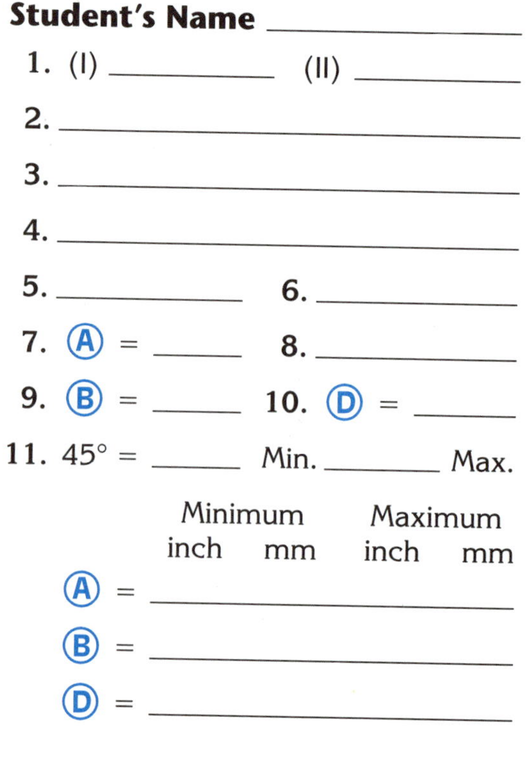
Student's Name ______________

1. (I) __________ (II) __________
2. ______________________
3. ______________________
4. ______________________
5. __________ 6. __________
7. (A) = ______ 8. __________
9. (B) = ______ 10. (D) = ______
11. 45° = ______ Min. ______ Max.

	Minimum inch	Minimum mm	Maximum inch	Maximum mm
(A) =				
(B) =				
(D) =				

11. Change the tolerances as follows: Two-place decimal dimensions $^{+.00}_{-.02}$; three-places, $^{+.000}_{-.0025"}$; millimeter dimensions $^{+.00}_{-.50}$; Angular dimensions $^{+.08°}_{-.25°}$. Then, compute the upper and lower limits for the 45° angle and dimensions (A), (B), and (D).

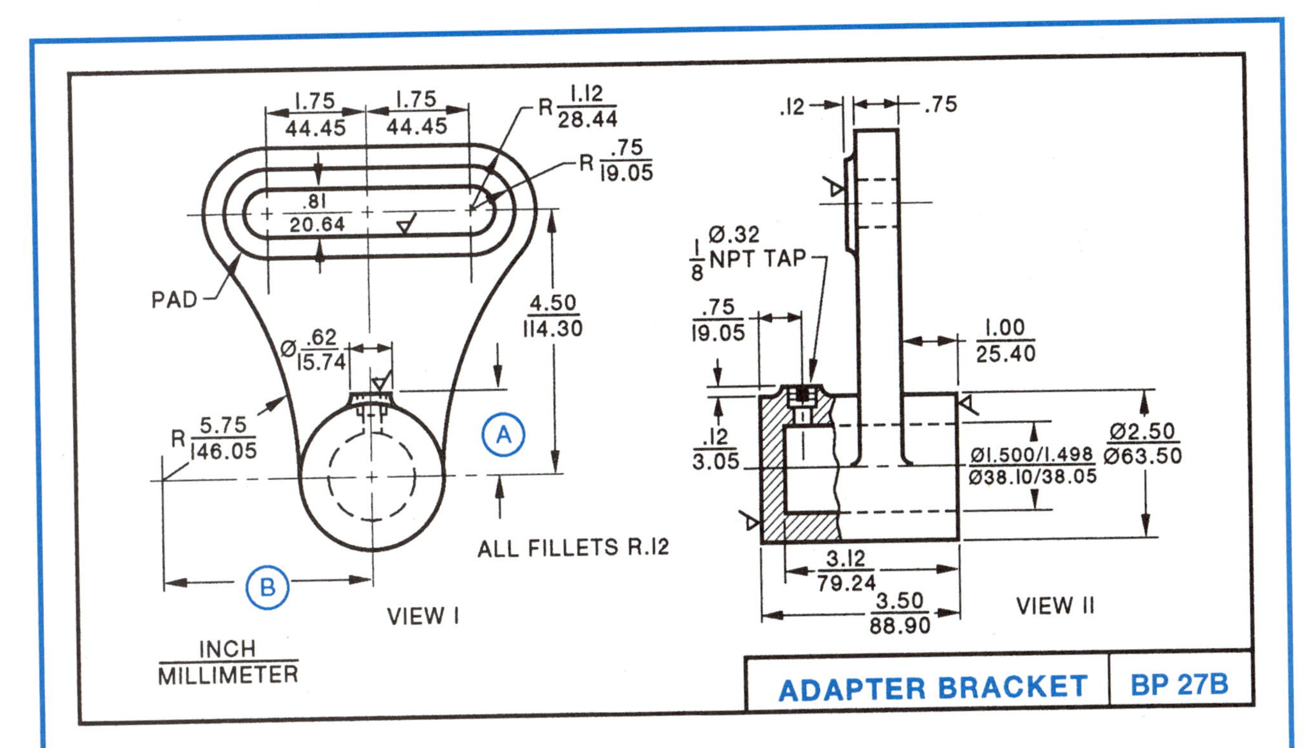

ADAPTER BRACKET (BP-27B)

Note: Compute all dimensions, where applicable, in Customary inch and SI metric (mm) values.

1. Name View I and View II (third-angle projection).
2. How many outside machined surfaces are indicated in View II?
3. Give the upper and lower limits of the bored hole.
4. How deep is the hole bored?
5. What is the drill size for the tapped hole?
6. What size pipe tap is used?
7. Give the center-to-center distance of the elongated slot.
8. Compute dimension Ⓐ.
9. Determine the overall width of the pad.
10. Compute dimension Ⓑ.
11. Add a ${}^{+.01''}_{-.00''}$ tolerance to all two-place decimal dimensions, and ${}^{+.20}_{-.00}$ mm to all metric dimensions. Then, determine the upper and lower limit dimensions for:
    - (a) The depth of the bored hole.
    - (b) The center-to-center distance between the elongated slot and the bored hole.
    - (c) The overall width of the bracket (View I).
    - (d) The overall bracket height.
    - (e) Dimensions Ⓐ and Ⓑ.

ASSIGNMENT B—UNIT 27

Student's Name ______________

1. (I) ______________ (II) ______________
2. ______________
3. Upper ________ " ________ mm
   Lower ________ " ________ mm
4. ________ " ________ mm
5. ______________
6. ______________
7. ________ " ________ mm
8. Ⓐ = ________ " ________ mm
9. ______________
10. Ⓑ = ______________

	Minimum inch	Minimum mm	Maximum inch	Maximum mm
11. (a)				
(b)				
(c)				
(d)				
(e) Ⓐ				
Ⓑ				

SECTION 7 Welding Drawings

Unit 28

SYMBOLS, REPRESENTATION, AND DIMENSIONING

WELDED STRUCTURES AND SYMBOLS

Many parts which were formerly cast in foundries are now being constructed by welding. Complete welding and machining information is conveyed from the part designer to the welder, machinist, and machine technician by graphic symbols. The symbols indicate the required type of weld, specific welding and machining dimensions, and other specifications.

TYPES OF WELDED JOINTS

There are five basic types of welded joints specified on drawings. Each type of joint is identified by the position of the parts to be joined together. Parts which are welded by using butt, corner, tee, lap, or edge-type joints are illustrated in figure 28-1.

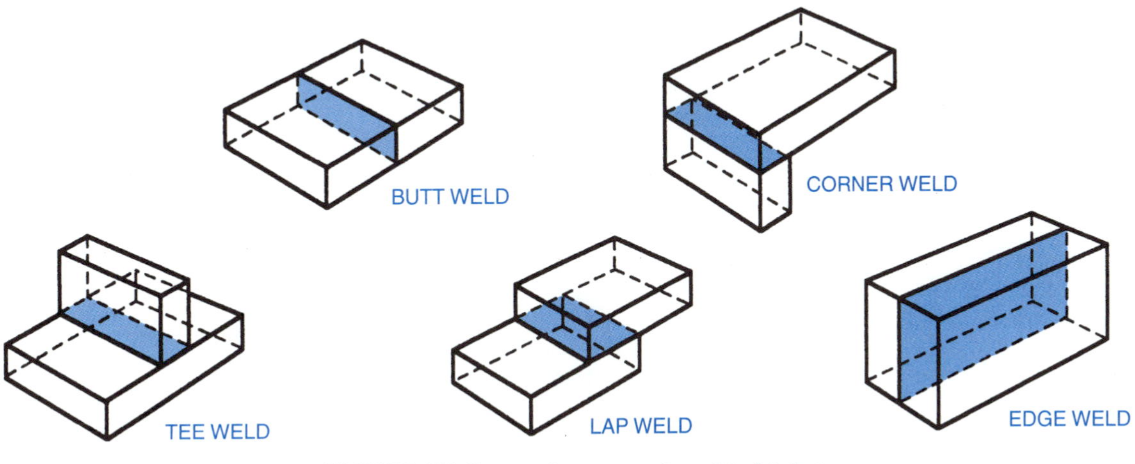

FIGURE 28-1 Basic types of welded joints.

ANSI SYMBOLS FOR ARC AND GAS WELDS

The most commonly used arc and gas welds for fusing parts are shown in figure 28-2. The welds are grouped as follows: (1) back or backing, (2) fillet, (3) plug and slot, and (4) groove. The groove welds are further identified as (A) square, (B) vee, (C) bevel, (D) **U**, (E) **J**, (F) flare, and (G) flare bevel.

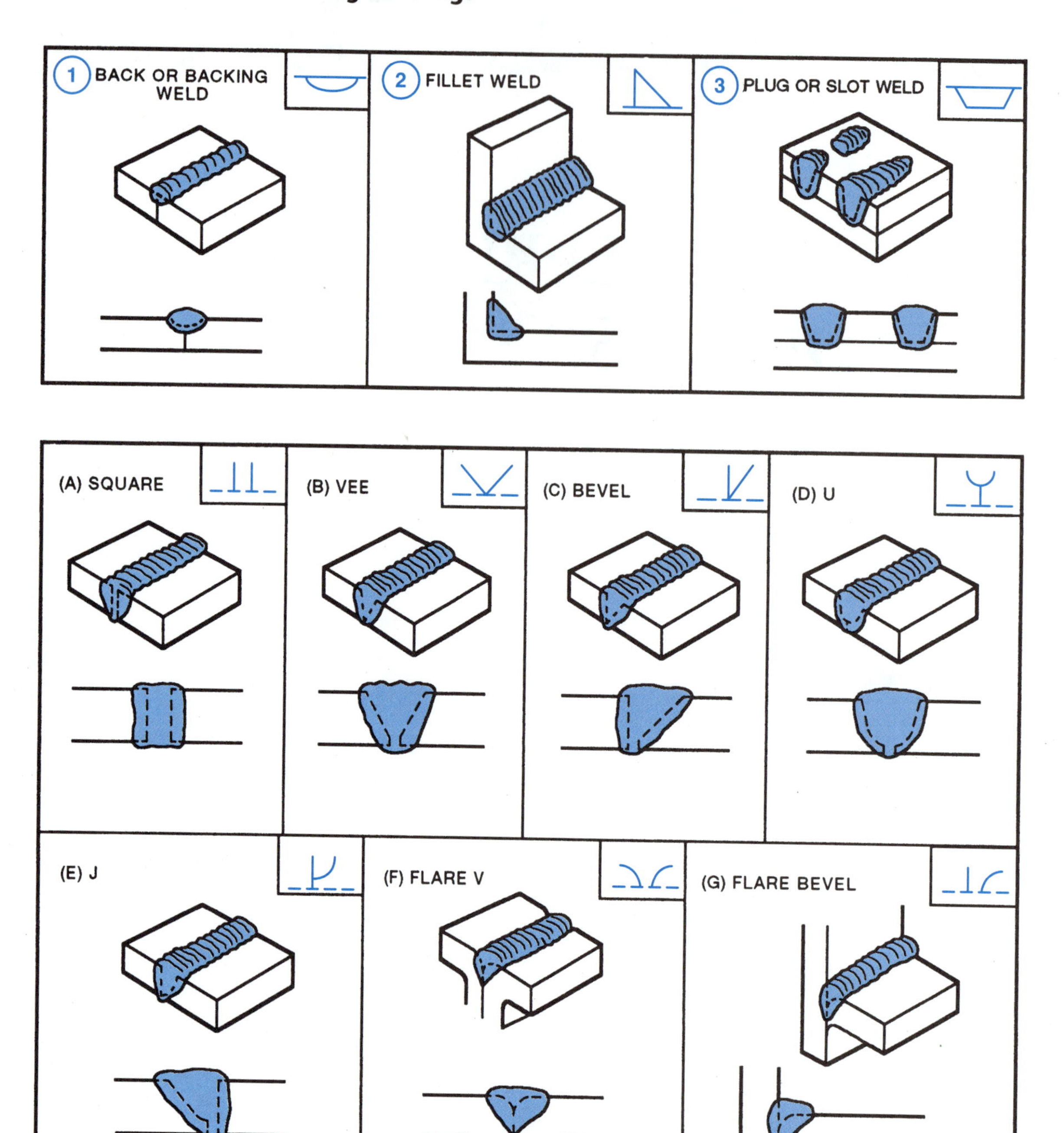

FIGURE 28-2 ANSI symbols and examples of common arc and gas welds.

RESISTANCE WELDING: TYPES AND SYMBOLS

Resistance welding is another method of fusing parts. The fusing temperature is produced in the particular area to be welded by applying force and passing electric current between two electrodes and the parts. Resistance welding does not require filler metal or fluxes. The symbols for general types of resistance welds are given in figure 28-3. These are *nonpreferred symbols.* Their replacement is recommended by using preferred symbols and including the process reference in the tail.

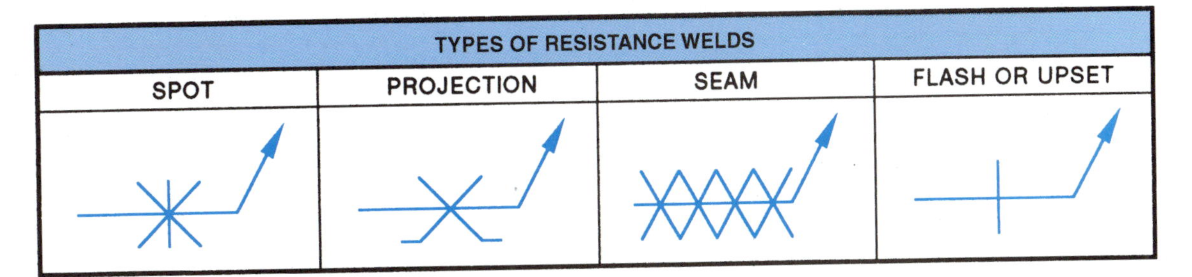

FIGURE 28-3 Nonpreferred symbols for resistance welds.

SUPPLEMENTARY WELD SYMBOLS

General supplementary weld symbols are shown in figure 28-4. These symbols convey additional information about the extent of welding, location, and contour of the weld bead. The contour symbols are placed above (**OTHER SIDE**) or below (**ARROW SIDE**) the weld symbol.

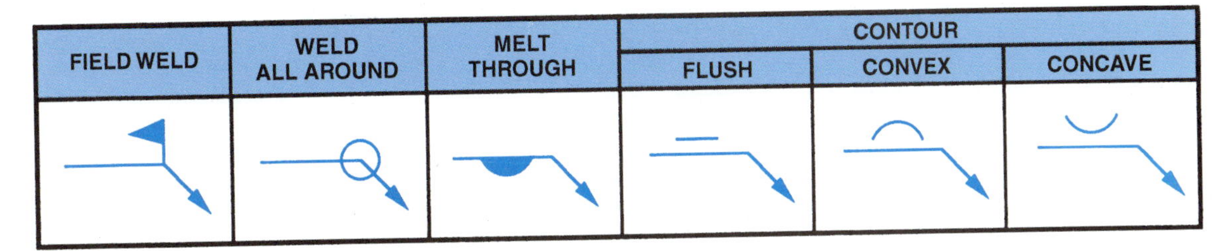

FIGURE 28-4 Supplementary weld symbols.

ANSI (AMERICAN WELDING SOCIETY) STANDARDS FOR SYMBOLS, REPRESENTATION ON DRAWINGS, AND DIMENSIONING

The main parts of the ANSI standard welding symbol (adopted in collaboration with the American Welding Society) are identified in figure 28-5.

The symbols used for parts ① through ⑨ are given in column (A), figure 28-6. A brief description of the function of each part follows in column (B). Examples of applications appear in column (C). ANSI standards are used on many metric drawings, pending the establishment of SI metric standards.

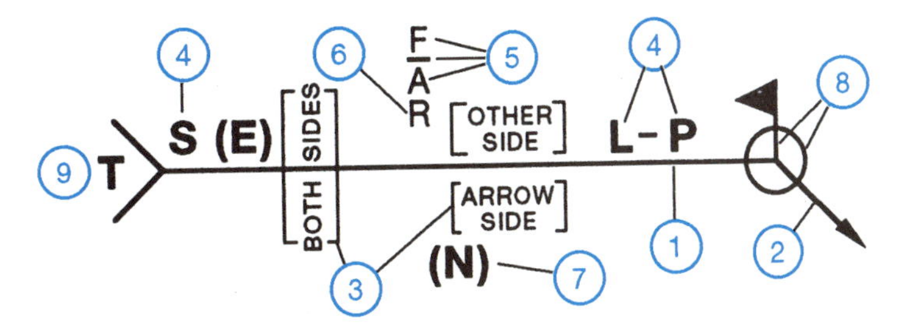

FIGURE 28-5 Parts of an ANSI standard welding symbol.

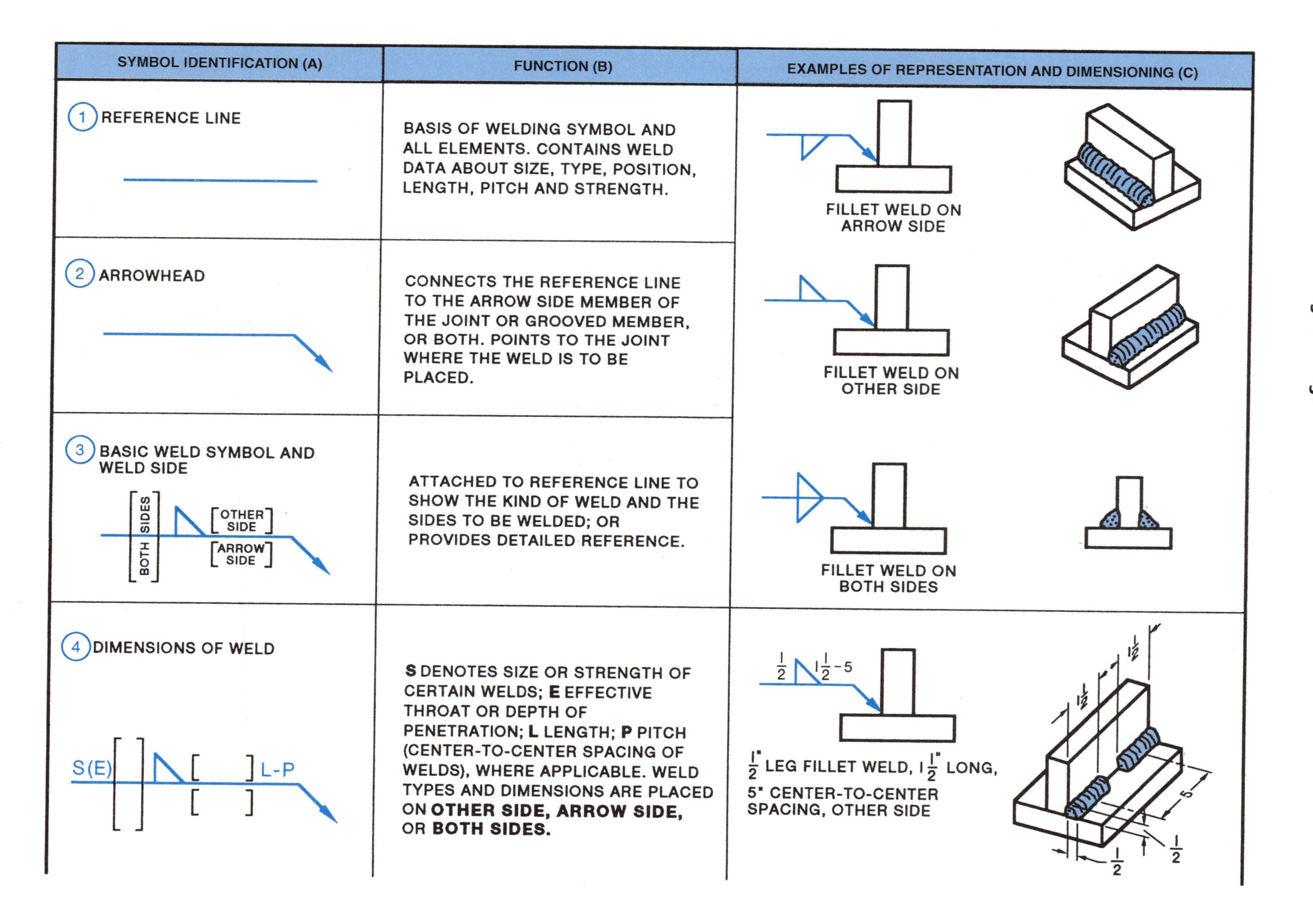

SYMBOL IDENTIFICATION (A)	FUNCTION (B)	EXAMPLES OF REPRESENTATION AND DIMENSIONING (C)
1 REFERENCE LINE	BASIS OF WELDING SYMBOL AND ALL ELEMENTS. CONTAINS WELD DATA ABOUT SIZE, TYPE, POSITION, LENGTH, PITCH AND STRENGTH.	FILLET WELD ON ARROW SIDE
2 ARROWHEAD	CONNECTS THE REFERENCE LINE TO THE ARROW SIDE MEMBER OF THE JOINT OR GROOVED MEMBER, OR BOTH. POINTS TO THE JOINT WHERE THE WELD IS TO BE PLACED.	FILLET WELD ON OTHER SIDE
3 BASIC WELD SYMBOL AND WELD SIDE BOTH SIDES; OTHER SIDE; ARROW SIDE	ATTACHED TO REFERENCE LINE TO SHOW THE KIND OF WELD AND THE SIDES TO BE WELDED; OR PROVIDES DETAILED REFERENCE.	FILLET WELD ON BOTH SIDES
4 DIMENSIONS OF WELD S(E) L-P	**S** DENOTES SIZE OR STRENGTH OF CERTAIN WELDS; **E** EFFECTIVE THROAT OR DEPTH OF PENETRATION; **L** LENGTH; **P** PITCH (CENTER-TO-CENTER SPACING OF WELDS), WHERE APPLICABLE. WELD TYPES AND DIMENSIONS ARE PLACED ON **OTHER SIDE, ARROW SIDE,** OR **BOTH SIDES.**	$\frac{1}{2}$ $1\frac{1}{2}$-5 $\frac{1}{2}$" LEG FILLET WELD, $1\frac{1}{2}$" LONG, 5" CENTER-TO-CENTER SPACING, OTHER SIDE $1\frac{1}{2}$ $1\frac{1}{2}$ 5 $\frac{1}{2}$ $\frac{1}{2}$

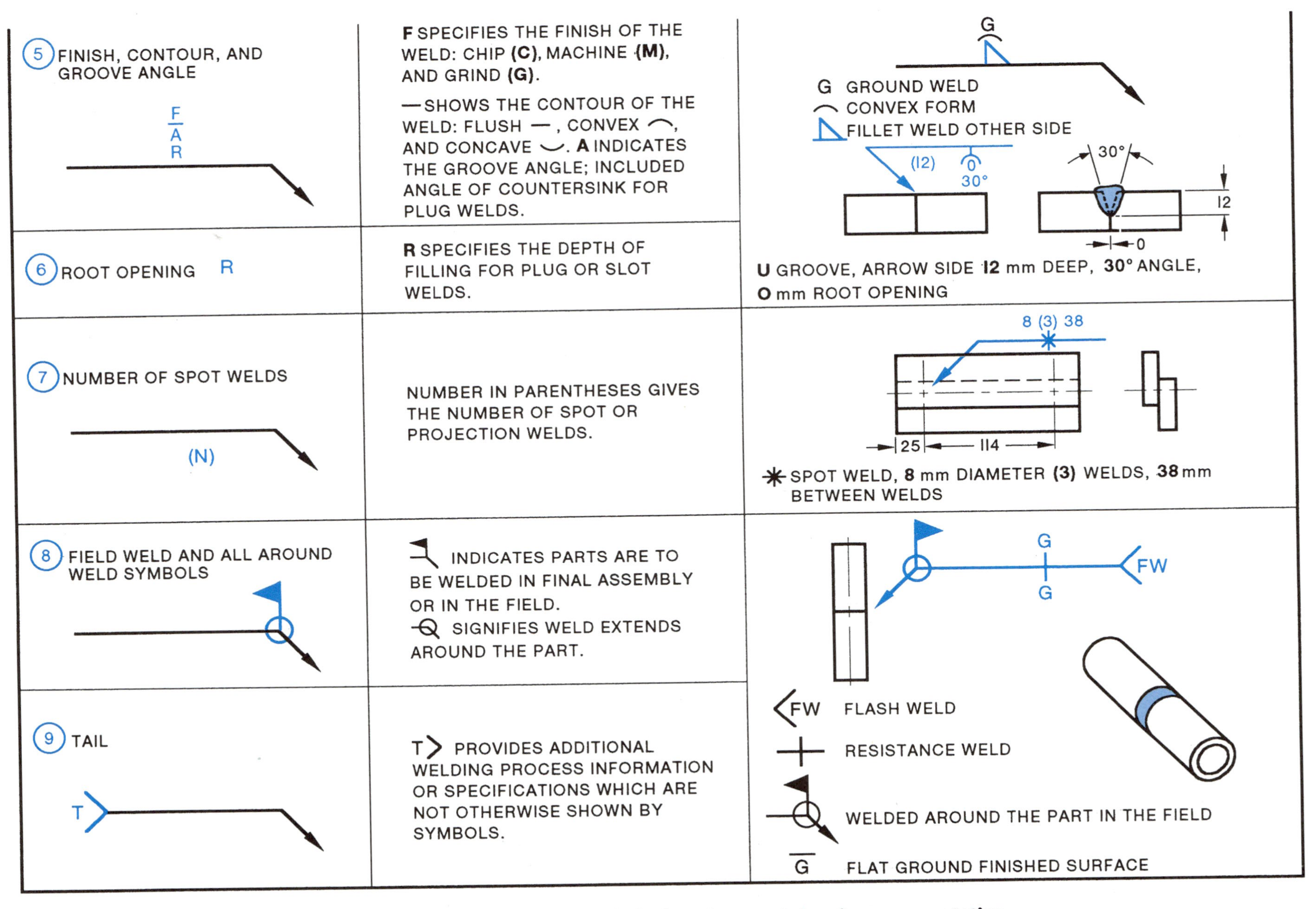

FIGURE 28-6 ANSI welding symbols, functions, and drawing representation.

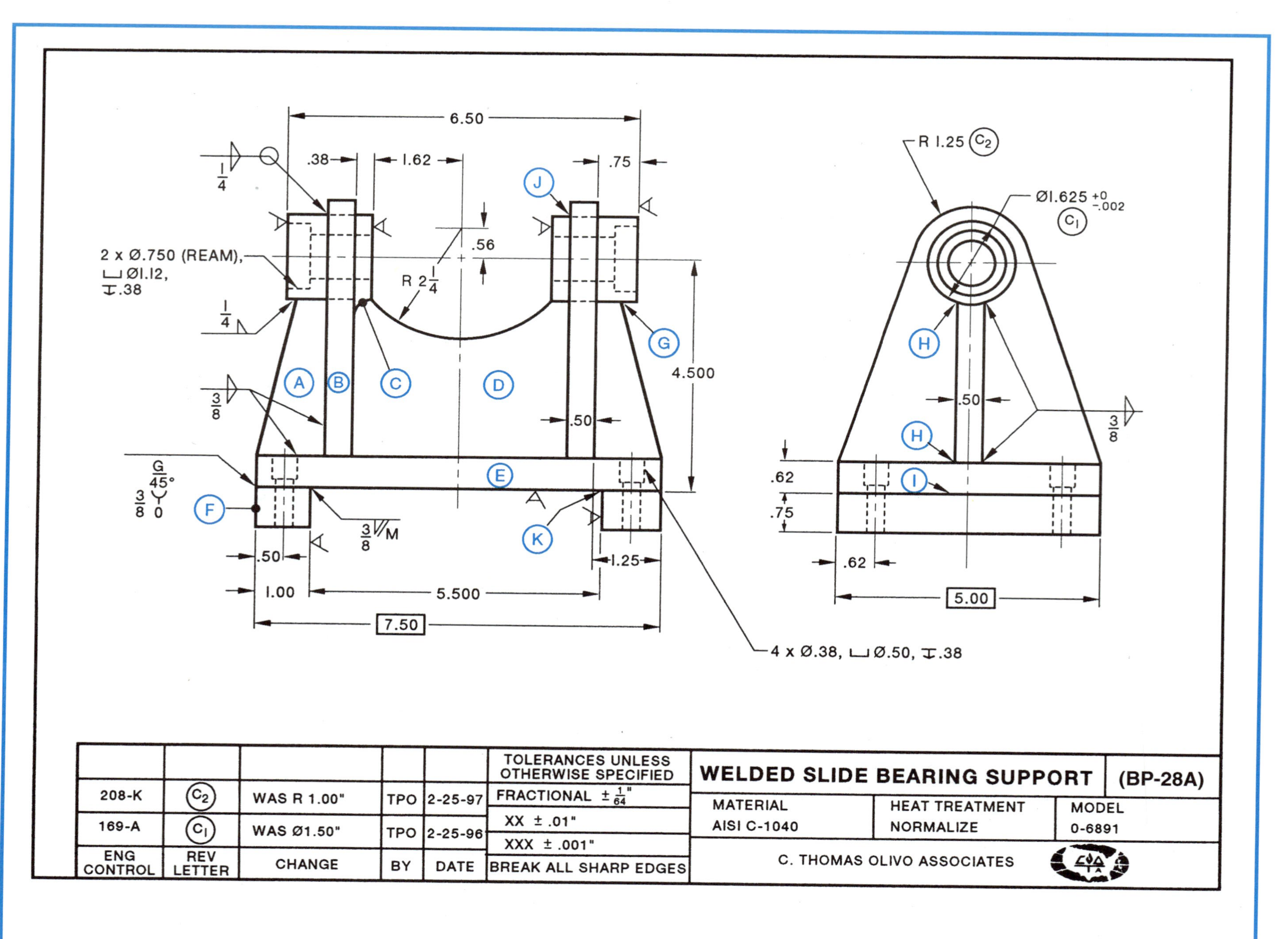
6.50
.38
1.62
.75
1/4
2 x Ø.750 (REAM), ⌴Ø1.12, ↧.38
.56
R 2 1/4
4.500
3/8
.50
G 45° 3/8 0
3/8 M
.50
1.00
5.500
1.25
7.50
R 1.25 C2
Ø1.625 +0 -.002 C1
.62
.75
5.00
4 x Ø.38, ⌴Ø.50, ↧.38
A
B
C
D
E
F
G
H
I
J
K
TOLERANCES UNLESS OTHERWISE SPECIFIED
FRACTIONAL ± 1/64"
XX ± .01"
XXX ± .001"
BREAK ALL SHARP EDGES
WELDED SLIDE BEARING SUPPORT
(BP-28A)
MATERIAL
AISI C-1040
HEAT TREATMENT
NORMALIZE
MODEL
0-6891
C. THOMAS OLIVO ASSOCIATES
208-K
C2
WAS R 1.00"
TPO
2-25-97
169-A
C1
WAS Ø1.50"
TPO
2-25-96
ENG CONTROL
REV LETTER
CHANGE
BY
DATE

WELDED SLIDE BEARING SUPPORT (BP-27A)

1. Give the nominal dimension for the overall length, depth (width), and height of the Bearing Support.
2. State what changes were made in dimensions (C1) and (C2) from the original.
3. Determine the upper and lower limit dimensions of the Shaft Bearing diameter.
4. a. Give the specifications for the reamed and recessed holes in the Shaft Bearings.
   b. Interpret each symbol and dimension. Include the tolerance for each diameter.
5. Prepare a Materials List with the following information:
   a. Give the required number of each part to fabricate the Bearing Support.
   b. Determine the size of the rectangular or round bar stock required for each part.
      Note: Convert each two-point decimal size to the equivalent fractional value, to the nearest 1/16".
   c. Give the code number for the steel to be used in the Bearing Support.
   d. Identify the heat treatment process.
6. a. Draw the ANSI symbols and values for welds (G) through (K).
   b. Interpret the meaning of each symbol and give the dimensions for welds (G) through (K).

ASSIGNMENT A—UNIT 28

Student's Name ______________________

1. L ________ D(W) ________ H ________
2. (C1) ____________ (C2) ____________
3. Upper ____________ Lower ____________
4. a. ______________________________
   b. ______________________________
5.

PART	NAME	NO. REQ'D (a)	SIZE (b)
(A)	END BRACE		
(B)	UPRIGHT BRACKET		
(C)	SHAFT BEARING		
(D)	CENTER BRACE		
(E)	BASE		
(F)	SLIDE BLOCKS		

c. ____________ d. ____________

6.

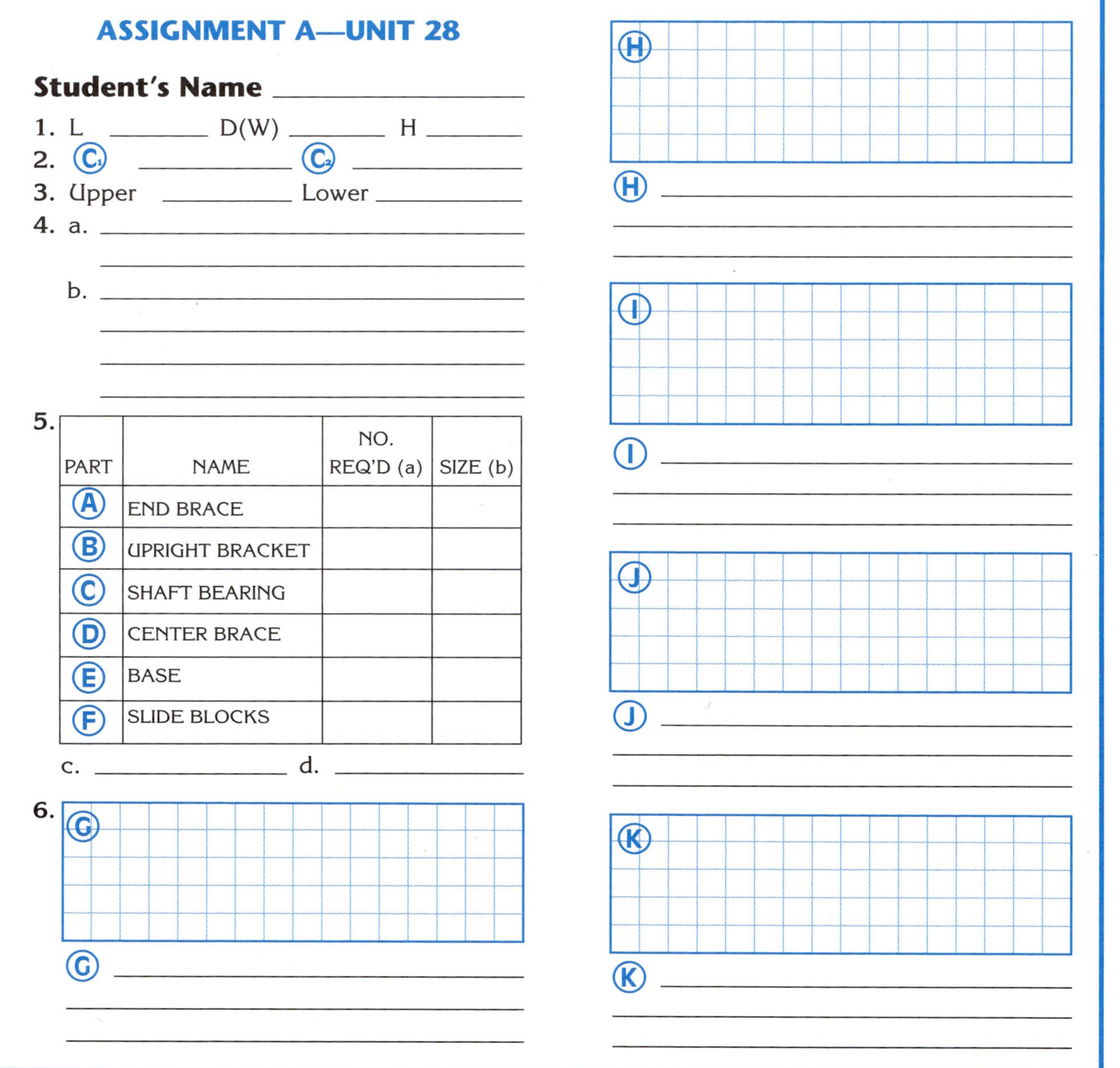

(G) ______________________________

(H) ______________________________

(I) ______________________________

(J) ______________________________

(K) ______________________________

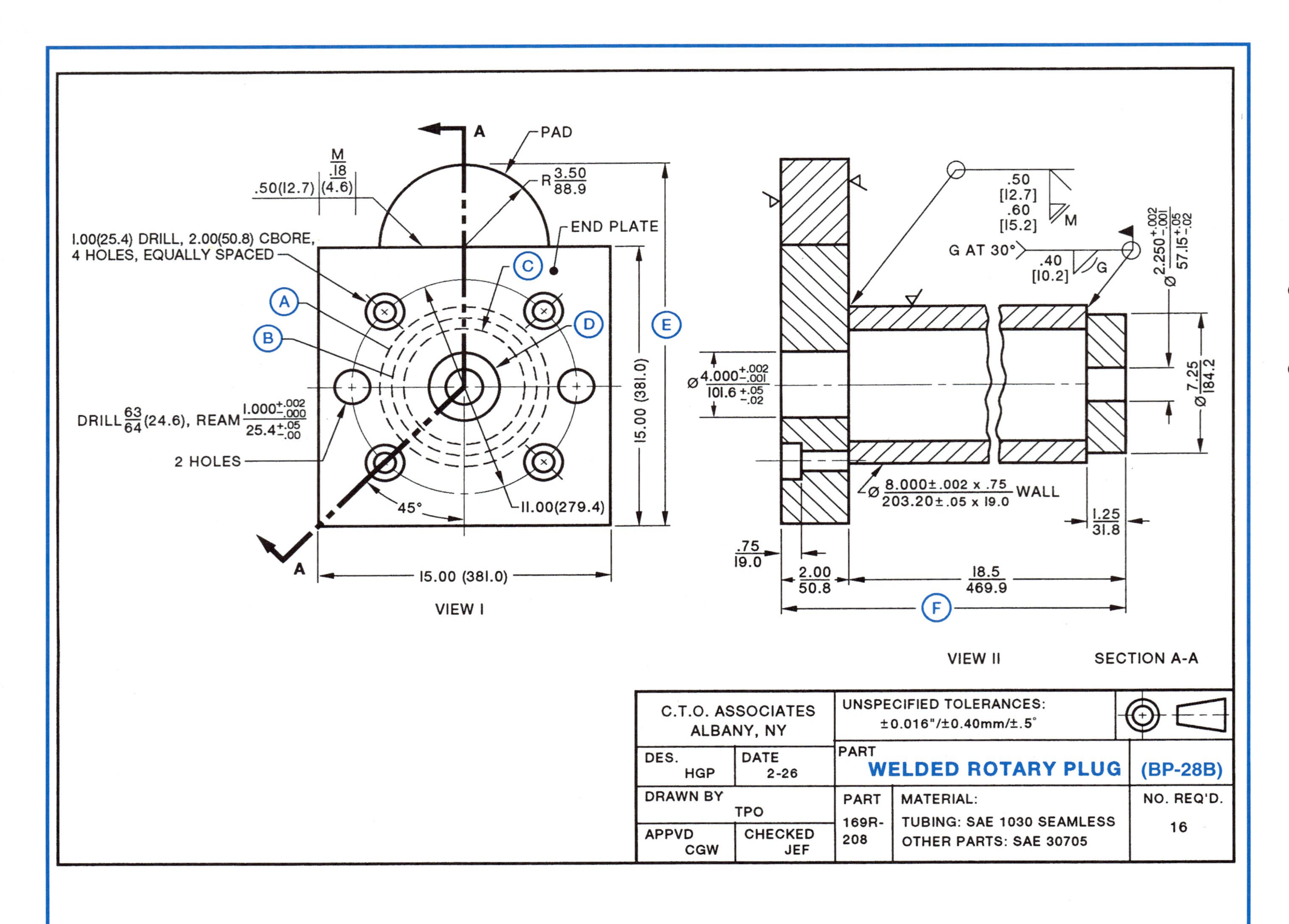

A
PAD
M .18
.50(12.7) (4.6)
R 3.50 88.9
END PLATE
1.00(25.4) DRILL, 2.00(50.8) CBORE, 4 HOLES, EQUALLY SPACED
A
B
C
D
E
DRILL 63/64 (24.6), REAM 1.000+.002 -.000 25.4+.05 -.00
2 HOLES
45°
11.00(279.4)
15.00 (381.0)
15.00 (381.0)
A
VIEW I
.50 [12.7] .60 [15.2]
M
G AT 30°
.40 [10.2]
G
Ø 2.250+.002 -.001 57.15+.05 -.02
Ø 4.000+.002 -.001 101.6+.05 -.02
Ø 7.25 184.2
Ø 8.000±.002 x .75 WALL 203.20±.05 x 19.0
1.25 31.8
.75 19.0
2.00 50.8
18.5 469.9
F
VIEW II
SECTION A-A
C.T.O. ASSOCIATES ALBANY, NY
UNSPECIFIED TOLERANCES: ±0.016"/±0.40mm/±.5°
DES. HGP
DATE 2-26
PART
WELDED ROTARY PLUG
(BP-28B)
DRAWN BY TPO
APPVD CGW
CHECKED JEF
PART 169R-208
MATERIAL:
TUBING: SAE 1030 SEAMLESS
OTHER PARTS: SAE 30705
NO. REQ'D. 16

WELDED ROTARY PLUG (BP-28B)

NOTE: All dimensions are to be given in both inch and millimeter values.

1. Name the two views.
2. List (a) the wall thickness of the tubing and (b) the material for the tube body and other parts.
3. Give the nominal diameters of (A), (B), and (C).
4. Determine the maximum and minimum diameters for the bored hole (D).
5. Establish the maximum overall height (E) and length (F).
6. Give the specifications for the reamed holes.
7. Identify (a) the minimum diameter and (b) the depth of the counterbored holes.
8. Specify the size of the welded pad.
9. Determine how many surfaces are to be machined.
10. Give the maximum dimensions of the circular end plate.
11. Indicate (a) the type of welded joint between the pad and the 2.00" thick plate, and (b) give the weld specifications.
12. Interpret the weld specifications for the tubing and square end plate.
13. Draw the ANSI symbols which describe the weld for the circular end plate and tubing body.
14. Interpret the symbols and dimensions for the circular end plate and body weld.

ASSIGNMENT B—UNIT 28

Student's Name ______

1. View I ______ View II ______
2. (a) Wall thickness ______
   (b) Materials. ______
   ______
3. (A) ______ nominal diameter
   (B) ______ nominal diameter
   (C) ______ nominal diameter
4. (D) Maximum diameter. ______
   Minimum diameter. ______
5. (E) Height ______ (F) Length ______
6. ______
   ______
   ______
7. (a) Minimum diameter ______
   (b) Minimum depth ______
8. ______
9. ______
10. ______
    ______
11. (a) Type ______
    (b) Specifications. ______
    ______
12. ______
    ______
    ______
    ______
13. ______
14. ______
    ______
    ______
    ______

SECTION 8 Computer Numerical Control (CNC) Fundamentals

Unit 29

DATUMS: ORDINATE AND TABULAR DIMENSIONING

COMPUTER NUMERICAL CONTROL

Machines controlled by data which are specially coded with numbers, letters, and special characters are operated by *numerical control (NC)*. This encoded information defines the path of the machine tool in accomplishing specific tasks. These instructions can be fed into the machine from several different sources—an operator, a computer, a punched tape, or a floppy disk. When the job changes, the instructions must be rewritten to accommodate these changes. This is a relatively easy job and this flexibility is the key to the growth of numerical control.

Actually, NC is a broad term that includes traditional approaches as well as more modern developments such as *computer numerical control (CNC)* and *direct numerical control (DNC)*.

The most popular application of NC is metal cutting, but NC is being successfully applied to other equipment, such as laser and other cutting and torching machines, tube bending machines, and punch presses. See Unit 30E for additional CNC information.

DATUMS

A *datum* is a reference feature from which dimensions are located and essential information is derived. A datum may be a point, line, plane, cylinder, or other exact feature. Datums are used in design, manufacturing, numerical control, and many other processes. Datums give location dimensions, furnish data for computations, and provide a reference source.

When a datum is specified, such as the two that are illustrated in figure 29-1, all features must be stated in relation to the datum reference. Different features of the part are located from the datum, not from other features. Datums must be clearly identified or easily recognizable. At least two, and often three, datums must be used to define or measure a part. Since all measurements are taken from datums (similar to those earlier used in base line dimensioning), errors in dimensioning are not cumulative.

ORDINATE DIMENSIONING AND NUMERICAL CONTROL MACHINE DRAWINGS

Ordinate dimensioning is a type of rectangular datum dimensioning. In ordinate dimensioning the dimensions are measured from two or three mutually related datum planes, figure 29-1. The datum planes are indicated as *zero coordinates*. Dimensions from the zero coordinates are represented on drawings as extension lines, without the use of dimension lines or arrowheads.

Specific features are located by the intersection of datum dimensions. Ordinate dimensioning is used when there would otherwise be a large number of dimensions and features and close tolerances are required. Datums and ordinate dimensioning are the foundation of numerically-controlled machine drawings and precision parts.

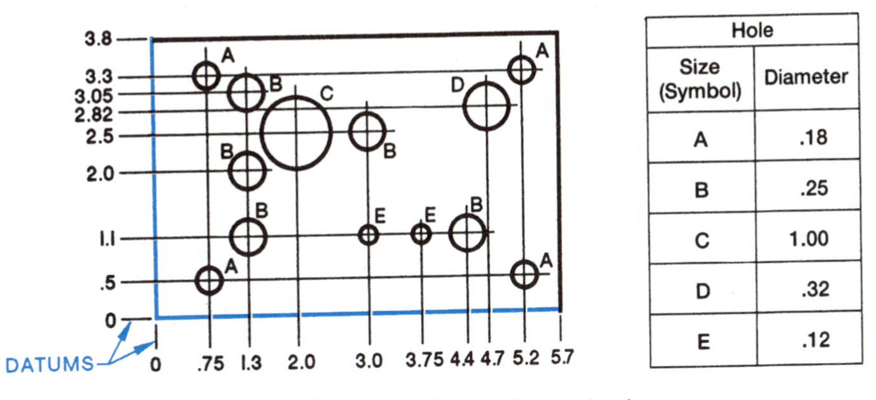

Hole	
Size (Symbol)	Diameter
A	.18
B	.25
C	1.00
D	.32
E	.12

FIGURE 29-1 Ordinate dimensioning.

TABULAR DIMENSIONING

Tabular dimensioning is another form of rectangular datum dimensioning. As the term implies, dimensions from intersecting datum planes are given in a table, figure 29-2. Tabular dimensioning helps to eliminate possible errors which could result from incorrectly reading a dimension when a large number are included on a drawing.

Values of **x** and **y** are measured along the respective datum lines from their intersection. The intersection is the origin of the datums. Tabular dimensioning is recommended where there are a great many repetitive features. These would make a dimensioned drawing difficult to read. Dimensions may also be given on a second or additional view for operations to be performed along **Z** or other axes.

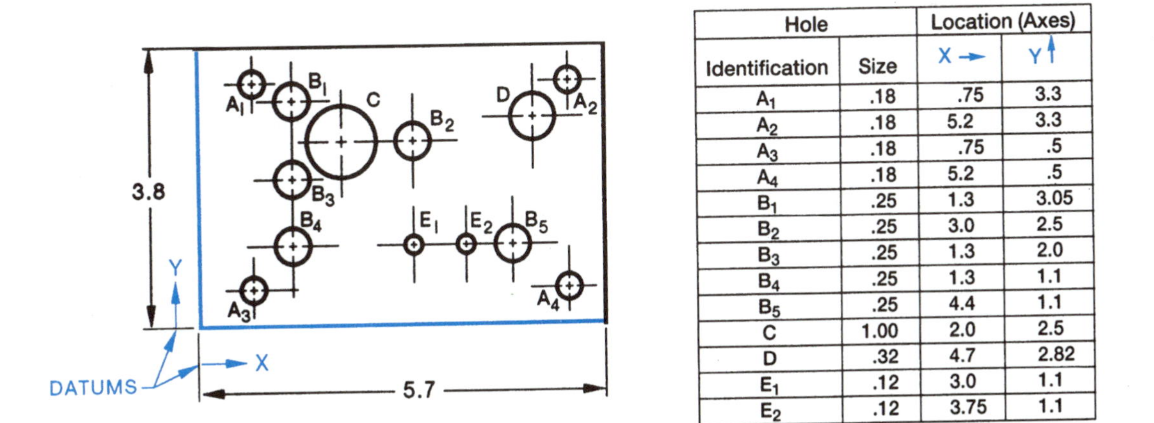

Hole		Location (Axes)	
Identification	Size	X →	Y ↑
A_1	.18	.75	3.3
A_2	.18	5.2	3.3
A_3	.18	.75	.5
A_4	.18	5.2	.5
B_1	.25	1.3	3.05
B_2	.25	3.0	2.5
B_3	.25	1.3	2.0
B_4	.25	1.3	1.1
B_5	.25	4.4	1.1
C	1.00	2.0	2.5
D	.32	4.7	2.82
E_1	.12	3.0	1.1
E_2	.12	3.75	1.1

FIGURE 29-2 Coordinate chart and tabular dimension drawing.

COORDINATE CHARTS

The dimensions of a coordinate and features that are to be fabricated on coordinates are given in a coordinate chart, figure 29-2. Only those dimensions that originate at the datums are found in such a chart. The coordinates may specify sufficient information to fabricate the entire feature or only a portion of it.

Parts or components may be represented on drawings by ordinate or tabular dimensioning in the English or metric system of measurement or in both systems.

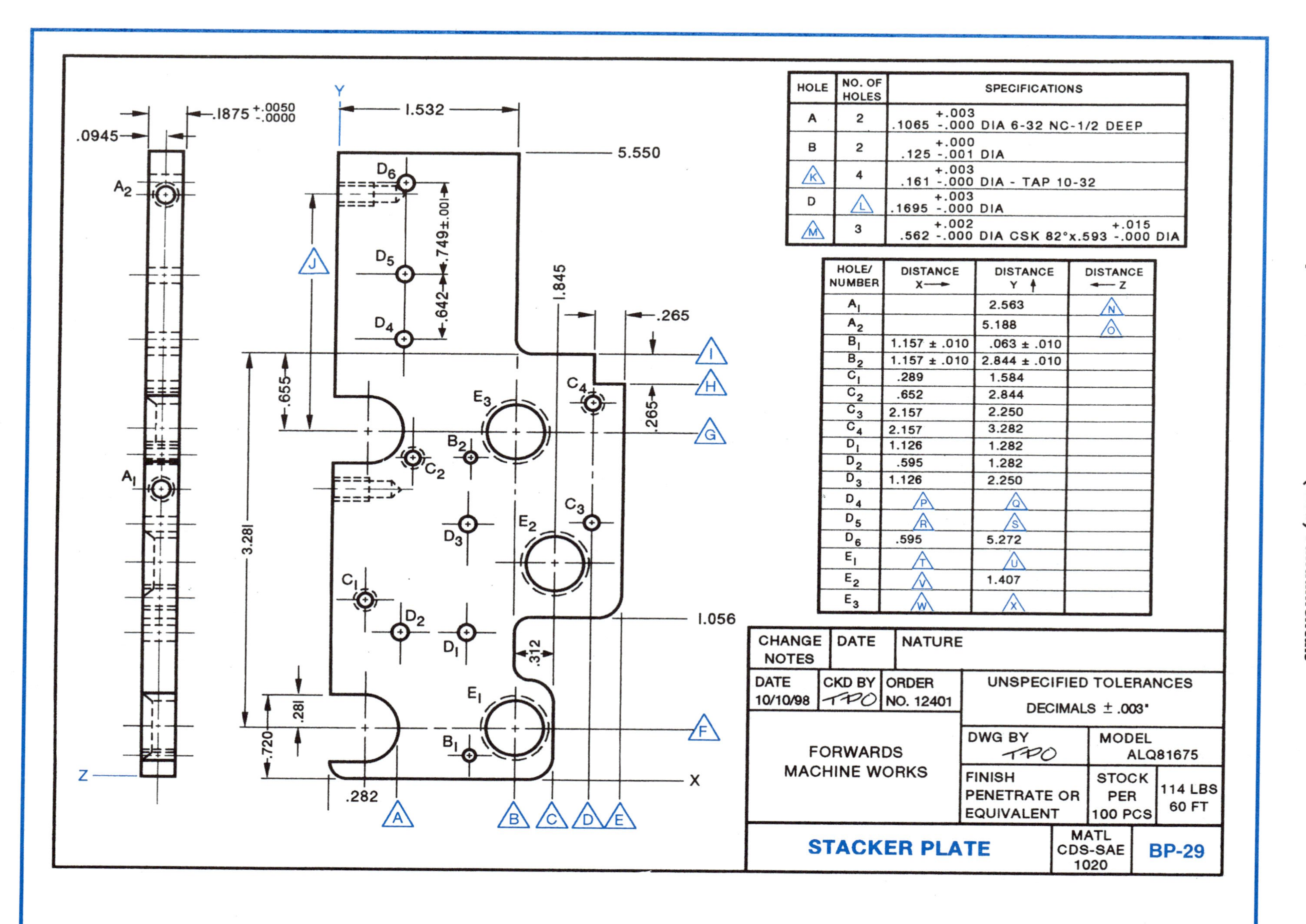

HOLE | NO. OF HOLES | SPECIFICATIONS
A | 2 | .1065 +.003 -.000 DIA 6-32 NC-1/2 DEEP
B | 2 | .125 +.000 -.001 DIA
K | 4 | .161 +.003 -.000 DIA - TAP 10-32
D | L | .1695 +.003 -.000 DIA
M | 3 | .562 +.002 -.000 DIA CSK 82°x.593 +.015 -.000 DIA
HOLE/NUMBER | DISTANCE X | DISTANCE Y | DISTANCE Z
A1 | | 2.563 | N
A2 | | 5.188 | O
B1 | 1.157 ± .010 | .063 ± .010 |
B2 | 1.157 ± .010 | 2.844 ± .010 |
C1 | .289 | 1.584 |
C2 | .652 | 2.844 |
C3 | 2.157 | 2.250 |
C4 | 2.157 | 3.282 |
D1 | 1.126 | 1.282 |
D2 | .595 | 1.282 |
D3 | 1.126 | 2.250 |
D4 | P | Q |
D5 | R | S |
D6 | .595 | 5.272 |
E1 | T | U |
E2 | V | 1.407 |
E3 | W | X |
CHANGE NOTES
DATE
NATURE
DATE 10/10/98
CKD BY TPO
ORDER NO. 12401
FORWARDS MACHINE WORKS
UNSPECIFIED TOLERANCES
DECIMALS ± .003"
DWG BY TPO
MODEL ALQ81675
FINISH PENETRATE OR EQUIVALENT
STOCK PER 100 PCS
114 LBS 60 FT
MATL CDS-SAE 1020
STACKER PLATE
BP-29
5.550
1.532
1.845
.265
.265
1.056
.312
.282
.281
.720
3.281
.655
.749±.001
.642
.1875 +.0050 -.0000
.0945
X
Y
Z
A
B
C
D
E
F
G
H
I
J
A1
A2
B1
B2
C1
C2
C3
C4
D1
D2
D3
D4
D5
D6
E1
E2
E3

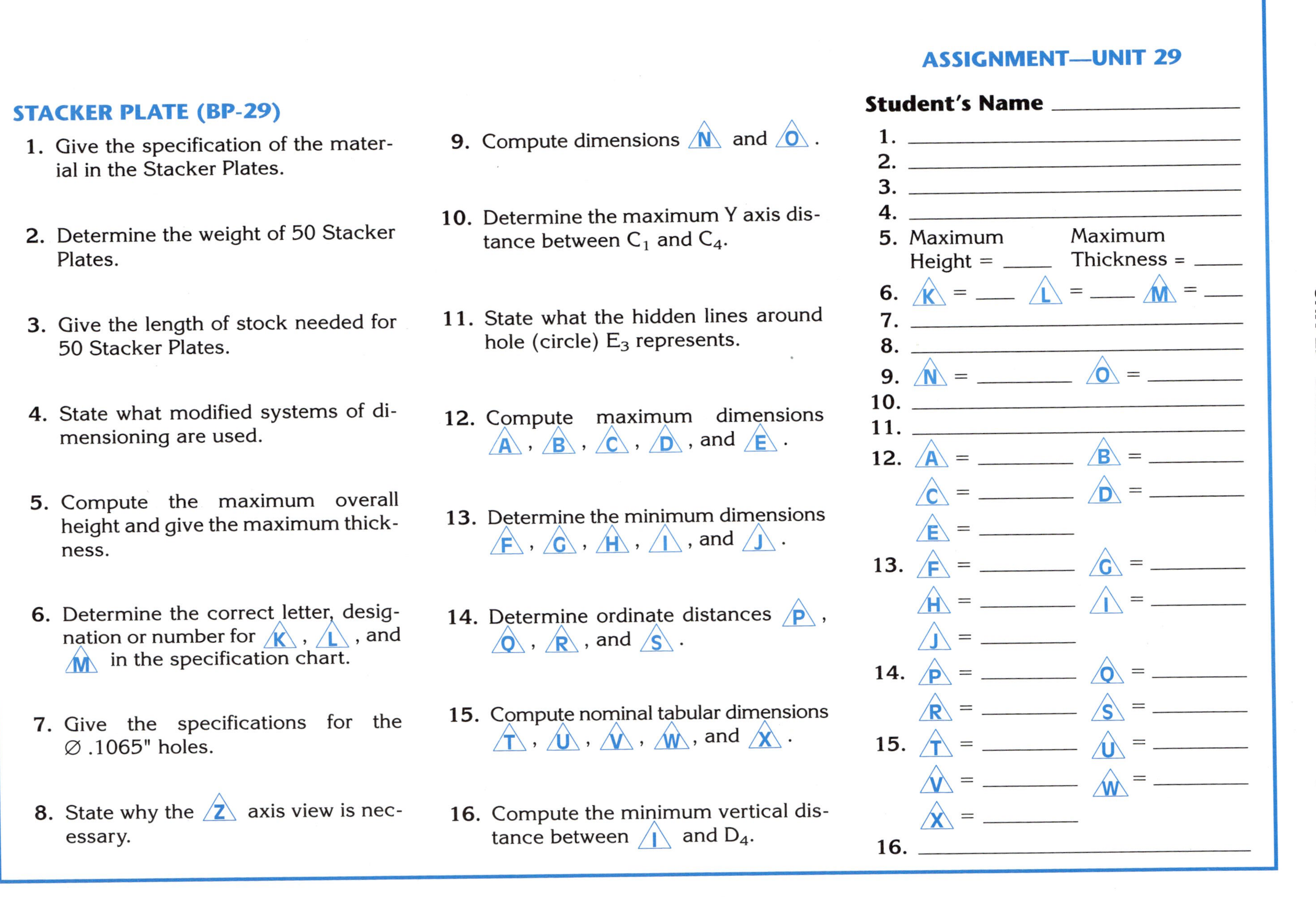

STACKER PLATE (BP-29)

1. Give the specification of the material in the Stacker Plates.
2. Determine the weight of 50 Stacker Plates.
3. Give the length of stock needed for 50 Stacker Plates.
4. State what modified systems of dimensioning are used.
5. Compute the maximum overall height and give the maximum thickness.
6. Determine the correct letter, designation or number for K, L, and M in the specification chart.
7. Give the specifications for the ∅ .1065" holes.
8. State why the Z axis view is necessary.
9. Compute dimensions N and O.
10. Determine the maximum Y axis distance between C_1 and C_4.
11. State what the hidden lines around hole (circle) E_3 represents.
12. Compute maximum dimensions A, B, C, D, and E.
13. Determine the minimum dimensions F, G, H, I, and J.
14. Determine ordinate distances P, Q, R, and S.
15. Compute nominal tabular dimensions T, U, V, W, and X.
16. Compute the minimum vertical distance between I and D_4.

ASSIGNMENT—UNIT 29

Student's Name ______________

1. ______________
2. ______________
3. ______________
4. ______________
5. Maximum Height = ____ Maximum Thickness = ____
6. K = ____ L = ____ M = ____
7. ______________
8. ______________
9. N = ________ O = ________
10. ______________
11. ______________
12. A = ________ B = ________
    C = ________ D = ________
    E = ________
13. F = ________ G = ________
    H = ________ I = ________
    J = ________
14. P = ________ Q = ________
    R = ________ S = ________
15. T = ________ U = ________
    V = ________ W = ________
    X = ________
16. ______________

SECTION 9 Geometric Dimensioning and Tolerancing

Unit 30

GEOMETRIC DIMENSIONING, TOLERANCING, AND DATUM REFERENCING

Geometric dimensioning and tolerancing (using datums) constitute a standardized ANSI system that complements more widely used conventional drafting room practices. Geometric dimensioning and tolerancing provides for the largest amount of permissible variation in form (shape and size) and position (location) for interchangeable parts.

Tolerances of form relate to straightness, flatness, roundness, cylindricity, perpendicularity, angularity, and parallelism. *Tolerances of position* relate to the location of such features as holes, grooves, stepped areas, and other details of a part.

Dimensioning and tolerancing, as covered in earlier units, relate to acceptable limits of deviation represented on drawings by using unilateral, bilateral, and high- and low-limit tolerancing. By contrast, geometric dimensioning and tolerancing provide for all characteristics of a part in terms of function, relationship to other mating parts, and to design, production, measurement, inspection, assembly, and operation.

TYPES, CHARACTERISTICS, AND SYMBOLS

The five basic types of tolerances are: (1) *form,* (2) *profile,* (3) *orientation,* (4) *location,* and (5) *runout.* Some tolerances relate to individual or related features of a part; others to a combination of individual and related features (figure 30-1A). Other characteristics are identified by *modifying symbols* (figure 30-1B) that are added following a tolerance value to indicate how the tolerance is modified. There are four basic modifying symbols: Ⓜ, Ⓛ, Ⓢ, and Ⓟ. *Maximum material condition* (MMC) implies that a tolerance, modified by the symbol Ⓜ, requires a part to be machined to a size limit where it contains the maximum amount of material. In other words, the MMC of a shaft of 0.750" diameter with $^{+0.002}_{-0.002}$ tolerancing contains the maximum amount of material when it measures 0.752". The *least material condition* (LMC), identified by the symbol Ⓛ, would be 0.748". The *regardless of feature size* (RFC) symbol Ⓢ indicates that a tolerance of form or position for any characteristic is maintained regardless of the actual size of a produced part. The *projected tolerance zone* symbol Ⓟ indicates that a tolerance (locational) zone is extended a specified distance.

DATUM REFERENCE FRAME AND PART FEATURES

To review, a *datum* identifies the origin of a dimensional relationship between a particular (designated) point or surface and a measurement. In geometric dimensioning and tolerancing, datums are identified in relation to a theoretical *reference frame.* The datum reference frame (similar to a drawing projection box) consists of three mutually perpendicular planes: *first datum plane, second datum plane,* and *third (tertiary) datum plane,* figure 30-2. The letters A, B, and C indicate the sequence (order of importance) of the datums; the arrowheads, the direction of the measurements.

(A) Geometric Tolerancing Characteristics and Symbols			
Features	**Type of Tolerance**	**Characteristic**	**Symbol**
Individual Features	Form	Straightness	⏤
		Flatness	⏥
		Roundness (Circularity)	○
		Cylindricity	⌭
Related Features	Orientation	Angularity	∠
		Perpendicularity	⊥
		Parallelism	//
	Location	Position	⌖
		Concentricity	◎
	Runout	Circular Runout	↗ ↗
		Total Runout	⌰ ⌰
Individual or Related Features	Profile	Profile of a Line	⌒
		Profile of a Surface	⌓

(B) Selected Modifying Terms and Symbols	
Term (Abbreviation)	**Symbol**
Maximum material condition (MMC)	Ⓜ
Least material condition (LMC)	Ⓛ
Regardless of feature size (RFC)	Ⓢ
Projected tolerance zone	Ⓟ

FIGURE 30-1 Geometric tolerancing characteristics, modifying terms, and symbols.

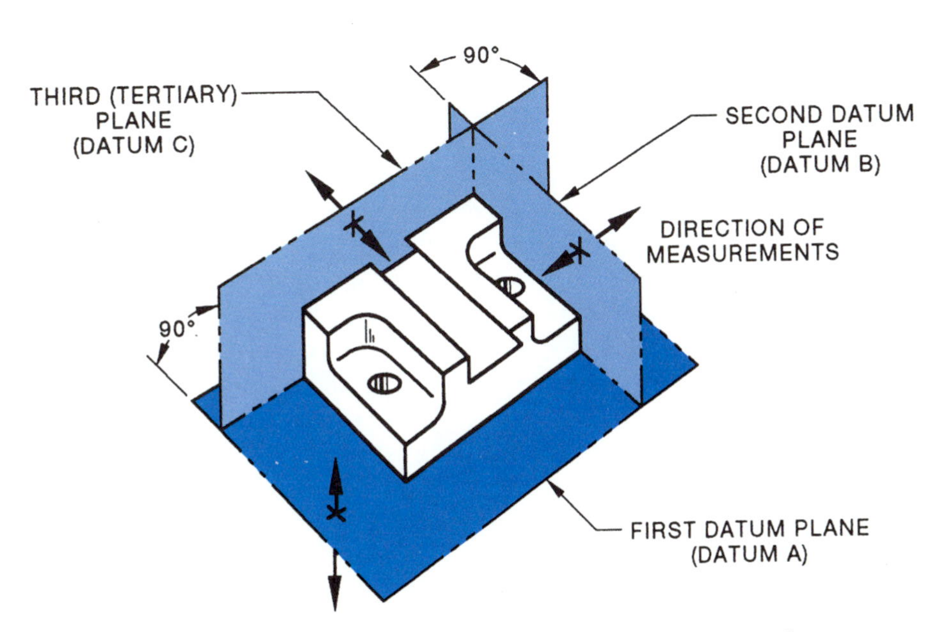

FIGURE 30-2 Datum planes, sequence of datums, and directions of measurements.

DATUM FEATURES OF CYLINDRICAL PARTS

Datums for cylindrical parts are generally related to three datum planes. Datum **A** identifies the base of the part. Datum **B** refers to the axis or center line. These datums are shown in figure 30-3. When used, Datum **C** is perpendicular to Datum **A** and parallel to Datum **B**.

DATUM FEATURE AND TARGET SYMBOLS

A *datum feature* symbol consists of a rectangular box or frame and the datum identifying letter (figure 30-4A). A dash (–) precedes and follows the letter. When tolerancing relates to two datums, the two letter symbols, separated by a dash, are contained in the datum feature frame.

A *datum target* symbol consists of a circle divided horizontally to form two half circles (figure 30-4B). A radial leader is used to reference the datum target symbol and contents to a specific surface, point, or line on a drawing. The size and location of the target area may be included using basic dimensioning practices.

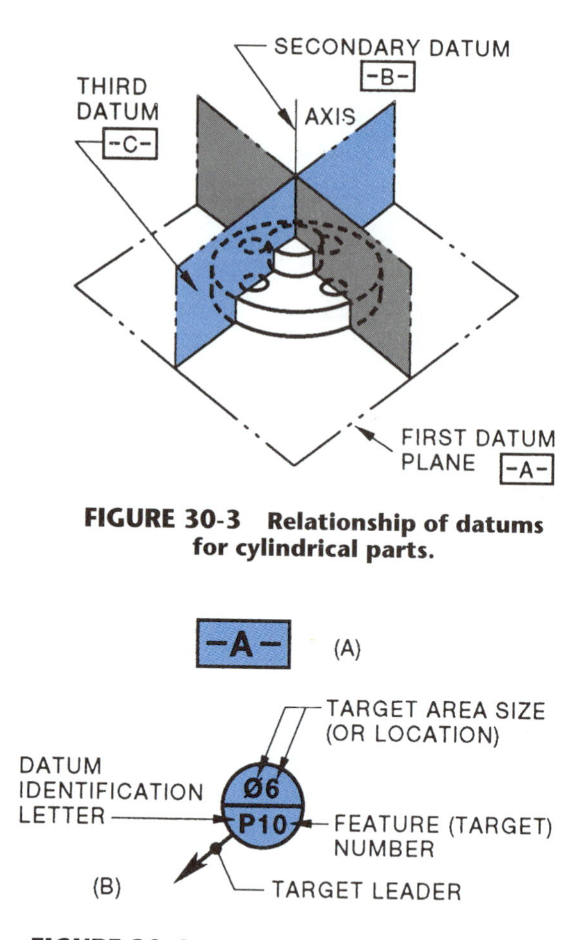

FIGURE 30-3 Relationship of datums for cylindrical parts.

FIGURE 30-4 Examples of datum feature and target control symbols.

FEATURE CONTROL FRAME

A *feature control frame* provides a convenient, accurate method for incorporating geometric characteristic symbols, tolerance values, modifying symbols, datum, and dimensioning and tolerancing information. The control frame may contain one or more compartments in which control information is identified following a specified sequence (figure 30-5).

The first entry in a feature control frame is the symbol of the geometric characteristic. The second entry provides the tolerancing dimension. When applicable, the symbol for material condition is the third entry. The tolerancing and material condition symbols are contained in one compartment of the frame. This information is followed by one or more additional compartments, depending on the number of datum reference letters (symbols) that are used. In instances when a datum is established from two datum features, both reference letters are separated by a dash and appear in a compartment, for example, | // | **0.05** | **A-B** |

Figure 30-5 provides a summary of how datums, as well as geometric dimensioning and tolerancing symbols and concepts, are contained in a feature control frame. Figure 30-6 illustrates the application of datum letters **-K-** and **-M-** and a feature control frame on an engineering drawing.

DIMENSION ORIGIN SYMBOL

A dimension line that has an arrowhead at each end indicates that the *origin point* of a dimension is not important. Therefore, the measurement may be taken from either end along the measurement line. When the part design requires the identification of the origin (reference point) for a specific measurement, it is necessary to show the origin of the dimension on an engineering drawing.

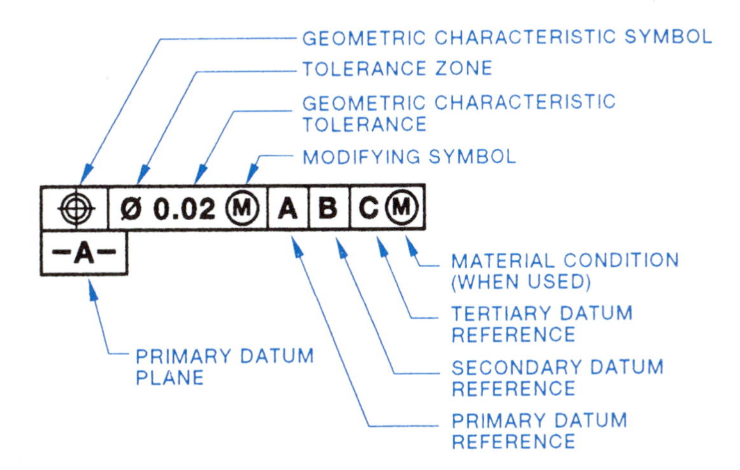

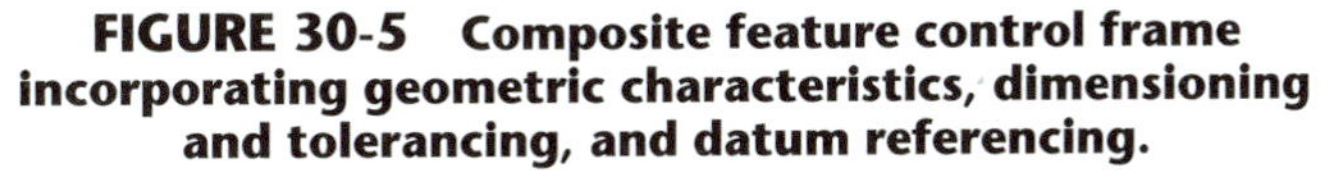

FIGURE 30-5 Composite feature control frame incorporating geometric characteristics, dimensioning and tolerancing, and datum referencing.

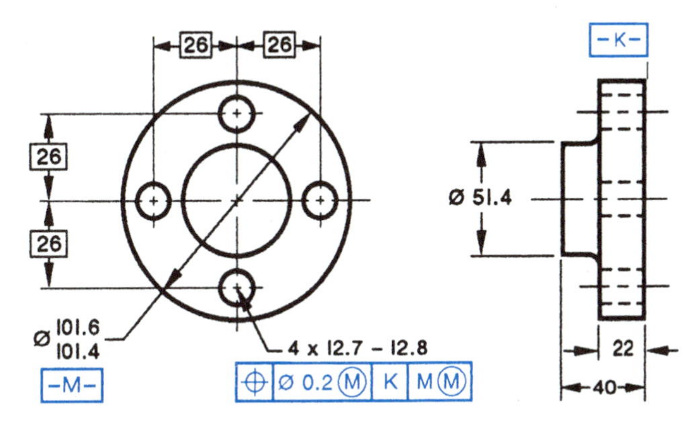

FIGURE 30-6 Application of feature control frame and datum letters on an engineering drawing.

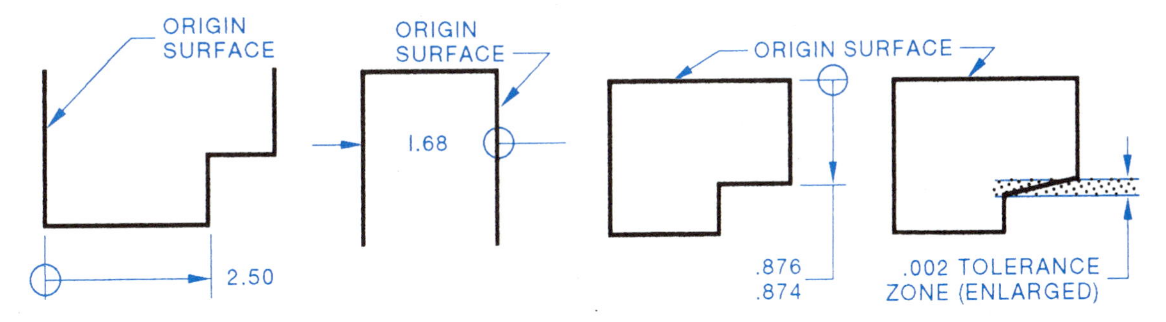

FIGURE 30-7 General applications of the dimension origin symbol.

A new ANSI *origin symbol* is used for this purpose. The original symbol consists of a circle around the extension line for a dimension at the reference (origin) point (⌖). Three general applications of the origin symbol are illustrated at A, B, and C, figure 30-7. The importance of using an origin symbol is emphasized at (C). Note that the tolerance (for design and manufacturing purposes) applies to the surface that is parallel to the origin surface.

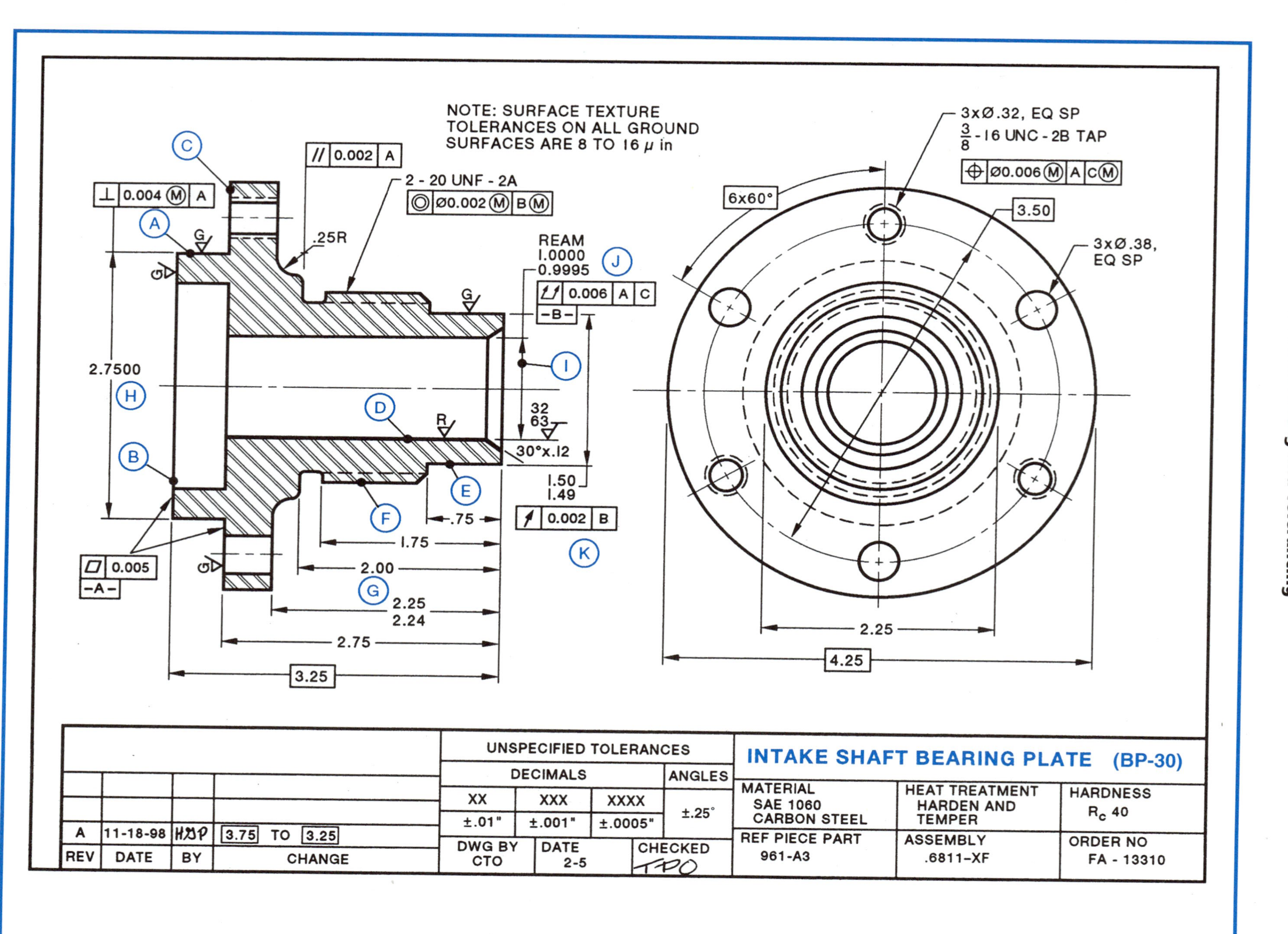
NOTE: SURFACE TEXTURE TOLERANCES ON ALL GROUND SURFACES ARE 8 TO 16 μ in
// 0.002 A
⊥ 0.004 Ⓜ A
2 - 20 UNF - 2A
◎ Ø0.002 Ⓜ B Ⓜ
.25R
REAM 1.0000 0.9995
0.006 A C
-B-
32 63
30°x.12
1.50 1.49
0.002 B
.75
1.75
2.00
2.25 2.24
2.75
3.25
2.7500
0.005
-A-
A
B
C
D
E
F
G
H
I
J
K
3xØ.32, EQ SP
3/8 - 16 UNC - 2B TAP
⊕ Ø0.006 Ⓜ A C Ⓜ
6x60°
3.50
3xØ.38, EQ SP
2.25
4.25
UNSPECIFIED TOLERANCES
DECIMALS
ANGLES
XX ±.01"
XXX ±.001"
XXXX ±.0005"
±.25°
INTAKE SHAFT BEARING PLATE (BP-30)
MATERIAL SAE 1060 CARBON STEEL
HEAT TREATMENT HARDEN AND TEMPER
HARDNESS R_c 40
REF PIECE PART 961-A3
ASSEMBLY .6811-XF
ORDER NO FA - 13310
DWG BY CTO
DATE 2-5
CHECKED TPO
A 11-18-98 HJP 3.75 TO 3.25
REV DATE BY CHANGE

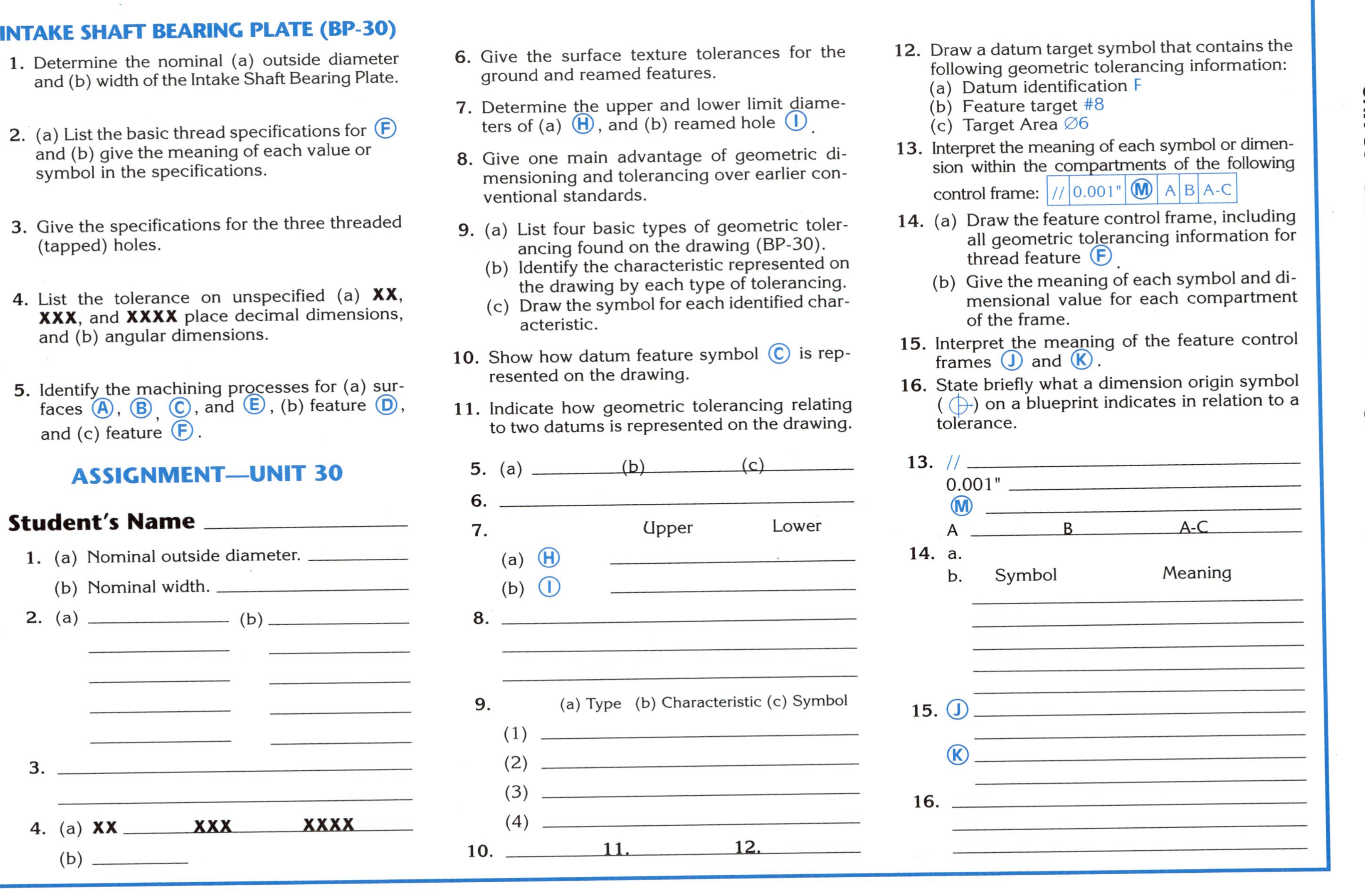

INTAKE SHAFT BEARING PLATE (BP-30)

1. Determine the nominal (a) outside diameter and (b) width of the Intake Shaft Bearing Plate.

2. (a) List the basic thread specifications for Ⓕ and (b) give the meaning of each value or symbol in the specifications.

3. Give the specifications for the three threaded (tapped) holes.

4. List the tolerance on unspecified (a) **XX**, **XXX**, and **XXXX** place decimal dimensions, and (b) angular dimensions.

5. Identify the machining processes for (a) surfaces Ⓐ, Ⓑ, Ⓒ, and Ⓔ, (b) feature Ⓓ, and (c) feature Ⓕ.

6. Give the surface texture tolerances for the ground and reamed features.

7. Determine the upper and lower limit diameters of (a) Ⓗ, and (b) reamed hole Ⓘ.

8. Give one main advantage of geometric dimensioning and tolerancing over earlier conventional standards.

9. (a) List four basic types of geometric tolerancing found on the drawing (BP-30).
   (b) Identify the characteristic represented on the drawing by each type of tolerancing.
   (c) Draw the symbol for each identified characteristic.

10. Show how datum feature symbol Ⓒ is represented on the drawing.

11. Indicate how geometric tolerancing relating to two datums is represented on the drawing.

12. Draw a datum target symbol that contains the following geometric tolerancing information:
    (a) Datum identification F
    (b) Feature target #8
    (c) Target Area ∅6

13. Interpret the meaning of each symbol or dimension within the compartments of the following control frame: | // | 0.001" Ⓜ | A | B | A-C |

14. (a) Draw the feature control frame, including all geometric tolerancing information for thread feature Ⓕ.
    (b) Give the meaning of each symbol and dimensional value for each compartment of the frame.

15. Interpret the meaning of the feature control frames Ⓙ and Ⓚ.

16. State briefly what a dimension origin symbol (⌖) on a blueprint indicates in relation to a tolerance.

ASSIGNMENT—UNIT 30

Student's Name ______________

1. (a) Nominal outside diameter. ______
   (b) Nominal width. ______

2. (a) ______ (b) ______
   ______ ______
   ______ ______
   ______ ______
   ______ ______

3. ______
   ______

4. (a) **XX** ______ **XXX** ______ **XXXX** ______
   (b) ______

5. (a) ______ (b) ______ (c) ______

6. ______

7. | | Upper | Lower |
   |---|---|---|
   | (a) Ⓗ | | |
   | (b) Ⓘ | | |

8. ______
   ______
   ______

9. (a) Type (b) Characteristic (c) Symbol
   (1) ______
   (2) ______
   (3) ______
   (4) ______

10. ______ 11. ______ 12. ______

13. // ______
    0.001" ______
    Ⓜ ______
    A ______ B ______ A-C ______

14. a.
    b. Symbol Meaning
    ______
    ______
    ______
    ______
    ______

15. Ⓙ ______
    ______
    Ⓚ ______
    ______

16. ______
    ______
    ______

SECTION 10 Computer Graphics Technology

Unit 31

COMPUTER-AIDED DRAFTING (CAD) AND DESIGN (CADD) AND ROBOTICS

Computer graphics broadly identifies drawings, three-dimensional displays, color animation, and other visual products drawn with the aid of a computer. *Computer-Aided Drafting (CAD)* is a computer graphics system. *CADD (Computer-Aided Design and Drafting)* is another computer graphics system.

CAPABILITIES OF CAD SYSTEMS

CAD systems utilize all of the principles and applications of drafting techniques, engineering data, and information generally found on drawings produced by conventional methods. CAD has the capability of generating lines of varying thickness, circles, arcs, ellipses, irregular curves, and polygons.

CAD scaled drawings may be enlarged or reduced and automatically dimensioned to provide necessary manufacturing information. Crosshatching for sectional views is produced by selecting the correct crosshatch lines for the part material, choosing the desired angle for the lines, and defining the area to be crosshatched. Single, multiple, and/or auxiliary views may be drawn using CAD. A completed drawing may be reproduced in black and white or color, stored in memory, recalled at any time, and edited for changes, corrections, or additions.

Details are generated by *zoning in* on a feature and *zoning out* to view the drawing in full. Notes and straight, slant, and architectural style letters and numbers may be selected and placed at desired locations on drawings.

CADD AND CAM

The functions generated by CADD are handled through a *central processing unit* (CPU) or computer. The computer includes a number of *integrated-circuit chips,* each of which controls a specific function. CADD, therefore, has added capability to perform complex mathematical processes, generate engineering data, and process commands to control tooling, manufacturing, architectural planning, electronic circuitry, and other automated production. For such applications, a CADD system is often interfaced with a *Computer-Aided Manufacturing (CAM)* system. The CAM system can be programmed to identify and control such manufacturing processes as drilling, boring, turning, milling, and grinding. It also provides for part inspection, testing, and assembly.

The CADD system translates the geometric model of a part or assembly, product information, and necessary engineering data into a computer language such as *Automatically Programmed Tools (APT)*. APT produces instructions for numerically controlled (NC), direct NC (DNC), and computer NC (CNC) machine tools.

CARTESIAN COORDINATE SYSTEM

The rectangular coordinate grid of the Cartesian coordinate system provides the foundation on which drawings produced by CAD are generated. The grid is formed by the intersecting of two planes at a right angle. The point of intersection is called the *origin, datum, zero reference point,* or *zero value.* The direction and line representing the horizontal plane form the **X** axis, and the vertical plane is the **Y** axis. Dimensions or values to the right of the origin (**0**) or vertical axis, and above the origin or the horizontal axis, are positive. Dimensions or values to the left of the origin or vertical axis, and below the origin or horizontal axis are negative (figure 31-1).

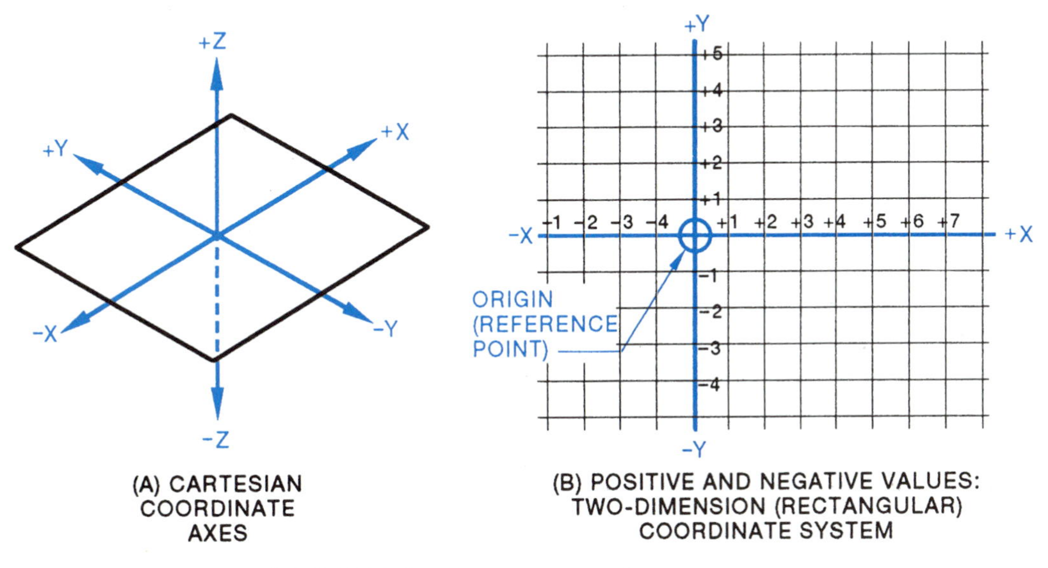

(A) CARTESIAN COORDINATE AXES

(B) POSITIVE AND NEGATIVE VALUES: TWO-DIMENSION (RECTANGULAR) COORDINATE SYSTEM

FIGURE 31-1 Cartesian coordinate system.

Lines are generated by locating two end points in relation to the **X** and **Y** axes. The appropriate keyboard command generates the line automatically (figure 31-2A). Similarly, circles are automatically drawn by locating the center and giving the radius, (figure 31-2B). Circles and arcs may also be generated by locating three points of a circle.

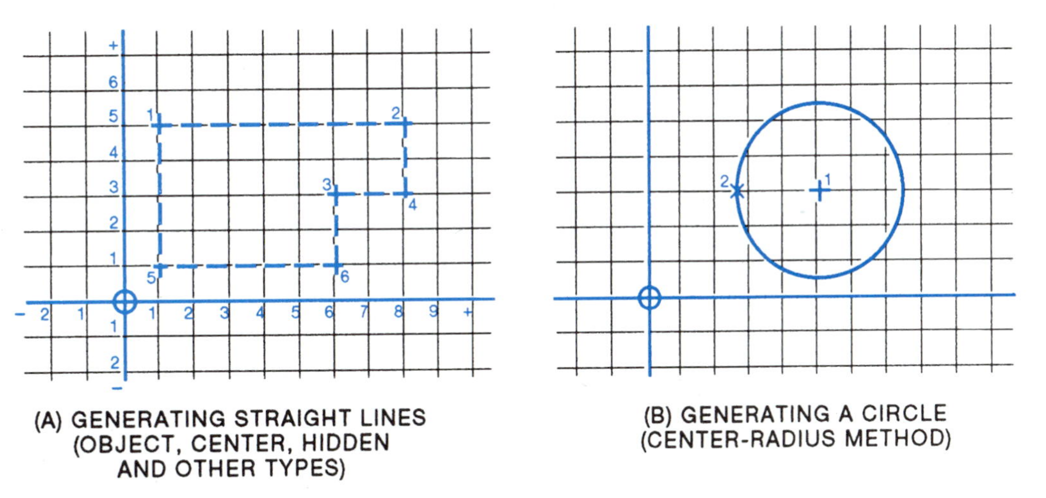

(A) GENERATING STRAIGHT LINES (OBJECT, CENTER, HIDDEN AND OTHER TYPES)

(B) GENERATING A CIRCLE (CENTER-RADIUS METHOD)

FIGURE 31-2 Locating points and generating lines and circles on a rectangular coordinate grid. Courtesy of Houston Instrument Division, AMETEK, Inc.

THREE-DIMENSIONAL COORDINATES

Every point on a CADD-generated drawing is defined by numerical locations and coordinate points. Since CADD uses a third axis (**Z**) and coordinates to add depth, pictorial three-dimensional (3-D) drawings are produced by inputting the three dimension data into a computer graphics system. Figure 31-3A illustrates a CADD drawing of a *3-D solid model.*

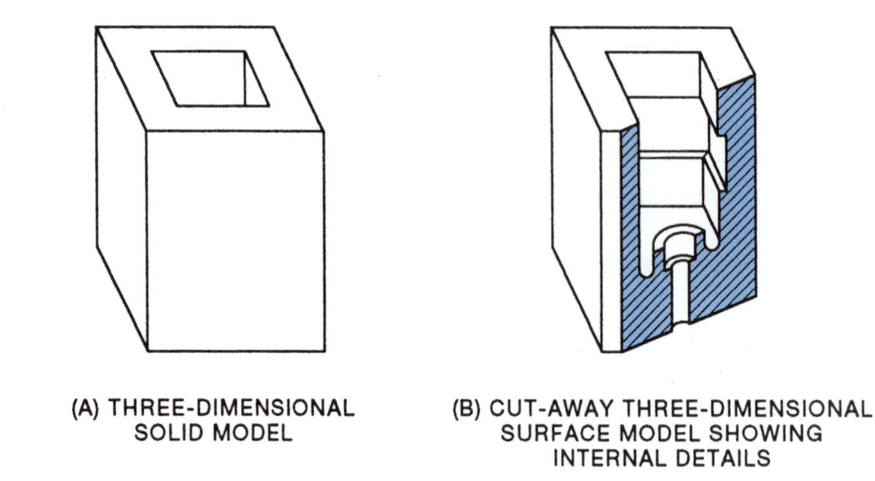

FIGURE 31-3 CAD-generated three-dimensional solid and cutaway models. Courtesy of Houston Instrument Division, AMETEK, Inc.

Different parts, sections, and surfaces may be rotated, cut away to view internal features, drawn, shaded or colored, and displayed on the *video display screen* (figure 31-3B). A color copy may then be produced by *pen plotters, electrostatic plotters, laser,* or other *printers.*

Three-dimensional drawings may also be created separately, and each part may be moved into its respective position in an assembly or exploded-parts drawing. Figure 31-4 shows a CADD pictorial assembly drawing produced from separately created drawings of the different parts.

FIGURE 31-4 CAD three-dimensional assembly drawing generated from separately drawn parts. Courtesy of McDonnell Douglas Manufacturing and Engineering Systems Company.

LIBRARY OF SYMBOLS

A typical drawing contains repeated symbols that are produced on conventional drawings using a template or by rubbing in place selected letters, symbols, or frames from transfer sheets. In CAD, once a symbol is designed, it may be stored, recalled, and re-used on a drawing by placing it at the required location.

INPUT AND OUTPUT DEVICES FOR CADD

Information and other graphic instructions are fed into the CADD computer by means of a keyboard. Corresponding to a typewriter, the *alphanumeric keyboard* contains letters, numbers, and other character keys. Another area of the keyboard permits graphic instructions to be entered in relation to **X**, **Y**, and **Z** coordinates. Additional *function keys* are included so that specific drafting and/or design functions may be programmed.

Some CADD systems include a complementary *function board.* Keys on the function board actuate still other particular commands or symbols. The term *menu* refers to a list of commands, symbols, and numbers for a particular CADD system. A *menu pad,* like a function board, provides yet another technique for the selection of commands to be processed through the computer. Data may also be entered into the computer graphics system by using a *sensitized graphics tablet* (called a *digitizer*) and a hand-held *stylus* or *puck.*

The video display screen, similar to a television screen, provides a common method of producing a drawing. When all information on the display screen (*softcopy*) is transmitted to the printer, a paper copy (*hardcopy*) is produced. The hardcopy print is complete with drawing data and other programmed information, and can be produced in single or multiple colors.

CADD AND ROBOTICS

CADD systems are used in robotics to perform the following tasks:

- Prepare design and working drawings of robot models.
- Store design and other drawings of parts to be manufactured in the CADD *database,* or produce three-dimensional graphic models.
- Represent graphically the *work cell* for specific robot functions in a 3-D model; including layout, design, and development specifications.
- Provide graphic feedback for study and correction of design and work flow problems.
- Simulate robot movements before the actual installation of a robot in manufacturing.

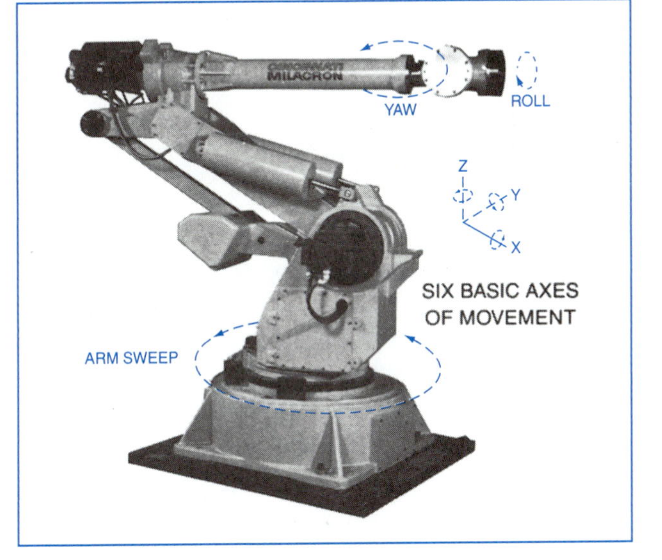

FIGURE 31-5 Six basic axes of movement of an industrial robot.

ROBOTICS: FUNCTIONS

Factory Automated Manufacturing (FAM) requires the use of a number of mechanical arms called *industrial robots.* A robot, similar to the one pictured in figure 31-5, may be programmed for the following tasks:

- Grip, press, hold, and accurately position workpieces to load and unload parts from work-positioning devices and conveyors (*transporters*) between machining stations.
- Perform similar processes in loading, unloading, and changing tools and accessories to tool storage drums (*matrices*) on machine tools.
- Execute programmed manufacturing, assembling, testing, and other processes.

Figure 31-6 provides an example of a computer-activated six-axis robot that is used for handling materials and parts from a conveyor system and loading and unloading a CNC *milling center* and a CNC *turning center* in a *manufacturing cell.* Robots are employed in both *handling applications* related to the movement of materials and *processor applications* in manipulating tools and equipment and performing work processes.

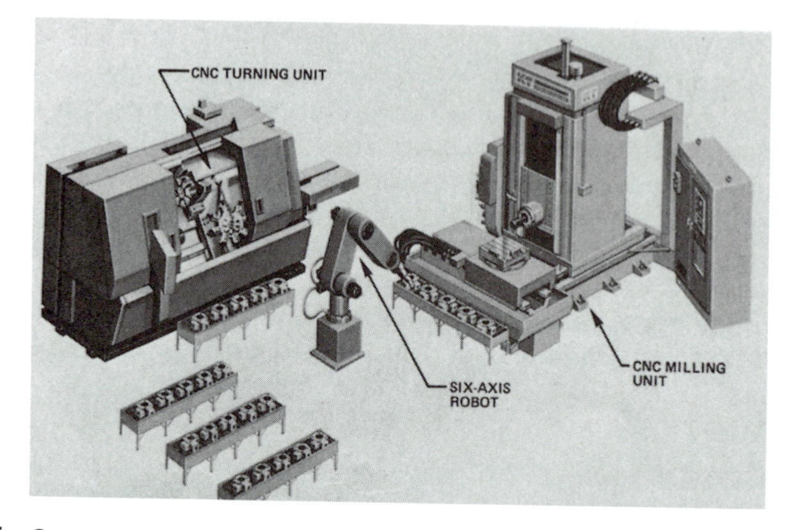

FIGURE 31-6 Computer-controlled six-axis robot handling workpieces between milling and turning machine tools. Courtesy of Kearney & Trecker Corporation.

INDUSTRIAL ROBOT DESIGN FEATURES

The three major components of a robot include: (1) the *robot hand* or *end effector,* (2) the *controller* in which program commands are recorded in memory and are programmed to repeat the motions, and (3) the *power unit.* A brief description of selected robot features follows.

End Effector. A robot generally has two sets of fingers. One set is used to load; the other, to unload. After loading and machining, the robot hand turns so the second set of fingers grips and removes the finished part.

Degrees of Freedom. This term identifies the number of axes of motion. For example, eight degrees of freedom means a robot has eight axes of motion.

Operating Reach and Envelope. A CADD graphic 3-D model of a robot performing a specific task includes an *inner envelope* and an *outer envelope* (figure 31-7). Tasks are performed at maximum speed with greater safety and positional accuracy, under maximum load within the inner envelope.

Loading and Unloading Design Features. Consideration is given in CADD/CAM systems involving robot applications to the following factors:

- Robot load capacity in terms of weight of the workpiece and grippers (fingers).
- Automatic control of increases or decreases in operating speeds according to work process requirements.
- Safe carrying capacity.
- Clearance of the arm, workpiece work stations, and processes to be performed.
- Positioning repeatability with accuracy.

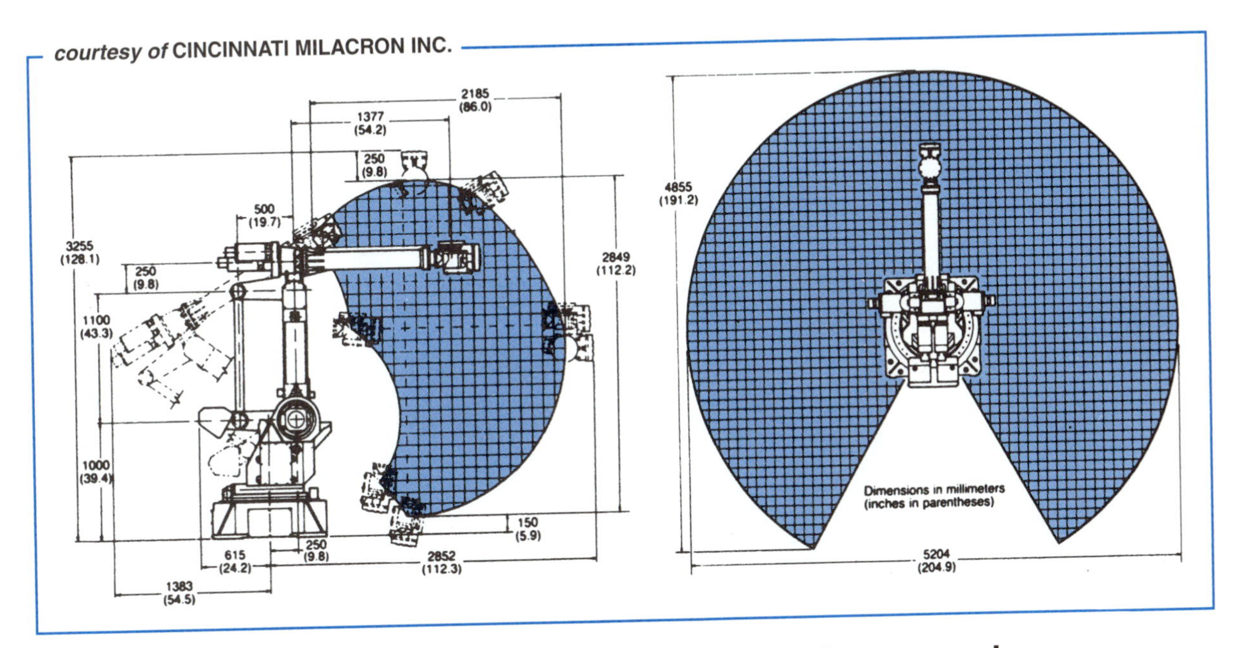

FIGURE 31-7 Robot work envelope describing the range, reach, degrees of freedom, and required floor space.

- Accommodating workpiece irregularities.
- Safe and accurate piloting of a workpiece into a fixture.
- Design of universal hands to accommodate the range of parts within a family-of-parts.

Figure 31-8 provides a graphic display of a flexible manufacturing system. FMS interlocks subsystems such as CADD, CAM, computer-directed rail-guided conveyors and robots for handling and for processor applications; turning, milling, grinding, and other CNC machining centers; measurement machine and quality control centers; and other manufacturing cell computer-actuated equipment.

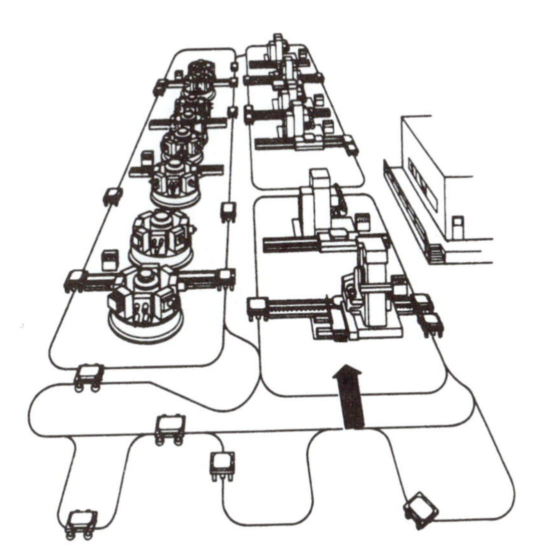

FIGURE 31-8 Model of computer-directed, random-order, flexible manufacturing system (FMS). Courtesy of Kearney & Trecker Corporation.

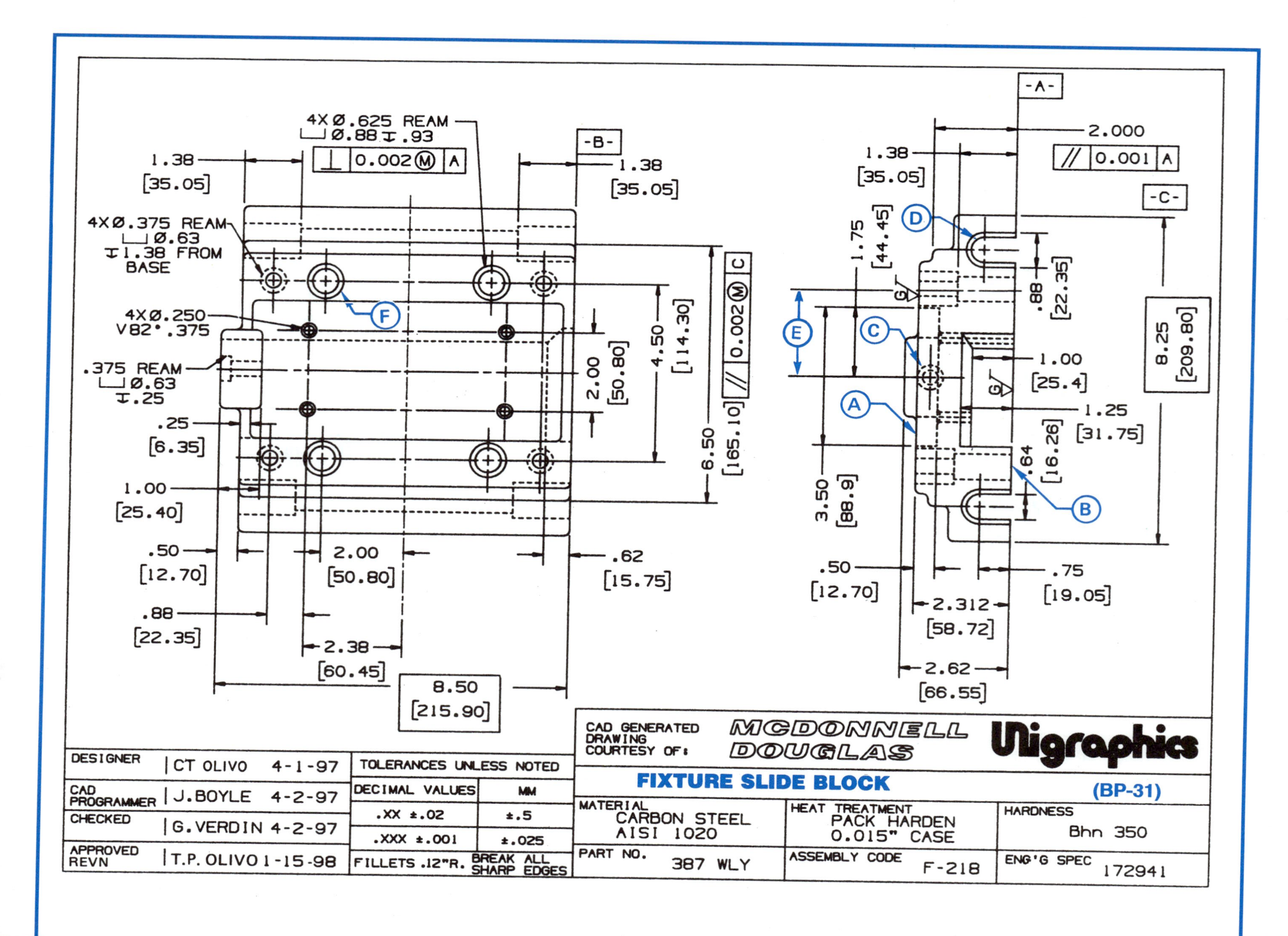
4XØ.625 REAM
Ø.88 ⊥.93
⊥ 0.002 Ⓜ A
-B-
-A-
-C-
2.000
// 0.001 A
1.38 [35.05]
4XØ.375 REAM
Ø.63
1.38 FROM BASE
4XØ.250
V82°.375
.375 REAM
Ø.63
.25
.25 [6.35]
1.00 [25.40]
.50 [12.70]
.88 [22.35]
2.00 [50.80]
2.38 [60.45]
8.50 [215.90]
.62 [15.75]
4.50 [114.30]
6.50 [165.10]
// 0.002 Ⓜ C
1.75 [44.45]
.88 [22.35]
1.00 [25.4]
1.25 [31.75]
.64 [16.26]
8.25 [209.80]
3.50 [88.9]
.50 [12.70]
.75 [19.05]
2.312 [58.72]
2.62 [66.55]
A
B
C
D
E
F
G
DESIGNER CT OLIVO 4-1-97
CAD PROGRAMMER J.BOYLE 4-2-97
CHECKED G.VERDIN 4-2-97
APPROVED REVN T.P. OLIVO 1-15-98
TOLERANCES UNLESS NOTED
DECIMAL VALUES
MM
.XX ±.02
±.5
.XXX ±.001
±.025
FILLETS .12"R.
BREAK ALL SHARP EDGES
CAD GENERATED DRAWING COURTESY OF:
MCDONNELL DOUGLAS
Unigraphics
FIXTURE SLIDE BLOCK
(BP-31)
MATERIAL CARBON STEEL AISI 1020
HEAT TREATMENT PACK HARDEN 0.015" CASE
HARDNESS Bhn 350
PART NO. 387 WLY
ASSEMBLY CODE F-218
ENG'G SPEC 172941

FIXTURE SLIDE BLOCK (BP-31)

1. Identify the method used to produce the drawing.
2. Give the nominal length, width, and thickness of the Fixture Slide Block.
3. Identify the two and three place decimal tolerances for unspecified inch and metric dimensions.
4. a. Identify the material used in the part.
   b. State the required depth of case hardening.
   c. Give the Brinell hardness number of the case.
5. a. Name the matching process for surfaces Ⓐ and Ⓑ.
   b. Determine the upper and lower decimal inch dimensional limits between these surfaces.
6. a. Write the specifications of the four 1/4" countersunk holes.
   b. Interpret the meaning of each dimension and symbol.
7. a. Give the positional (centerline) dimensions for detail Ⓒ.
   b. Give specifications for this detail.
   c. Interpret each dimensional value and symbol.
8. a. List the two nominal radii for Ⓓ in inches and millimeters.
   b. State the maximum and minimum depth of Ⓓ in inches.
   c. Identify the geometric tolerancing allowance on the center to center distance between the slots Ⓓ.
   d. Translate the meaning of the geometric dimensioning requirements given in the tolerancing frame.
9. Determine the maximum and minimum center to center dimension Ⓔ ("/mm).
10. a. Give complete dimensioning and geometric tolerancing requirements for holes Ⓕ.
    b. Translate the tolerancing frame information.
11. List three basic CAD system capabilities.
12. State two functions which distinguish CAD from CADD.
13. Identify a major function that is served by interlocking CADD and CAM.
14. a. Identify the mathematical system that forms the basis for CAD.
    b. Make a sketch and label the following: (1) point of origin, (2) **X** axis, (3) **Y** axis, and (4) positive (+) and negative (−) numbered line values.
15. Provide a general definition of the function of industrial robots.
16. Describe briefly the meaning of the following tasks in relation to robots.
    a. Simulating robot movements, and
    b. Drawing of inner and outer envelopes.

ASSIGNMENT—UNIT 31

Student's Name ________________

1. ________________
2. L ______ W ______ T ______
3. XX ________________
   XXX ________________
4. (a) ______ (b) ______ (c) ______
5. (a) ______ (b) Upper ______
   Lower ______
6. (a)

   (b) ________________

7. (a) ________________
   (b) ________________
   (c) ________________
8. (a) Inch ______ mm ______
   (b) Max. ______ Min. ______
   (c)
   (d) ________________
9. Max. ______ Inch ______ mm
   Min. ______ Inch ______ mm
10. (a) ________________
    (b) ________________

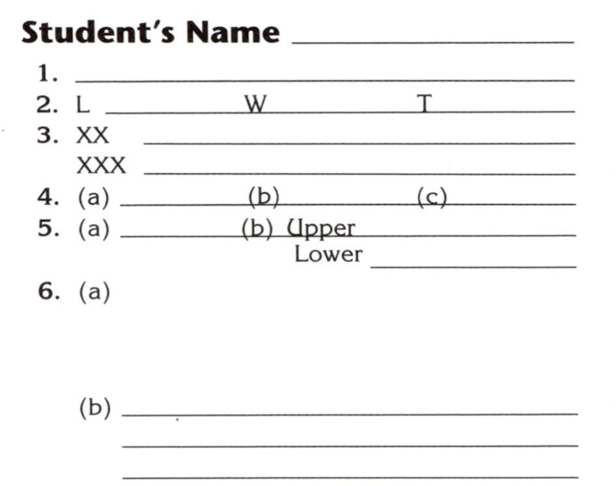

11. (1) ________________
    (2) ________________
    (3) ________________
12. (1) ________________
    (2) ________________
13. ________________
14. (a) ________________
    (b)
15. ________________
16. (a) ________________
    (b) ________________

PART 2

TECHNICAL SKETCHING

SECTION 11 Sketching Lines and Basic Forms

Unit 32

SKETCHING HORIZONTAL, VERTICAL, AND SLANT LINES

THE VALUE OF SKETCHING

Many parts or assembled units may be described clearly and adequately by one or more freehand sketches. Sketching is another way of conveying ideas rather than a method of making perfect completed drawings.

Technicians, therefore, in addition to being able to interpret drawings and blueprints accurately and easily, are often required to make sketches. By sketching additional views not found on blueprints, a tradesperson can study the part thoroughly to understand the process required to fabricate or machine it. Since the only tools needed are a soft pencil and any available paper, sketches may be made conveniently at any time or place.

SKETCHING LINES FREEHAND

The development of correct skills in sketching lines freehand is more essential for the beginner than speed. After the basic principles of sketching are learned and skill is acquired in making neat and accurate sketches, then stress should be placed on speed.

Although shop sketches are made on the job and special pencils are not required, a medium soft to medium (HB, No. 2) lead pencil with a cone-shaped point will produce the best results. For fine lines, use a fairly sharp point, for heavier lines, round the point more, figure 32-1. A sharp point may be kept longer by rotating the pencil. Adjustable pencil holders are often used in drafting/design rooms, shops, and laboratories. The holders accommodate 0.05, 0.07 and 0.09 mm diameter leads for drawing fine or heavy lines, using HB lead.

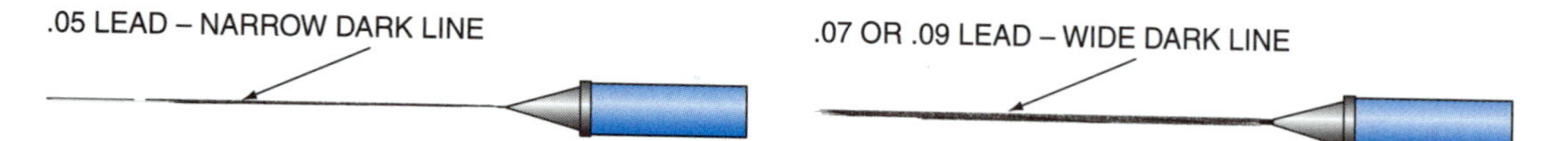

FIGURE 32-1 Sharpness of pencil point influences line weight.

SKETCHING HORIZONTAL LINES

The same principles of drafting which apply to the making of a mechanical drawing and the interpretation of blueprints are used for making sketches.

In sketching, the pencil is held 3/4" to 1" from the point so the lines to be drawn may be seen easily and there is a free and easy movement of the pencil. A free arm movement makes it possible to sketch smooth, neat lines as compared with the rough and inaccurate lines produced by a finger and wrist movement.

In planning the first horizontal line on a sketch, place two points on the paper to mark the beginning and end of the line, figure 32-2A. Place the pencil point on the first dot. Then, with a free arm movement and with the eyes focused on the right point, draw the line from left to right, figure 32-2B.

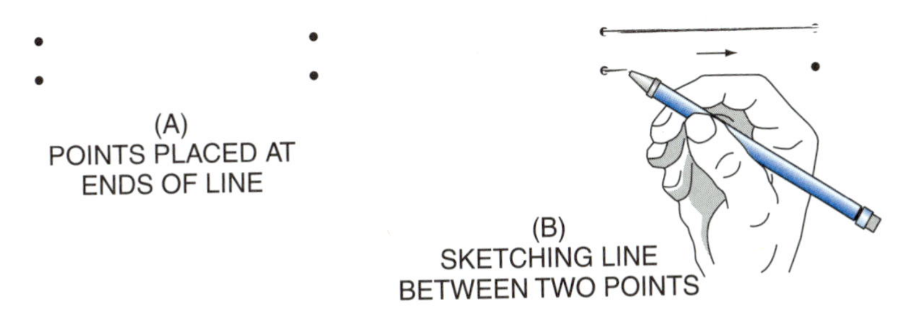

FIGURE 32-2 Sketching horizontal lines.

Examine the line for straightness, smoothness, and weight. If the line is too light, a softer pencil or a more rounded point may be needed. On long lines, extra dots are often placed between the start and finish points of the line. These intermediate dots are used as a guide for drawing long straight lines.

SKETCHING VERTICAL LINES

The same techniques for holding the pencil and using a free arm movement apply in sketching vertical lines. Dots again may be used to indicate the beginning and end of the vertical line. Start the line from the top and move the pencil downward as shown in Figure 32-3. For long lines, if possible, the paper may be turned to a convenient position.

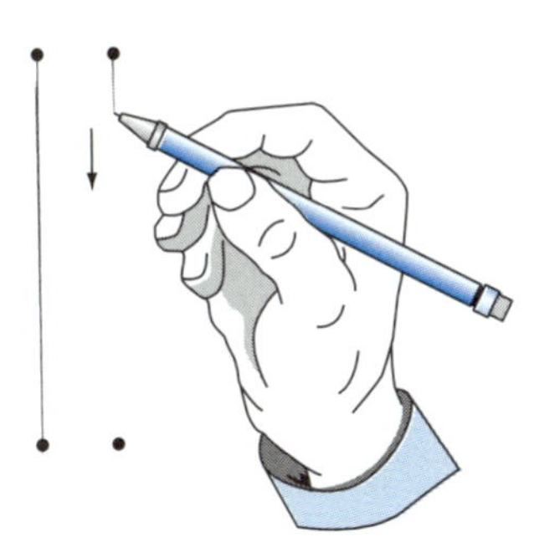

FIGURE 32-3 Sketching vertical lines.

FIGURE 32-4 Sketching slant lines.

SKETCHING SLANT LINES

The three types of straight lines which are widely used in drawing and sketching are: (1) horizontal, (2) vertical, and (3) inclined or slant lines.

The slant line may be drawn either from the top down or from the bottom up. The use of dots at the starting and stopping points for slant lines is helpful to the beginner. Slant lines, figure 32-4, are produced with the same free arm movement used for horizontal and vertical lines.

USING GRAPH PAPER

Many industries recommend the use of graph paper for making sketches. Graph paper is marked to show how many squares are included in each inch. The combination of light ruled lines at a fixed number per inch makes it possible to draw neat sketches which are fairly accurate in size. Throughout this text, squared or isometric graph sections are included for each sketching assignment. Other ruled section papers are available with diametric, oblique, or perspective lines.

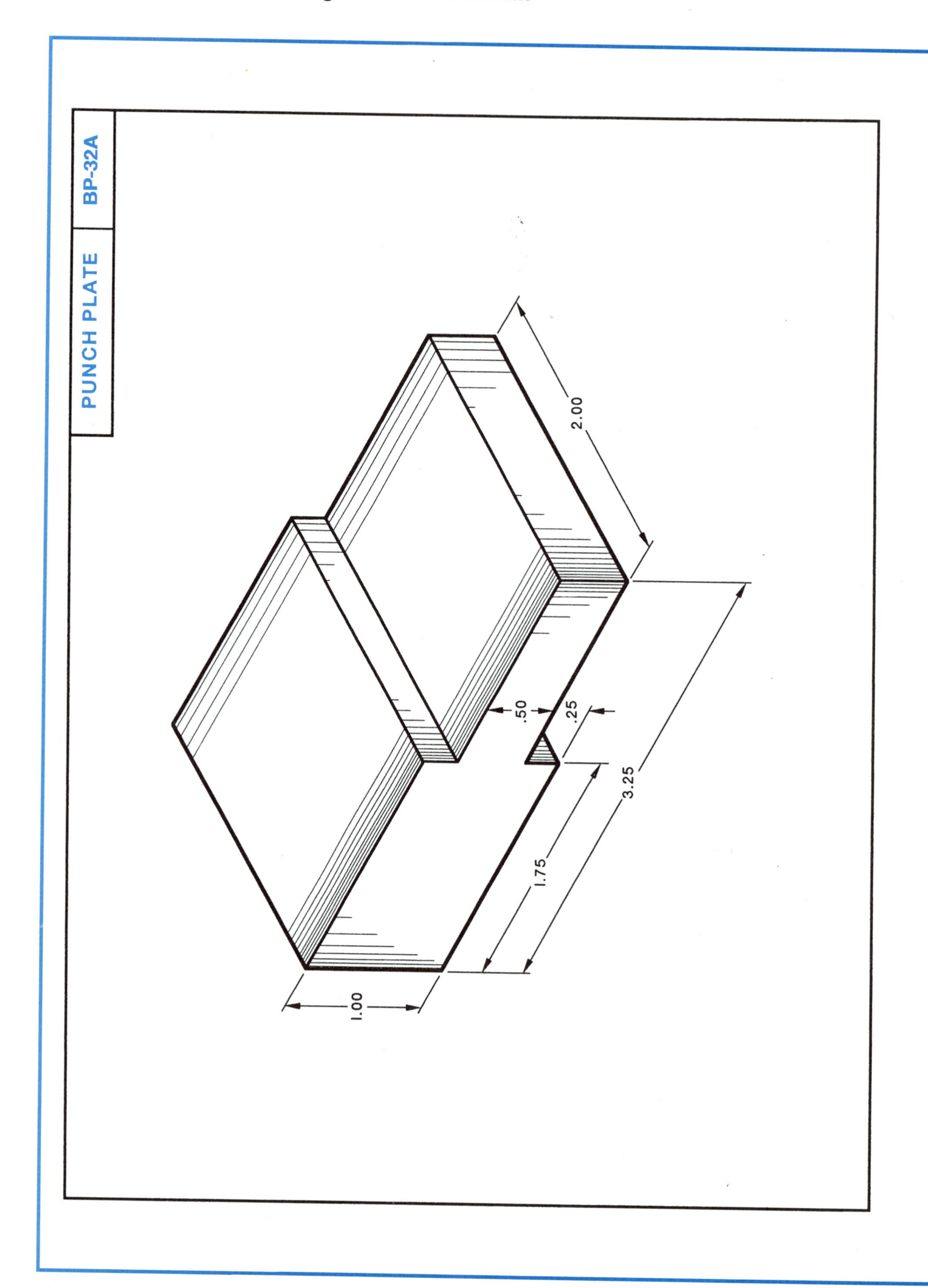
PUNCH PLATE
BP-32A
2.00
3.25
.50
.25
1.75
1.00

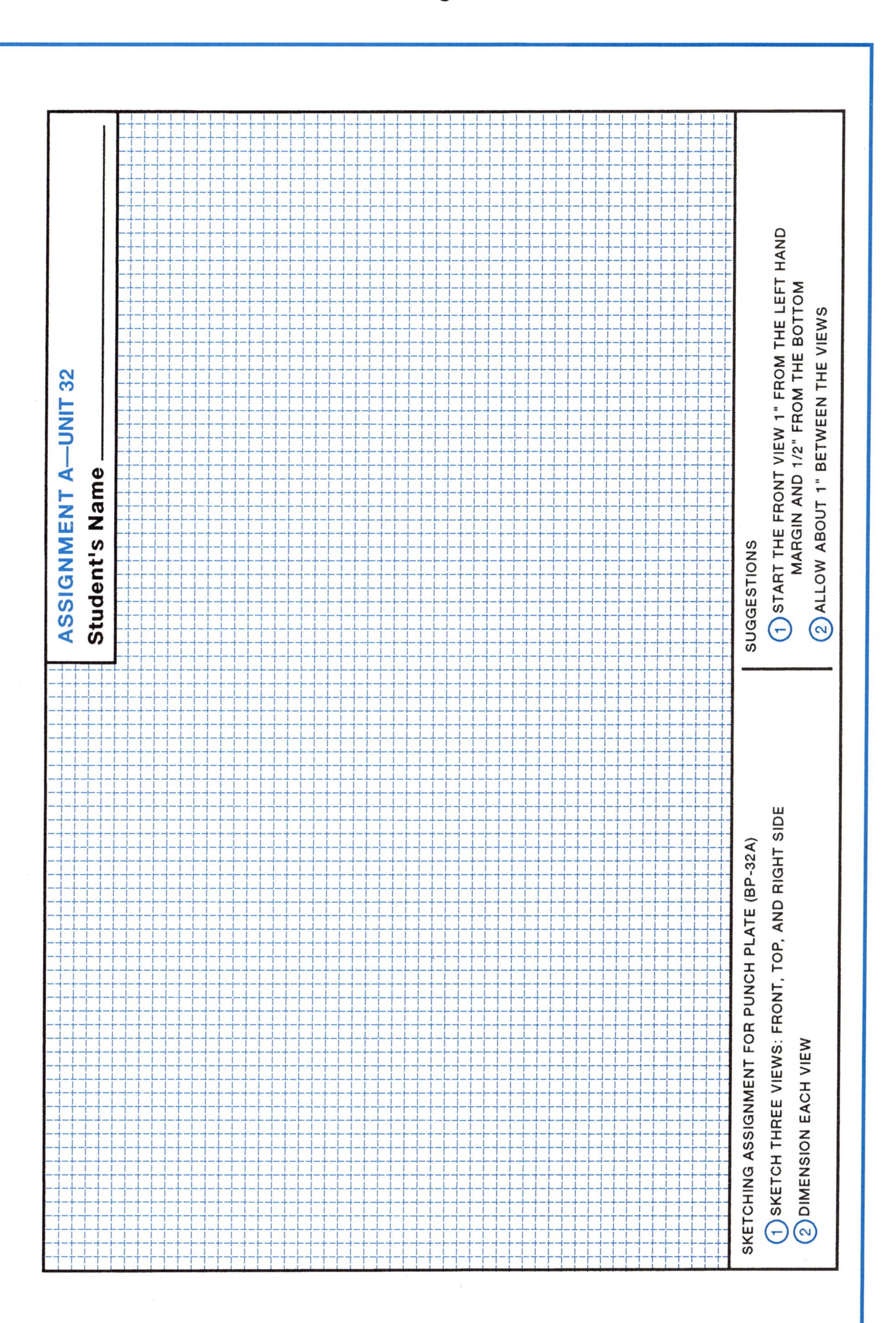
ASSIGNMENT A—UNIT 32
Student's Name
SKETCHING ASSIGNMENT FOR PUNCH PLATE (BP-32A)
① SKETCH THREE VIEWS: FRONT, TOP, AND RIGHT SIDE
② DIMENSION EACH VIEW
SUGGESTIONS
① START THE FRONT VIEW 1" FROM THE LEFT HAND MARGIN AND 1/2" FROM THE BOTTOM
② ALLOW ABOUT 1" BETWEEN THE VIEWS

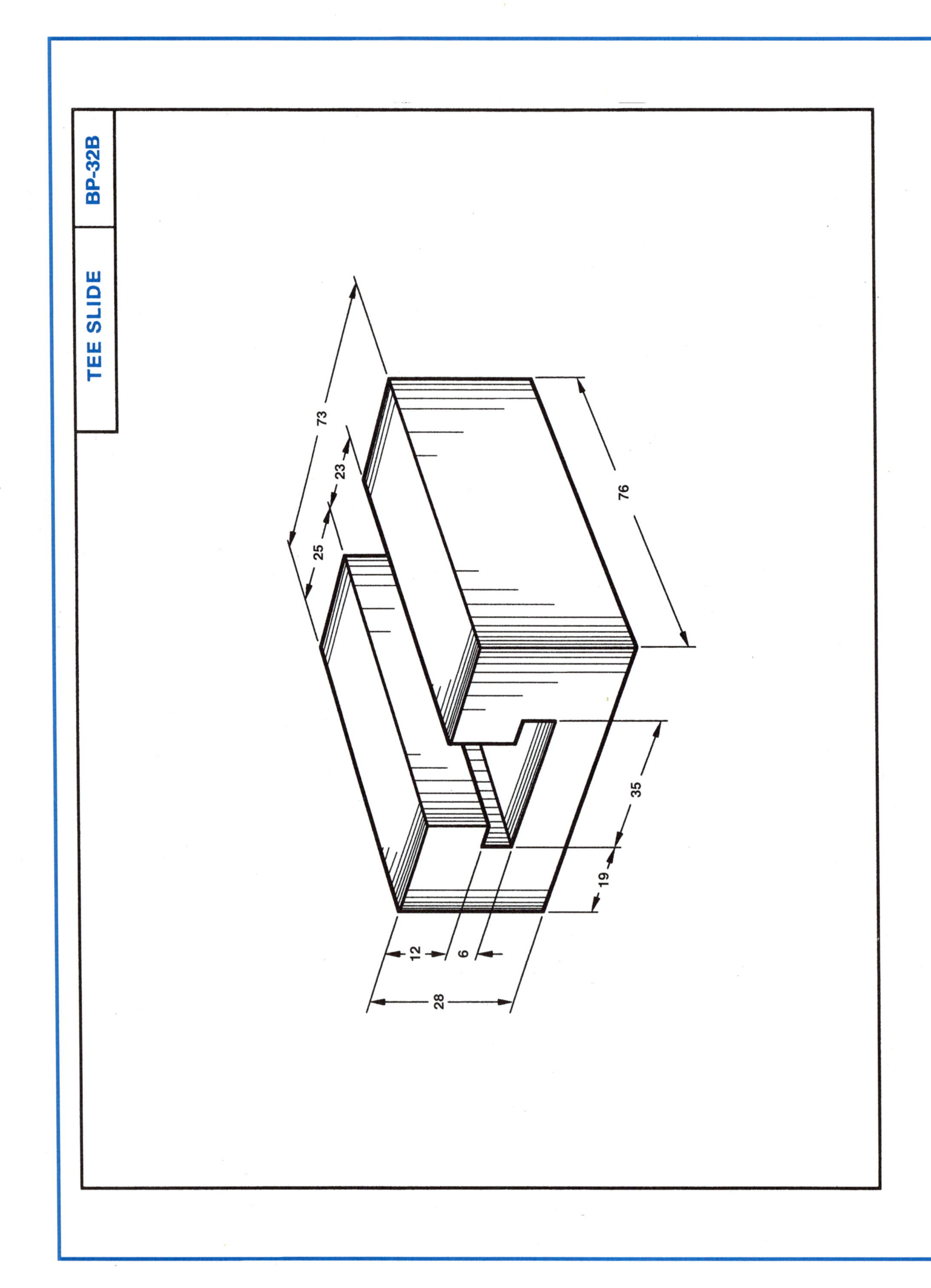
TEE SLIDE
BP-32B
73
23
25
76
35
19
12
6
28

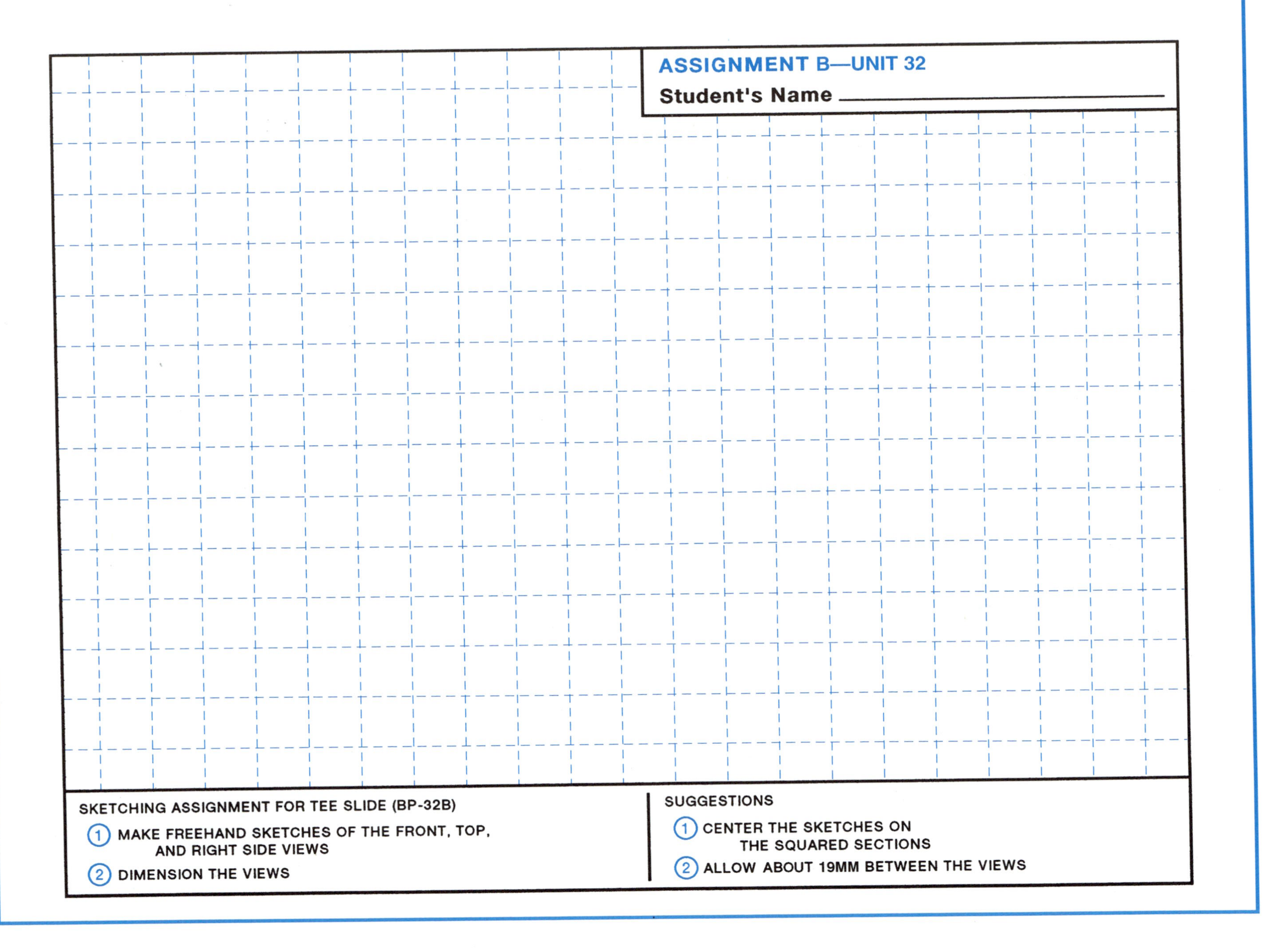
ASSIGNMENT B—UNIT 32
Student's Name
SKETCHING ASSIGNMENT FOR TEE SLIDE (BP-32B)
1 MAKE FREEHAND SKETCHES OF THE FRONT, TOP, AND RIGHT SIDE VIEWS
2 DIMENSION THE VIEWS
SUGGESTIONS
1 CENTER THE SKETCHES ON THE SQUARED SECTIONS
2 ALLOW ABOUT 19MM BETWEEN THE VIEWS

Unit 33

SKETCHING CURVED LINES AND CIRCLES

ARCS CONNECTING STRAIGHT LINES

Often, when describing a part, it is necessary to draw both straight and curved lines. When part of a circle is shown, the curved line is usually called an *arc*.

When an arc must be drawn so that it connects two straight lines (in other words, is tangent to them), it may be sketched easily if five basic steps are followed, figure 33-1.

STEP 1 Extend the straight lines so they intersect.

STEP 2 Step off the same distance from the intersection (center) of both lines.

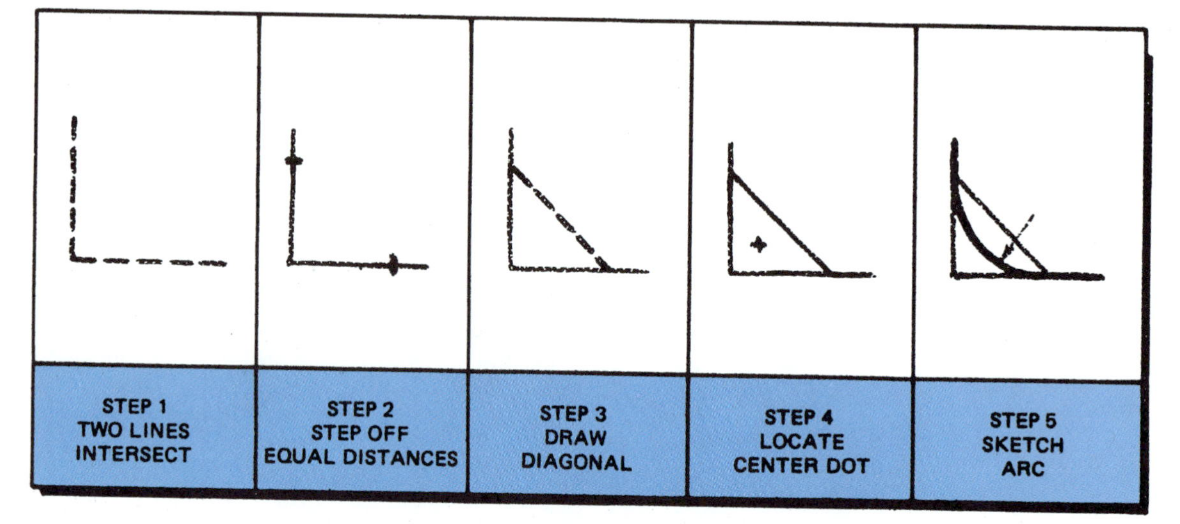

FIGURE 33-1 Sketching an arc freehand.

STEP 3 Draw the diagonal line through these two points to form a triangle.

STEP 4 Place a dot in the center of the triangle.

STEP 5 Start at one of the lines and sketch an arc which runs through the dot and ends on the other line. Darken the arc and erase all unnecessary lines.

ARCS CONNECTING STRAIGHT AND CURVED LINES

Two other line combinations using straight lines and arcs are also very common. The first is the case of an arc connecting a straight line with another arc; the second is where an arc connects two other arcs. The same five basic steps are used as illustrated in figure 33-2.

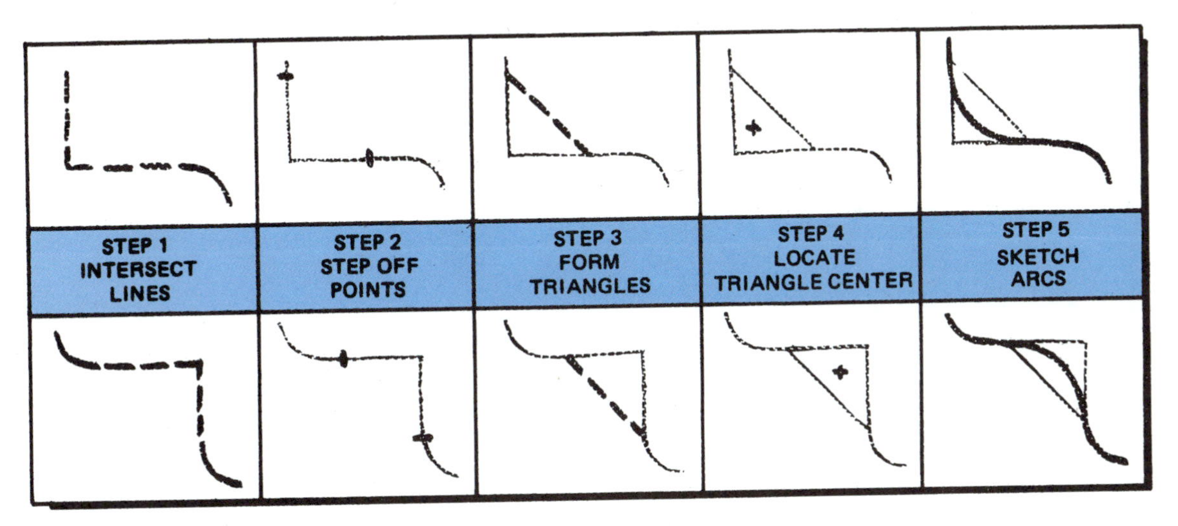

FIGURE 33-2 Sketching arcs in combination with other arcs.

SKETCHING CIRCLES

Shop sketches also require the drawing of circles or parts of a circle. While there are many ways to draw a circle freehand, a well-formed circle can be sketched by following five simple steps which are described and illustrated in figure 33-3.

STEP 1 Lay out the vertical and horizontal center lines in the correct location. Measure off half the diameter of the circle on each side of the two center lines.

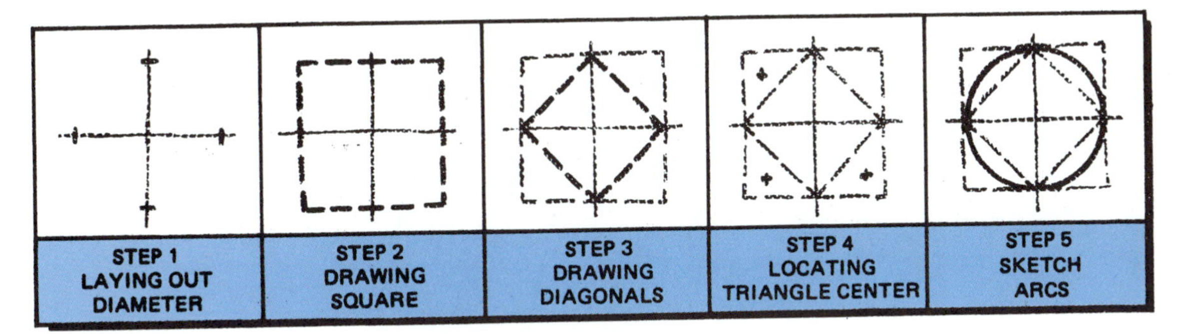

FIGURE 33-3 Sketching a circle.

STEP 2 Draw lightly two vertical and two horizontal lines passing through the points marked off on the center lines. These lines, properly drawn, will be parallel and will form a square.

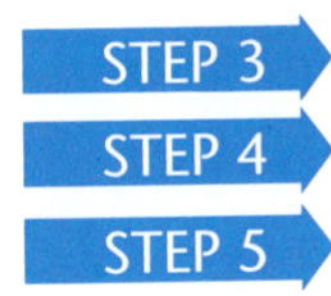

STEP 3 Draw diagonal lines from each corner of the square to form four triangles.

STEP 4 Place a dot in the center of each triangle through which an arc is to pass.

STEP 5 Sketch the arc for one quarter of the circle. Start at one center line and draw the arc through the dot to the next center line. Continue until the circle is complete. Darken the circle and erase all the guide lines to simplify the reading of the sketch.

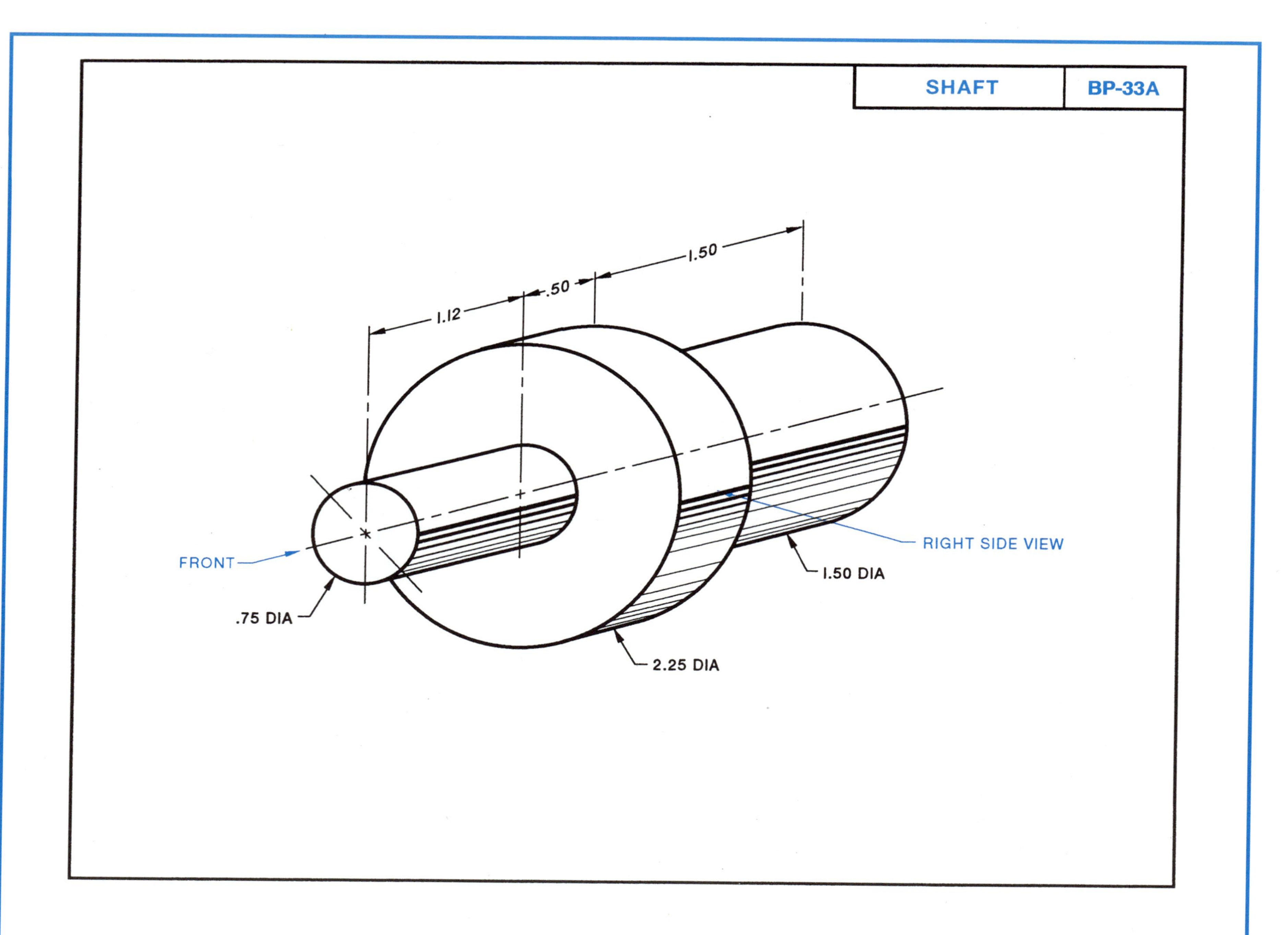
SHAFT
BP-33A
1.12
.50
1.50
FRONT
.75 DIA
2.25 DIA
1.50 DIA
RIGHT SIDE VIEW

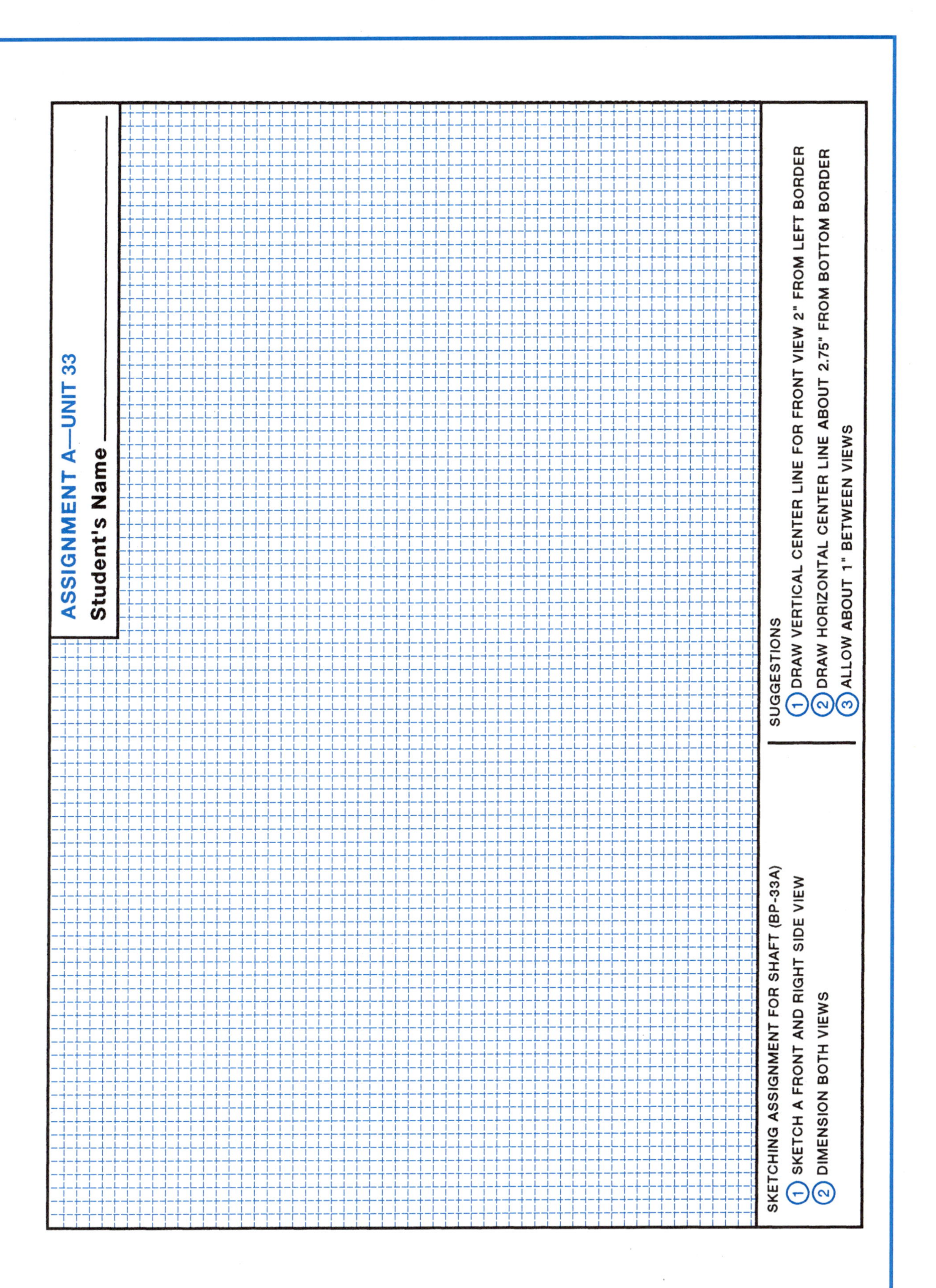
ASSIGNMENT A—UNIT 33
Student's Name
SKETCHING ASSIGNMENT FOR SHAFT (BP-33A)
1 SKETCH A FRONT AND RIGHT SIDE VIEW
2 DIMENSION BOTH VIEWS
SUGGESTIONS
1 DRAW VERTICAL CENTER LINE FOR FRONT VIEW 2" FROM LEFT BORDER
2 DRAW HORIZONTAL CENTER LINE ABOUT 2.75" FROM BOTTOM BORDER
3 ALLOW ABOUT 1" BETWEEN VIEWS

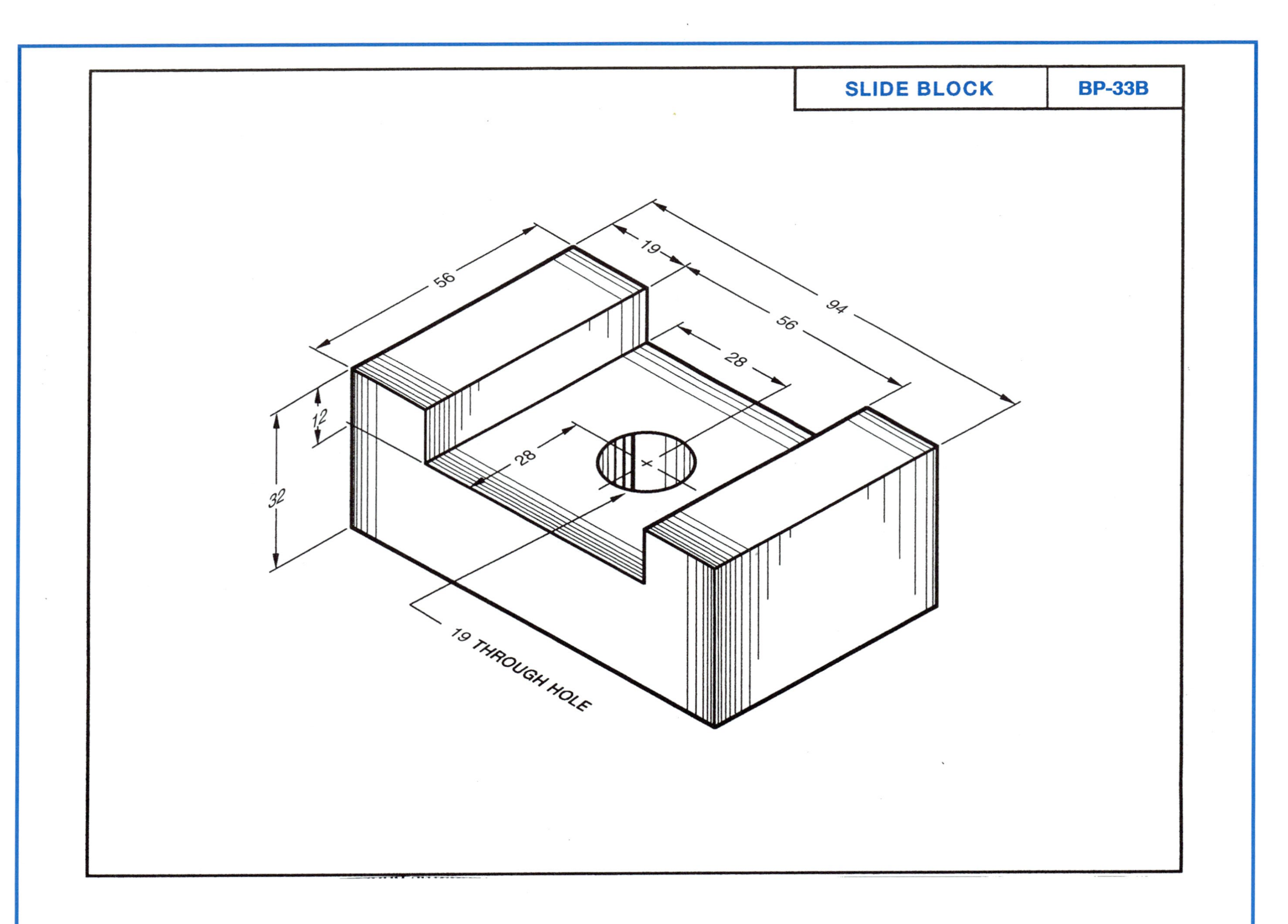
SLIDE BLOCK
BP-33B
56
19
94
56
28
12
32
28
19 THROUGH HOLE

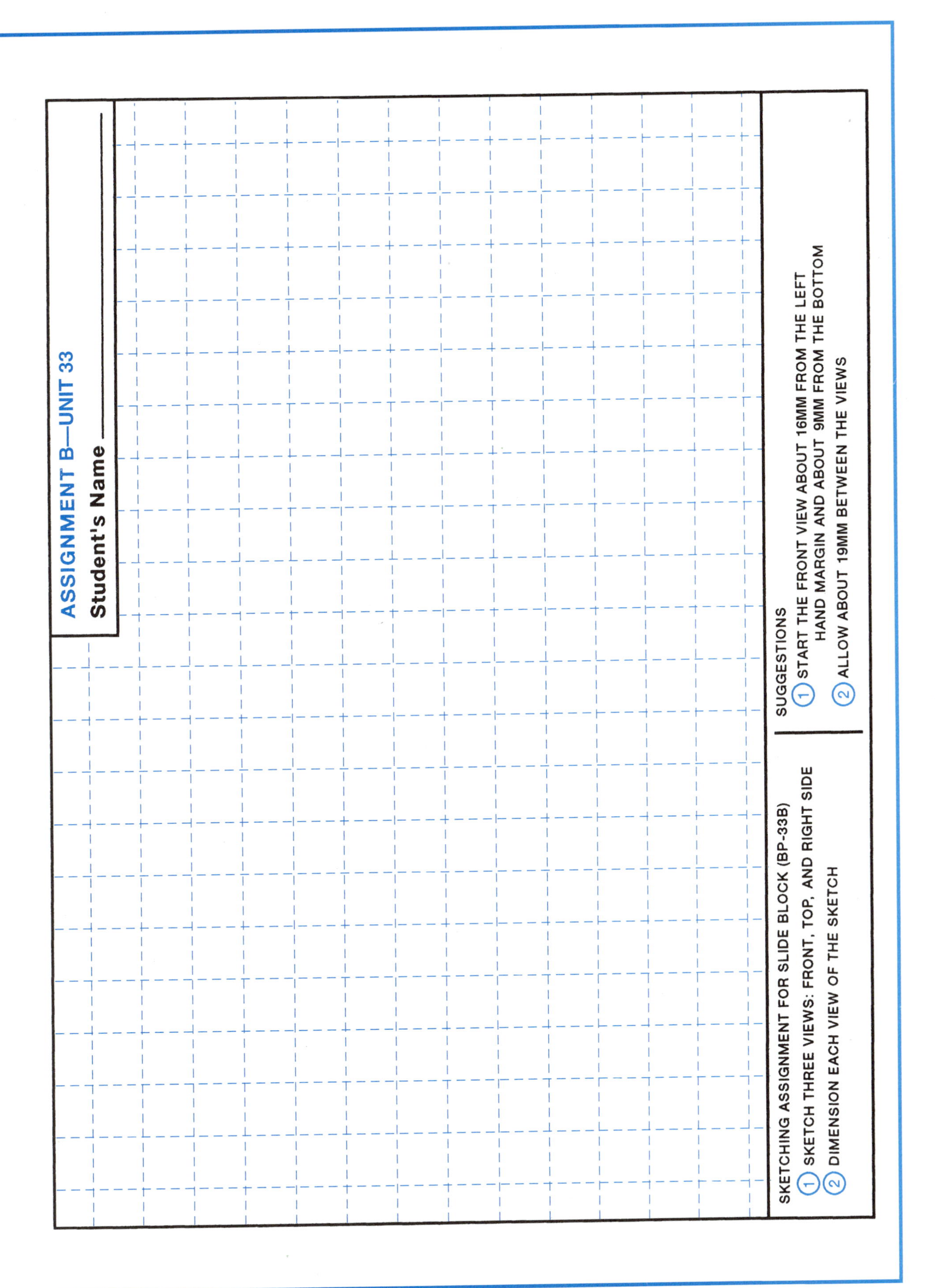
ASSIGNMENT B—UNIT 33
Student's Name
SKETCHING ASSIGNMENT FOR SLIDE BLOCK (BP-33B)
1 SKETCH THREE VIEWS: FRONT, TOP, AND RIGHT SIDE
2 DIMENSION EACH VIEW OF THE SKETCH
SUGGESTIONS
1 START THE FRONT VIEW ABOUT 16MM FROM THE LEFT HAND MARGIN AND ABOUT 9MM FROM THE BOTTOM
2 ALLOW ABOUT 19MM BETWEEN THE VIEWS

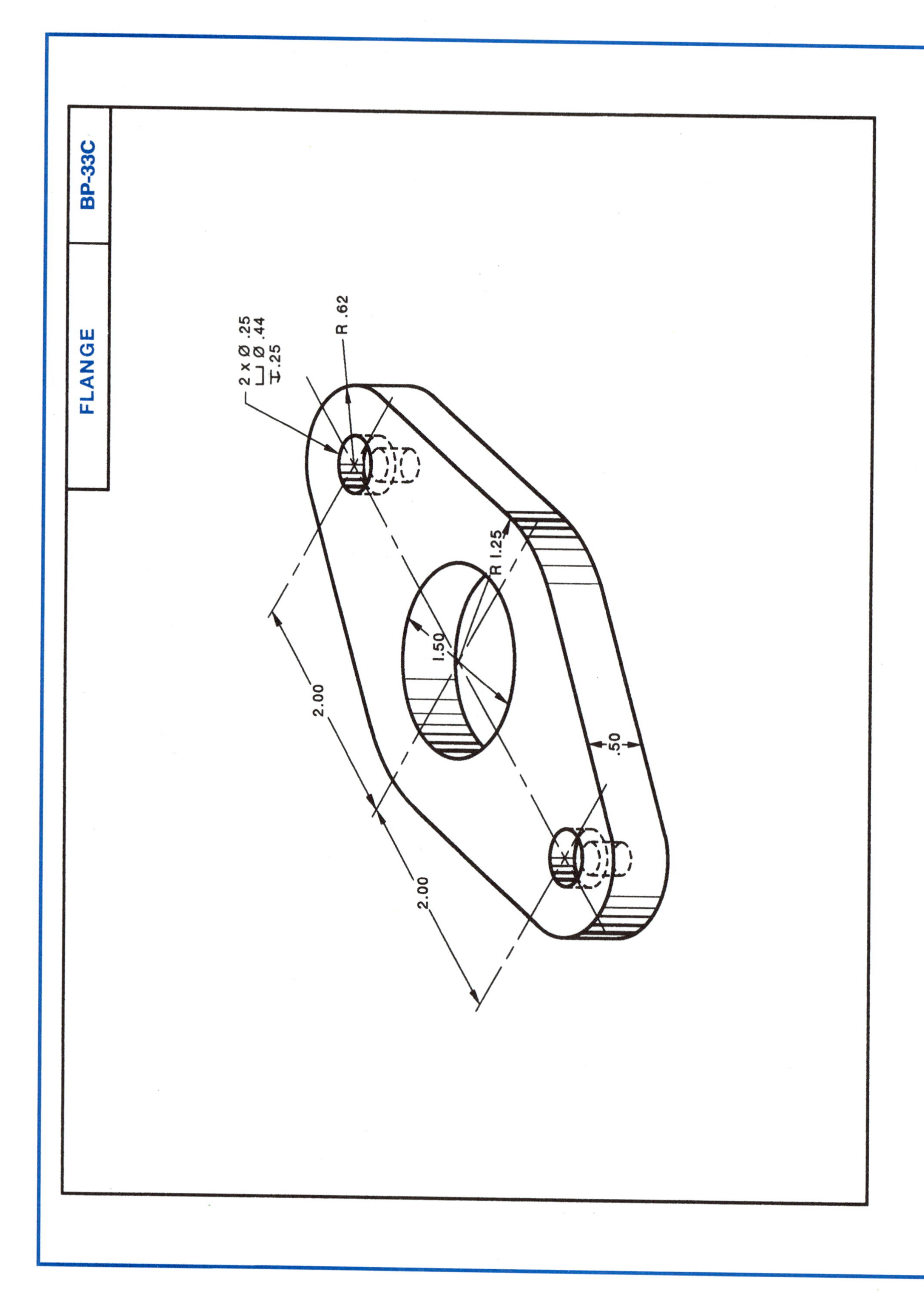
FLANGE
BP-33C
2 x Ø .25
⌴ Ø .44
↧ .25
R .62
R 1.25
1.50
2.00
2.00
.50

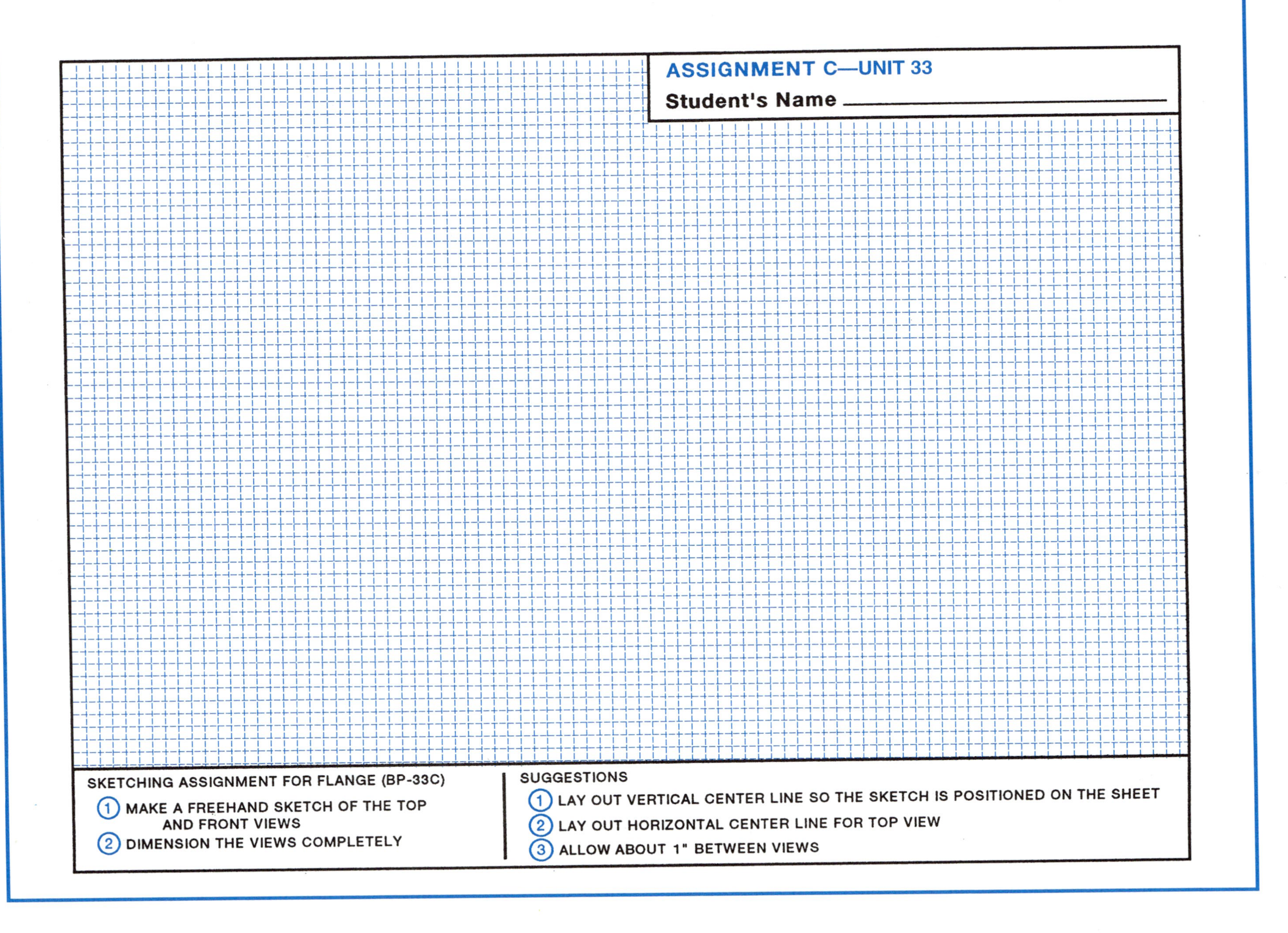
ASSIGNMENT C—UNIT 33
Student's Name
SKETCHING ASSIGNMENT FOR FLANGE (BP-33C)
1 MAKE A FREEHAND SKETCH OF THE TOP AND FRONT VIEWS
2 DIMENSION THE VIEWS COMPLETELY
SUGGESTIONS
1 LAY OUT VERTICAL CENTER LINE SO THE SKETCH IS POSITIONED ON THE SHEET
2 LAY OUT HORIZONTAL CENTER LINE FOR TOP VIEW
3 ALLOW ABOUT 1" BETWEEN VIEWS

Unit 34

SKETCHING IRREGULAR SHAPES

Parts that are irregular in shape often look complicated to sketch. However, the object may be drawn easily if it is first visualized as a series of square and rectangular blocks. Then, by using straight, slant, and curved lines in combination with each other, it is possible to draw in the squares and rectangles the exact shape of the part.

The Shaft Support shown in figure 34-1 is an example of a machine part that can be sketched easily by *blocking*. The support must first be thought of in terms of basic squares and rectangles. After these squares and rectangles are determined, the step-by-step procedures which are commonly used to simplify the making of a sketch of an irregularly-shaped piece are applied. These step-by-step procedures are described next and their application is shown in figure 34-2.

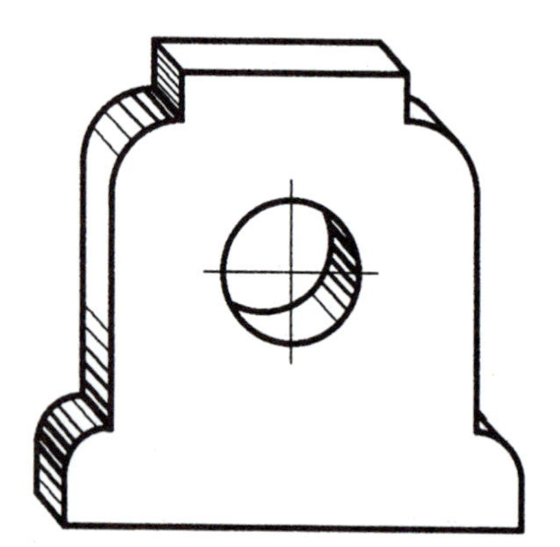

FIGURE 34-1 Example of irregularly-shaped part to be sketched.

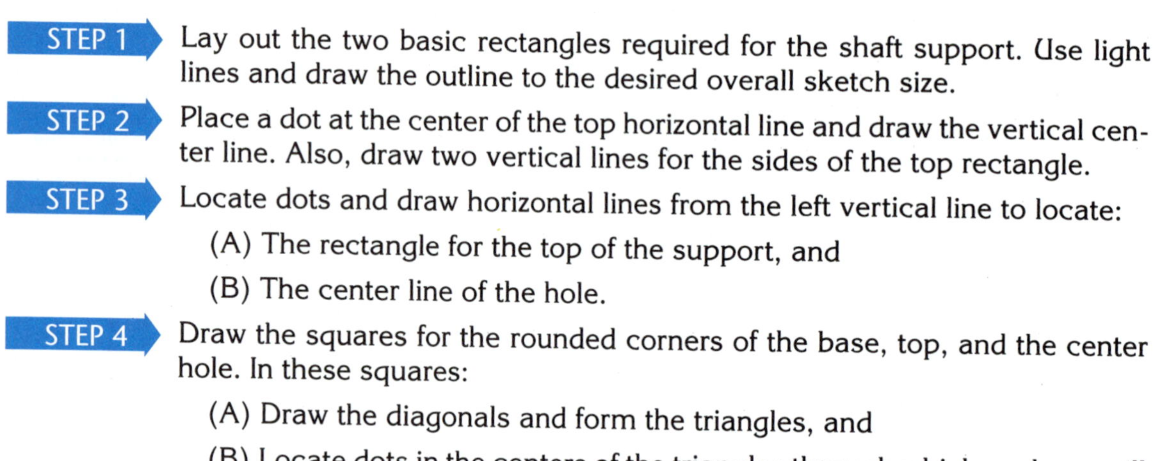

STEP 1 Lay out the two basic rectangles required for the shaft support. Use light lines and draw the outline to the desired overall sketch size.

STEP 2 Place a dot at the center of the top horizontal line and draw the vertical center line. Also, draw two vertical lines for the sides of the top rectangle.

STEP 3 Locate dots and draw horizontal lines from the left vertical line to locate:

(A) The rectangle for the top of the support, and

(B) The center line of the hole.

STEP 4 Draw the squares for the rounded corners of the base, top, and the center hole. In these squares:

(A) Draw the diagonals and form the triangles, and

(B) Locate dots in the centers of the triangles through which each arc will pass.

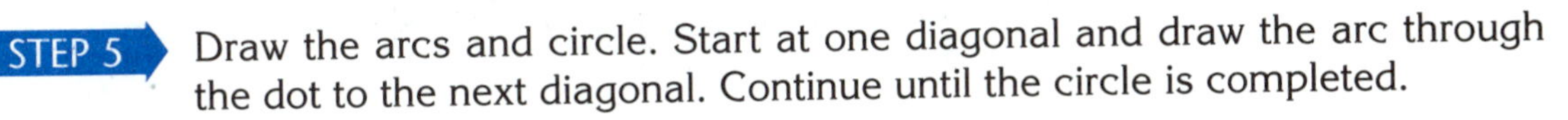

STEP 5 Draw the arcs and circle. Start at one diagonal and draw the arc through the dot to the next diagonal. Continue until the circle is completed.

STEP 6 Draw the side view if necessary.

STEP 7 Darken all object lines and dimension. Erase those lines used in construction that either do not simplify the sketch or are not required to interpret the sketch quickly and accurately.

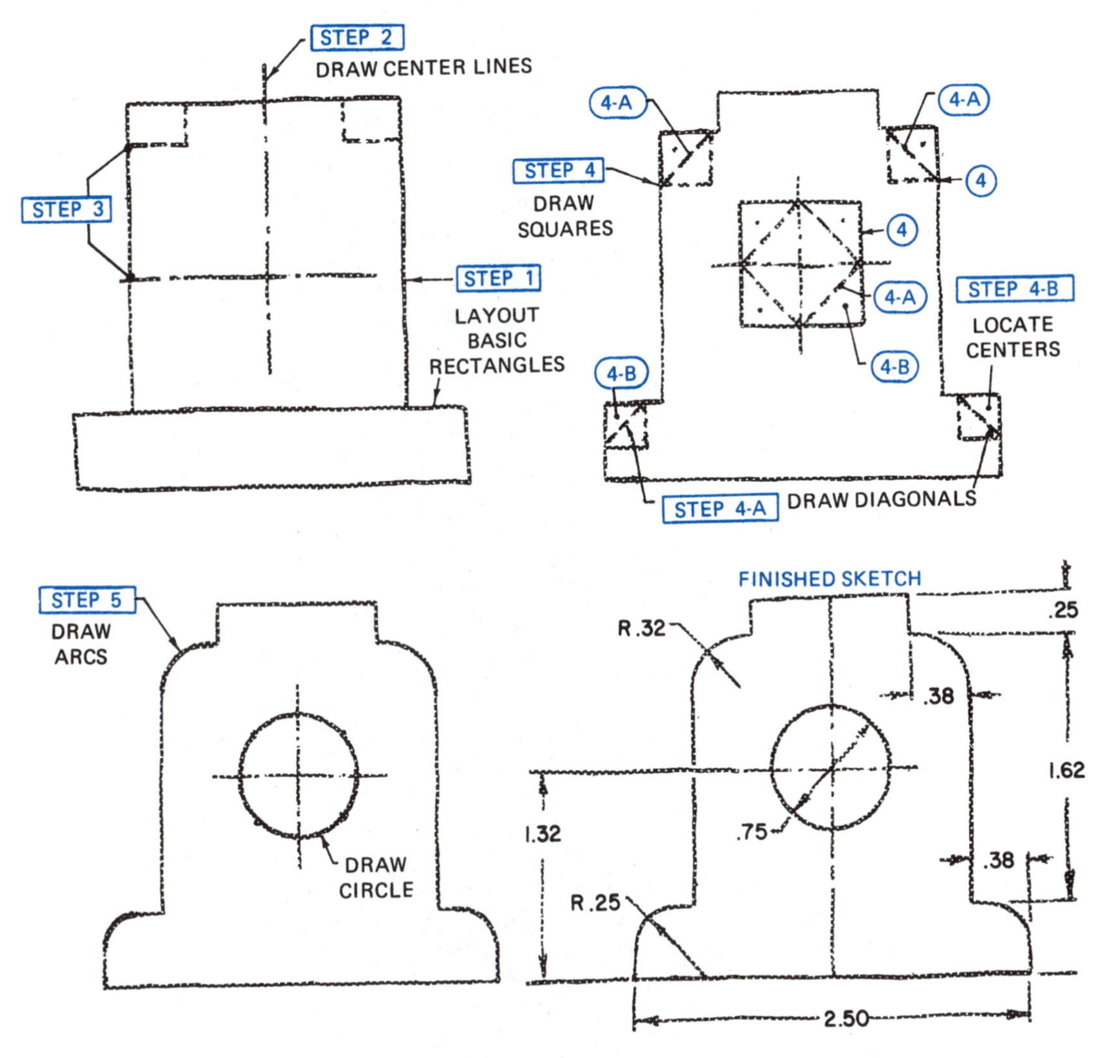

FIGURE 34-2 Sketching an irregular shape.

The techniques described in this unit have been found by tested experience to be essential in training the beginner to sketch accurately.

As skill is developed in drawing straight, slant and curved lines freehand in combination with each other, some of the steps in sketching irregularly-shaped parts may be omitted and the process shortened.

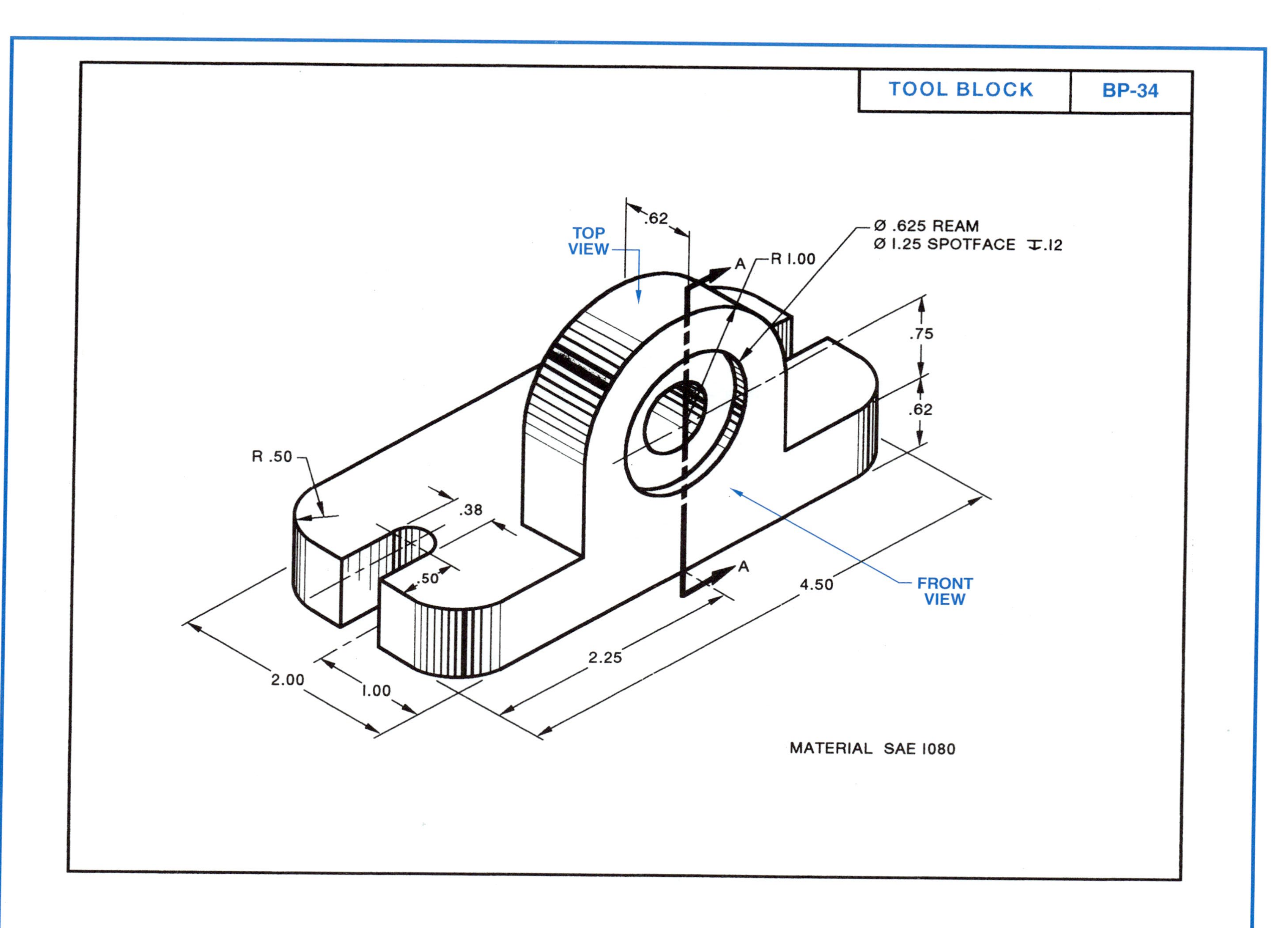
TOOL BLOCK
BP-34
.62
TOP VIEW
A
R 1.00
Ø .625 REAM
Ø 1.25 SPOTFACE ↧.12
.75
.62
R .50
.38
.50
A
4.50
FRONT VIEW
2.00
1.00
2.25
MATERIAL SAE 1080

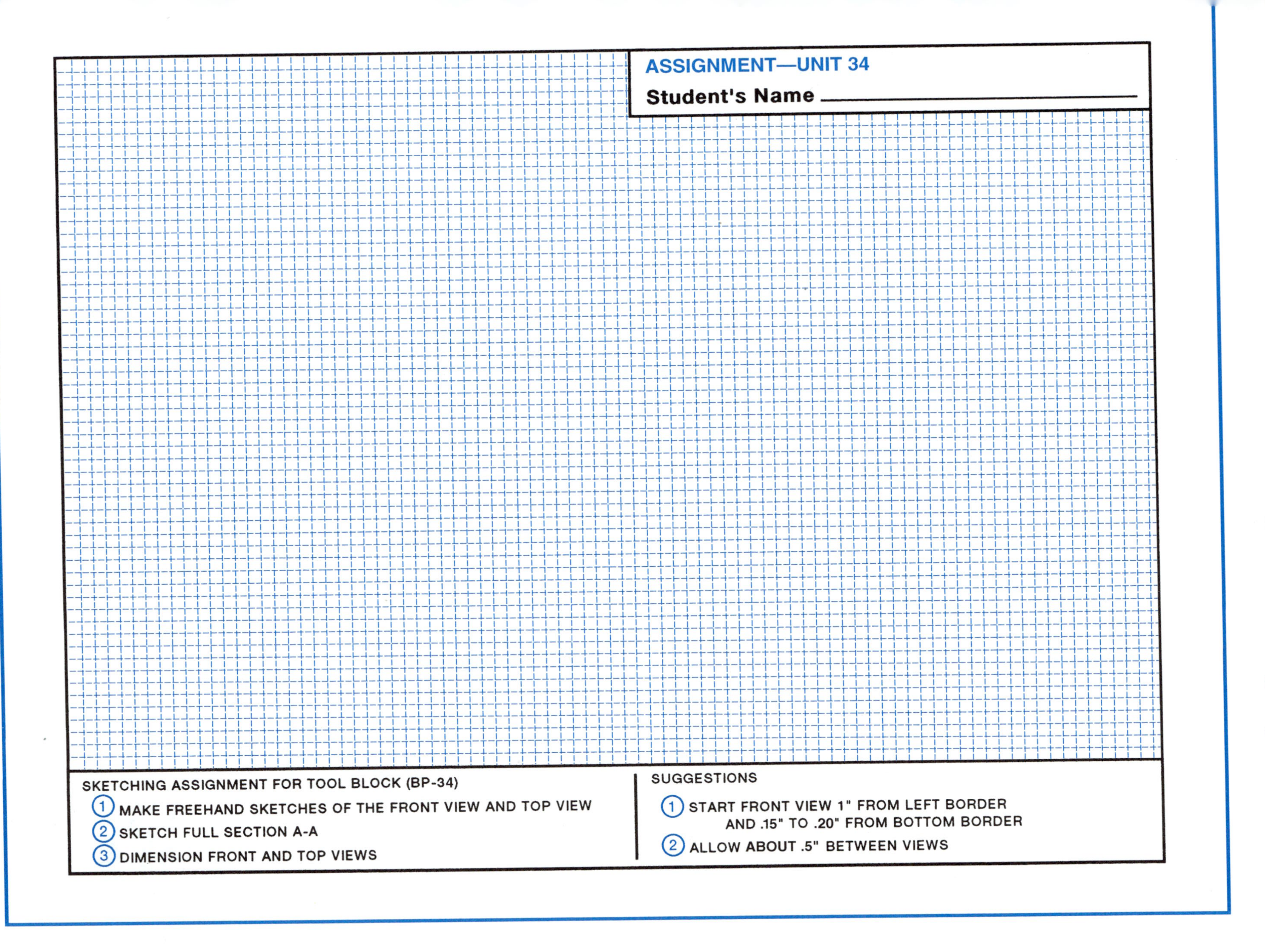
ASSIGNMENT—UNIT 34
Student's Name
SKETCHING ASSIGNMENT FOR TOOL BLOCK (BP-34)
1 MAKE FREEHAND SKETCHES OF THE FRONT VIEW AND TOP VIEW
2 SKETCH FULL SECTION A-A
3 DIMENSION FRONT AND TOP VIEWS
SUGGESTIONS
1 START FRONT VIEW 1" FROM LEFT BORDER AND .15" TO .20" FROM BOTTOM BORDER
2 ALLOW ABOUT .5" BETWEEN VIEWS

Unit 35

SKETCHING FILLETS, RADII, AND ROUNDED CORNERS AND EDGES

Wherever practical, sharp corners are rounded and sharp edges are eliminated. This is true of fabricated and hardened parts where a sharp corner reduces the strength of the object. It is also true of castings where sharp edges and corners may be difficult to produce, figure 35-1.

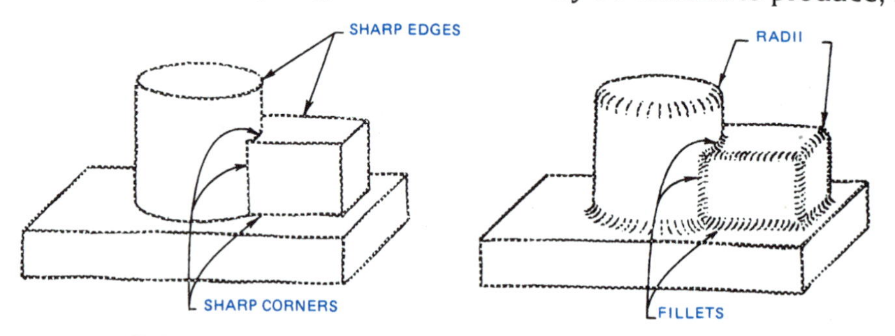

FIGURE 35-1 Eliminating sharp edges and corners.

Where an outside edge is rounded, the convex edge is called a *round edge* or *radius*. A rounded inside corner is known as a *fillet*.

SKETCHING A FILLET OR RADIUS

Regardless of whether a fillet or radius is required, the steps for sketching each one are identical. The step-by-step procedure is illustrated in figure 35-2.

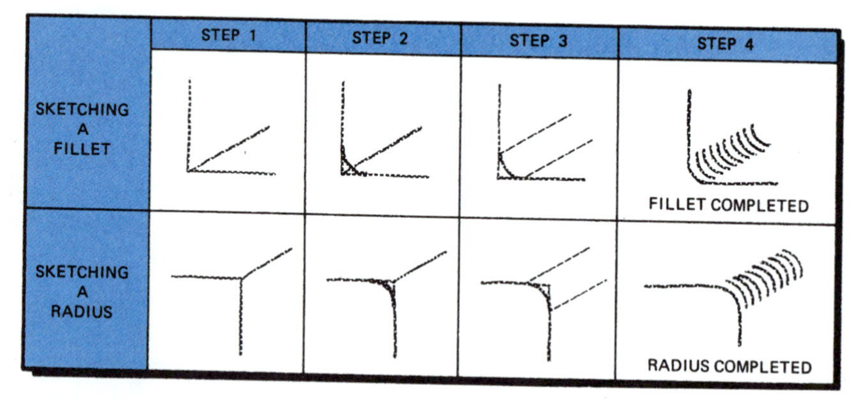

FIGURE 35-2 Steps in sketching a fillet or radius.

STEP 1 Sketch the lines which represent the edge or corner.

STEP 2 Draw the arc to the required radius.

STEP 3 Draw two lines parallel to the edge line from the two points where the arc touches the straight lines.

STEP 4 Sketch a series of curved lines across the work with the same arc as the object line. These curved lines start at one of the parallel lines and terminate at the other end.

SKETCHING RADII AND FILLETS

The method of sketching corners at which radii or fillets come together is shown in the three steps in figure 35-3.

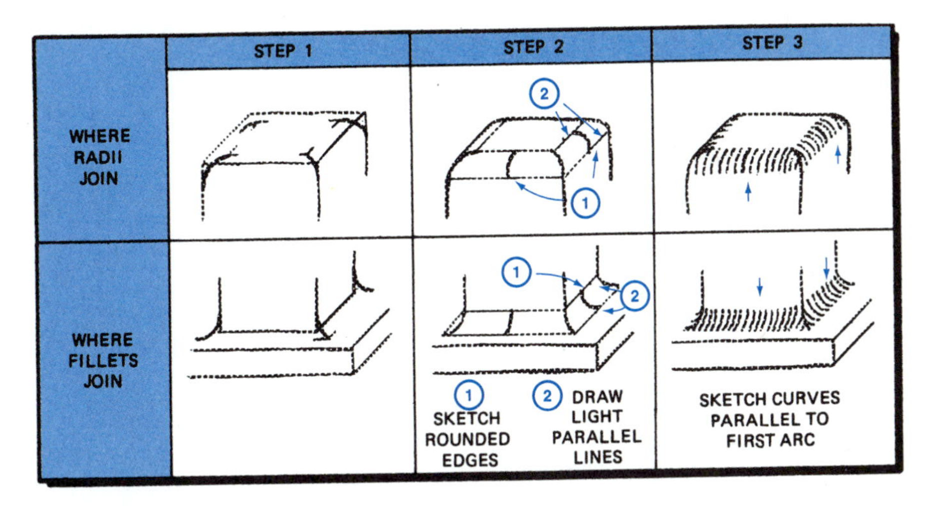

FIGURE 35-3 Sketching rounded corners and edges.

SKETCHING CORNERS ON CIRCULAR PARTS

The direction of radius of fillet lines changes on circular objects at a center line. The curved lines tend to straighten as they approach the center line and then slowly curve in the opposite direction beyond that point. The direction of curved lines for outside radii of round parts is shown in figure 35-4A. The curved lines for the intersecting corners are shown in figure 35-4B.

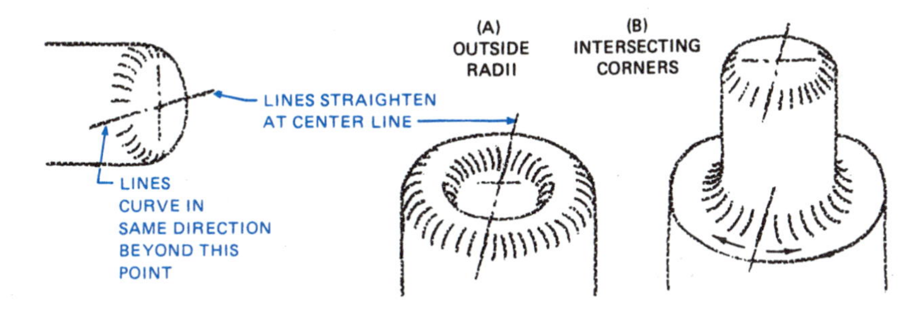

FIGURE 35-4 Direction of curved lines on round parts.

BASIC SHADING/RENDERING TECHNIQUES

Shading (sometimes referred to as *rendering*) is often used on drawings that have hard-to-read part features. Where required, as a general rule, the draftsperson uses the least amount of shading possible. Shading may be produced freehand, generated by computer, or created by cutting out cross hatch, dot, or other shade line patterns from an acetate sheet and pressing the shade material on a drawing.

Finely-spaced, thin, straight lines and/or concentric finely-spaced curved lines (or narrow, solid bands) are used for shading. Figure 35-5 shows how shade lines and solid shade bands clearly identify the end radius, shoulder fillet, cylindrical body, and keyseat features of the part.

FIGURE 35-5 Shading with straight and curved lines and narrow, solid bands.

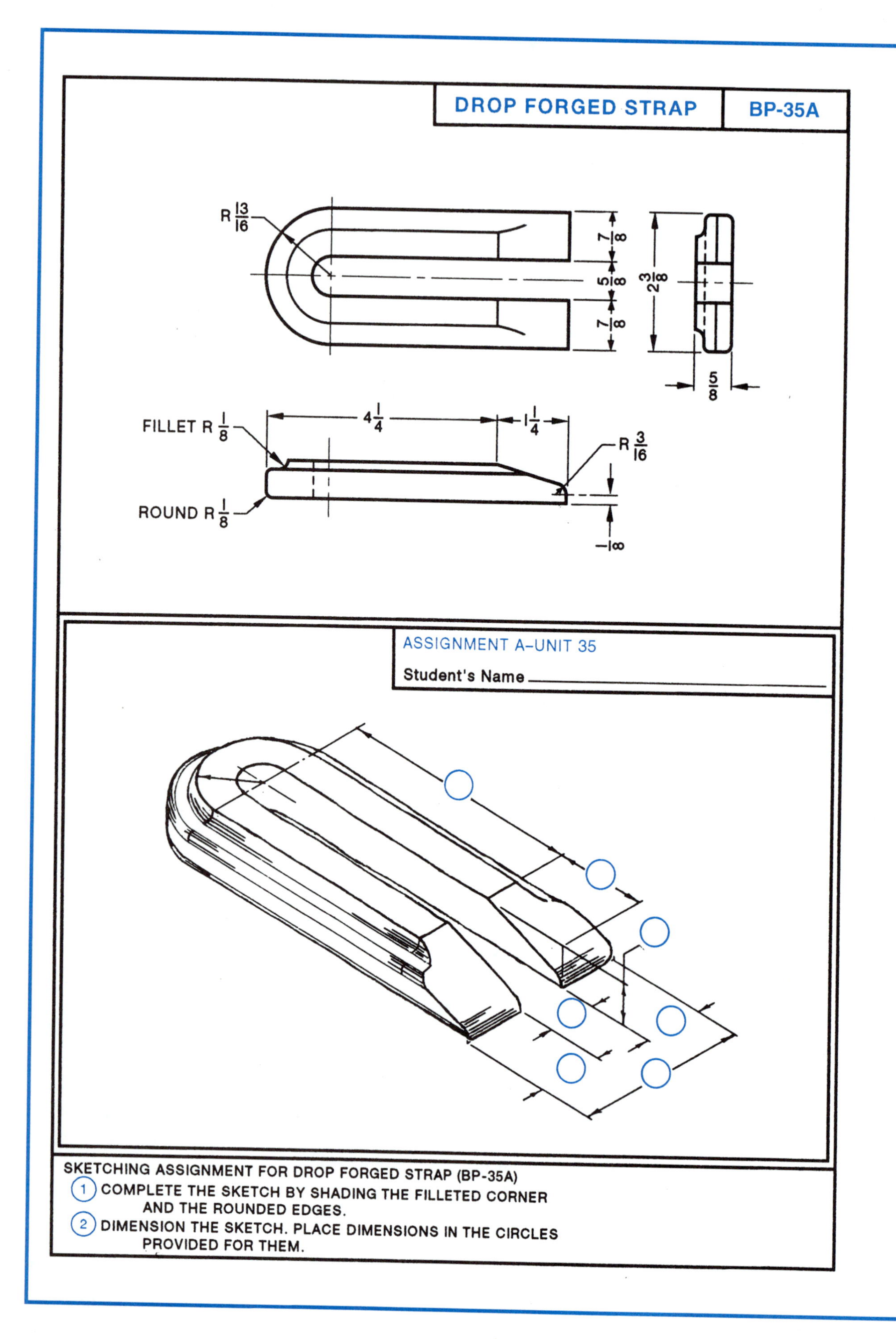
DROP FORGED STRAP
BP-35A
R 13/16
7/8
5/8
7/8
2 3/8
5/8
FILLET R 1/8
4 1/4
1 1/4
R 3/16
ROUND R 1/8
1/8
ASSIGNMENT A–UNIT 35
Student's Name
SKETCHING ASSIGNMENT FOR DROP FORGED STRAP (BP-35A)
1 COMPLETE THE SKETCH BY SHADING THE FILLETED CORNER AND THE ROUNDED EDGES.
2 DIMENSION THE SKETCH. PLACE DIMENSIONS IN THE CIRCLES PROVIDED FOR THEM.

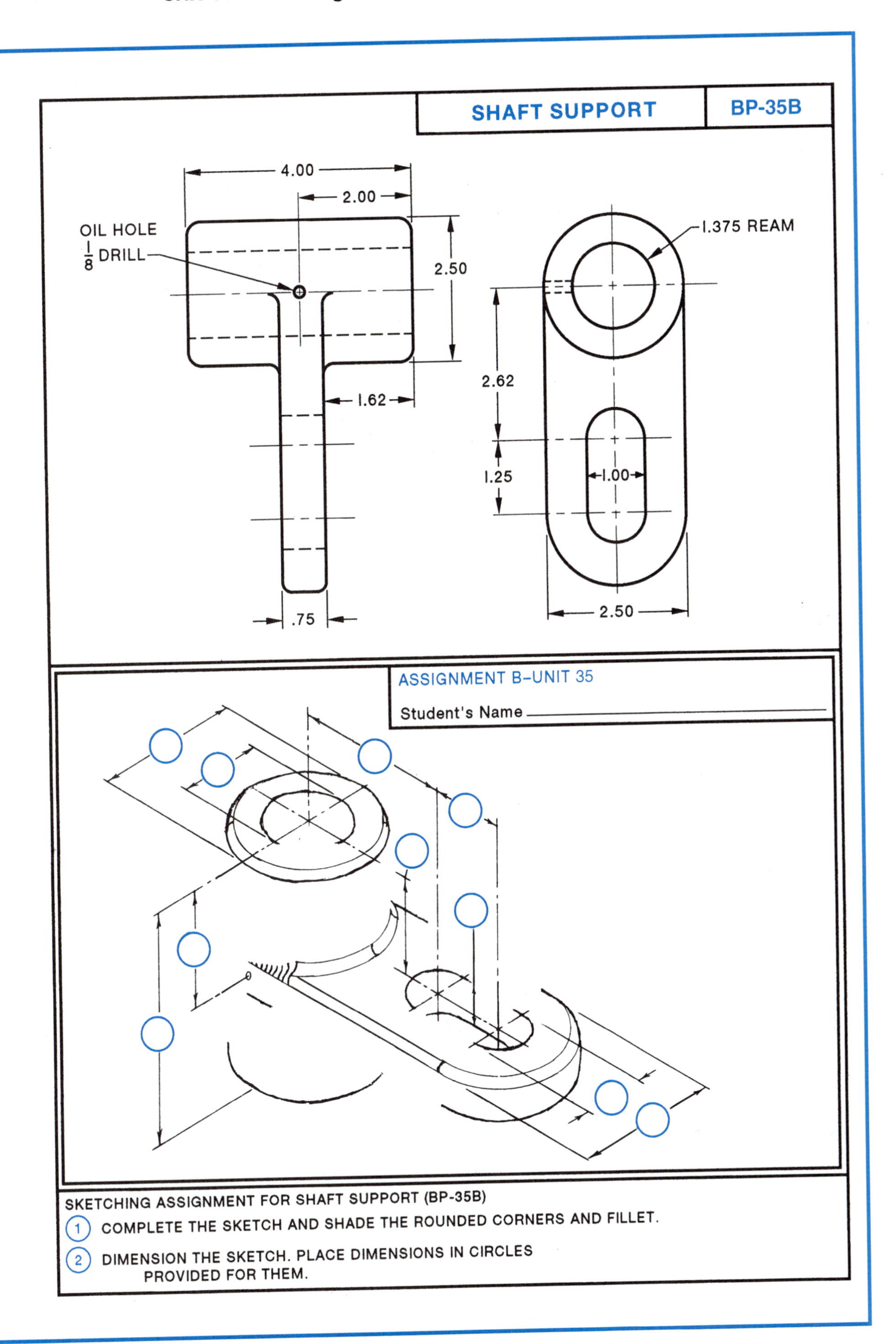
SHAFT SUPPORT
BP-35B
4.00
2.00
OIL HOLE
1/8 DRILL
2.50
1.62
.75
1.375 REAM
2.62
1.25
1.00
2.50
ASSIGNMENT B–UNIT 35
Student's Name
SKETCHING ASSIGNMENT FOR SHAFT SUPPORT (BP-35B)
1 COMPLETE THE SKETCH AND SHADE THE ROUNDED CORNERS AND FILLET.
2 DIMENSION THE SKETCH. PLACE DIMENSIONS IN CIRCLES PROVIDED FOR THEM.

SECTION 12 Freehand Lettering

Unit 36

FREEHAND VERTICAL LETTERING

The true shape of a part or mechanism may be described accurately on a drawing by using combinations of lines and views. Added to these are the lettering, which supplies additional information, and the dimensions. For exceptionally accurate work and to standardize the shape and size of the letters, the lettering is done with a guide or the lettering may be produced by computer and then added to a drawing. In most other cases, the letters are formed freehand.

An almost square style letter known as *Gothic lettering* is very widely used because it is legible and the individual letters are simple enough to be made quickly and accurately. Gothic letters may be either vertical (straight) or inclined (slant), and upper or lower case. Lower case letters are rarely used on manufacturing drawings, but are used at times with construction drawings.

FORMING UPPER CASE LETTERS

The shape of each vertical letter (both upper and lower case) and each number will be discussed in this unit. All of these letters and numbers are formed by combining vertical, horizontal, slant, and curved lines. The upper case letters should be started first as they are easiest to make.

Very light guide lines should be used to keep the letters straight and of uniform height. A soft pencil is recommended for lettering because it is possible to guide the pencil easily to form good letters.

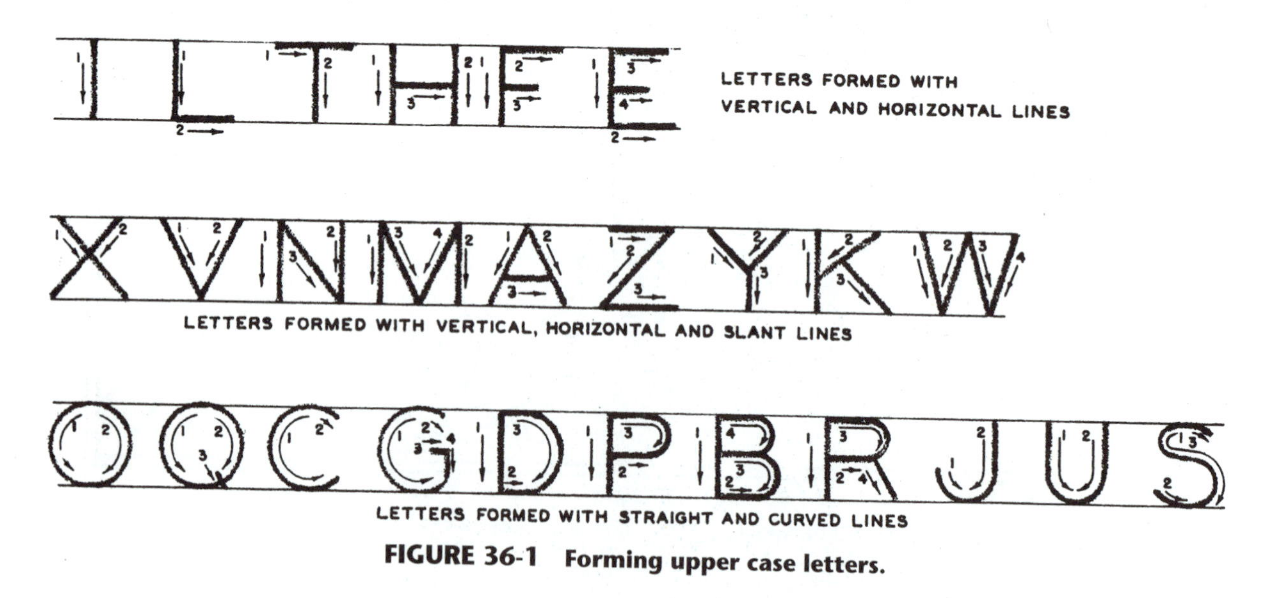

FIGURE 36-1 Forming upper case letters.

The shape of each vertical letter and number is given in figures 36-1, 36-2, and 36-3. The fine (light) lines with arrows which appear with every letter give the direction and number of the strokes needed to form the letter. The small numbers indicate the sequence of the strokes. Note that most letters are narrower than they are long. Lettering is usually done with a single stroke to keep the line weight of each letter uniform.

FIGURE 36-2 Forming vertical numerals.

SPACING LETTERS

To achieve good lettering, attention must be given to the proper spacing between letters, words, and lines. Words and lines that are either condensed and run together or spread out are difficult to read, cause inaccuracies, and detract from an otherwise good drawing.

The space between letters should be about one-fourth the width of a regular letter. For example, the slant line of the letter **A** should be one-quarter letter width away from the top line of the letter **T**. Judgment must be used in the amount of white space left between letters so that it is as equal as possible. The letters will then look in balance and will be easy to read.

Between words, a space two-thirds the full width of a normal letter should be used. Lines of lettering are easiest to read when a space of from one-half to the full height of the letters is left between the lines.

LOWER CASE VERTICAL LETTERS

Lower case vertical letters are formed in a manner different from that used to form upper case letters. The main portion, or body, of most Gothic lower case letters is approximately the same width and height. The parts that extend above or below the body are one-half the height of the body. The shape of each vertical lower case letter and the forming of the letters are indicated in figure 36-3.

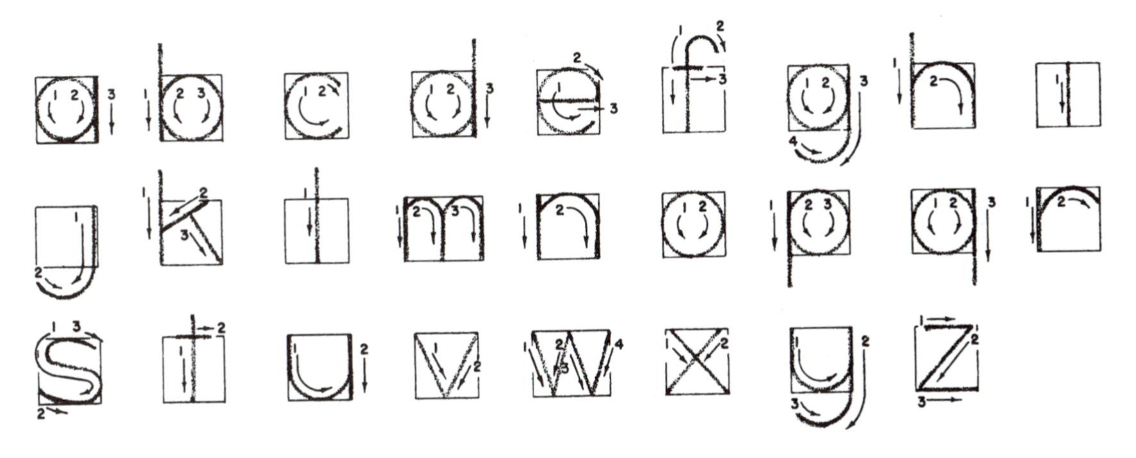

FIGURE 36-3 Forming vertical lower-case letters.

The spacing between lower case letters and words is the same as for capitals. The spacing between lines should be equal to the height of the body of the lower case letter to allow for the lines extending above and below the body of certain letters.

LETTERING FRACTIONS

Since fractions are important, they must not be subordinate to any of the lettering. The height of each number in the numerator and denominator must be at least two-thirds the height of a whole number. The dividing line of the fraction is in the center of the whole number. There must be a space between the numerator, the dividing line, and the denominator so that neither number touches the line, figure 36-4.

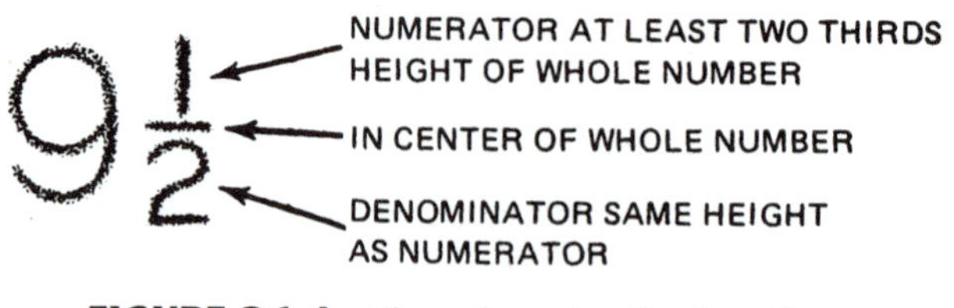

FIGURE 36-4 Size of parts of a fraction.

Lettering for Microfilm and Photocopies

Lettering used on drawings which are to be microfilmed or photocopied require special attention. The National Microfilm Association has developed an adaptation of the single-stroke Gothic characters for general use and increased legibility in drawing reproduction. The letters are known as *microfont.* Figure 36-5 shows only the vertical style. Many companies that use microfilming continue to use standard letters, large and dense enough to clearly reproduce.

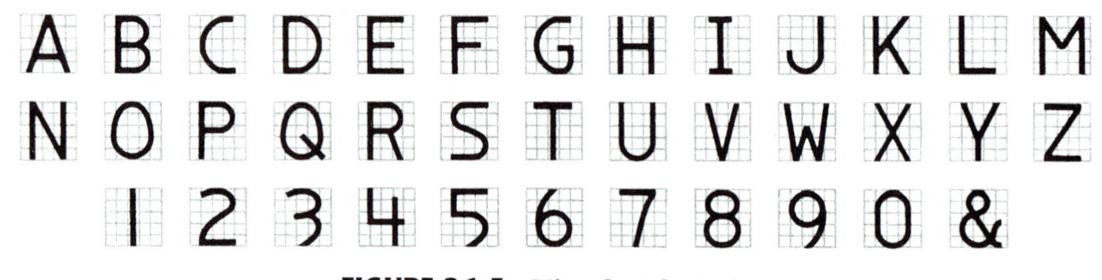

FIGURE 36-5 Microfont lettering

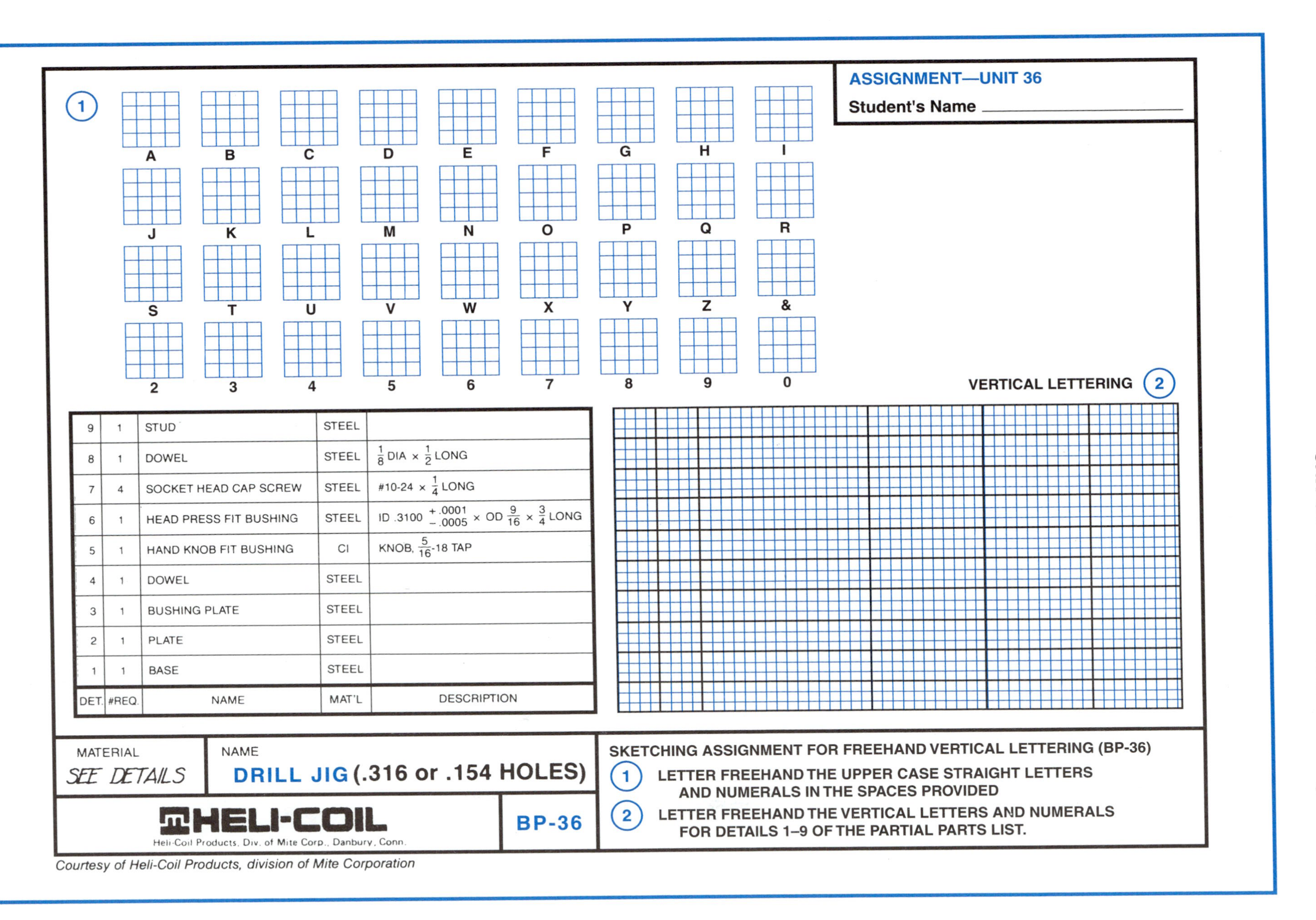

ASSIGNMENT—UNIT 36

Student's Name ____________

① A B C D E F G H I
J K L M N O P Q R
S T U V W X Y Z &
2 3 4 5 6 7 8 9 0

VERTICAL LETTERING ②

DET.	#REQ.	NAME	MAT'L	DESCRIPTION
9	1	STUD	STEEL	
8	1	DOWEL	STEEL	$\frac{1}{8}$ DIA × $\frac{1}{2}$ LONG
7	4	SOCKET HEAD CAP SCREW	STEEL	#10-24 × $\frac{1}{4}$ LONG
6	1	HEAD PRESS FIT BUSHING	STEEL	ID .3100 $^{+.0001}_{-.0005}$ × OD $\frac{9}{16}$ × $\frac{3}{4}$ LONG
5	1	HAND KNOB FIT BUSHING	CI	KNOB, $\frac{5}{16}$-18 TAP
4	1	DOWEL	STEEL	
3	1	BUSHING PLATE	STEEL	
2	1	PLATE	STEEL	
1	1	BASE	STEEL	

MATERIAL SEE DETAILS

NAME DRILL JIG (.316 or .154 HOLES)

HELI-COIL

Heli-Coil Products, Div. of Mite Corp., Danbury, Conn.

BP-36

SKETCHING ASSIGNMENT FOR FREEHAND VERTICAL LETTERING (BP-36)

① LETTER FREEHAND THE UPPER CASE STRAIGHT LETTERS AND NUMERALS IN THE SPACES PROVIDED

② LETTER FREEHAND THE VERTICAL LETTERS AND NUMERALS FOR DETAILS 1–9 OF THE PARTIAL PARTS LIST.

Courtesy of Heli-Coil Products, division of Mite Corporation

Unit 37

FREEHAND INCLINED LETTERING

Inclined letters and numbers are used on many drawings as they can be formed with a very natural movement and a slant similar to that used in everyday writing. The shape of the inclined or slant letter is the same as that of the vertical or straight letter except that circles and parts of circles are elliptical and the axis of each letter is at an angle, figure 37-1. Slant letters may be formed with the left hand using the same techniques of shaping and spacing as are used by the right hand. The only difference is that for some letters and numerals the strokes are reversed.

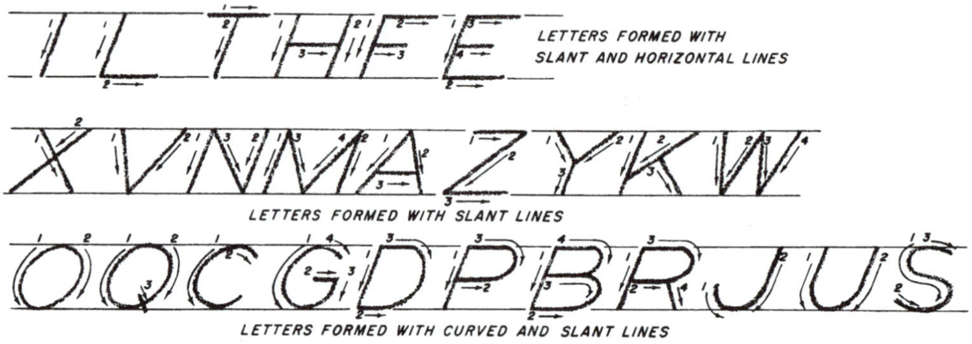

LETTERS FORMED WITH SLANT AND HORIZONTAL LINES

LETTERS FORMED WITH SLANT LINES

LETTERS FORMED WITH CURVED AND SLANT LINES

FIGURE 37-1 **Forming upper case slant letters.**

FORMING SLANT LETTERS AND NUMERALS

The angle of slant of the letters and numerals may vary, depending on individual preference. Lettering at an angle of from 60 to 75 degrees is practical, as it is easy to read and produce. Light horizontal guide lines to assure the uniform height of letters are recommended for the beginner. Angle guide lines assist the beginner in keeping all the letters shaped correctly and spaced properly. The direction of the strokes for each letter and numeral, and the spacing between letters, words, and lines, are the same as for straight letters. The shape of each upper case slant letter is illustrated in figure 37-1; each numeral is shown in figure 37-2; and each lower case slant letter is shown in figure 37-3.

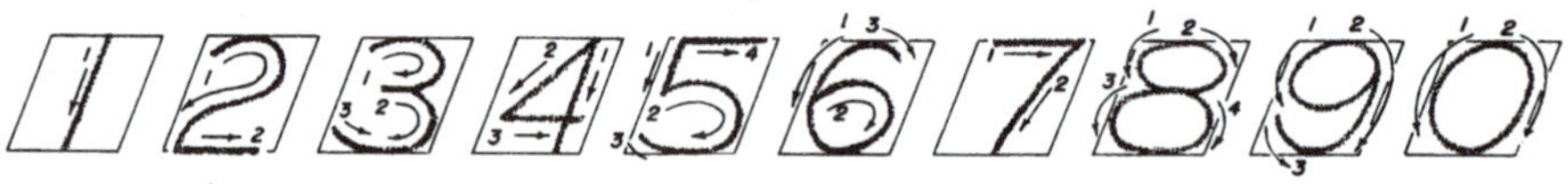

FIGURE 37-2 **Forming slant numerals.**

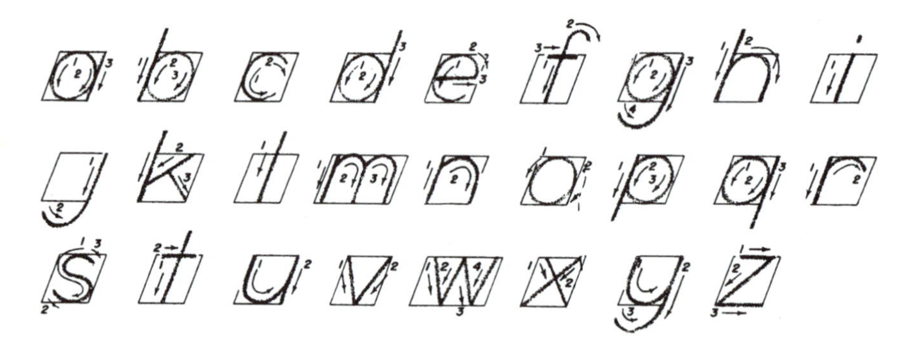

FIGURE 37-3 **Forming lower case slant letters.**

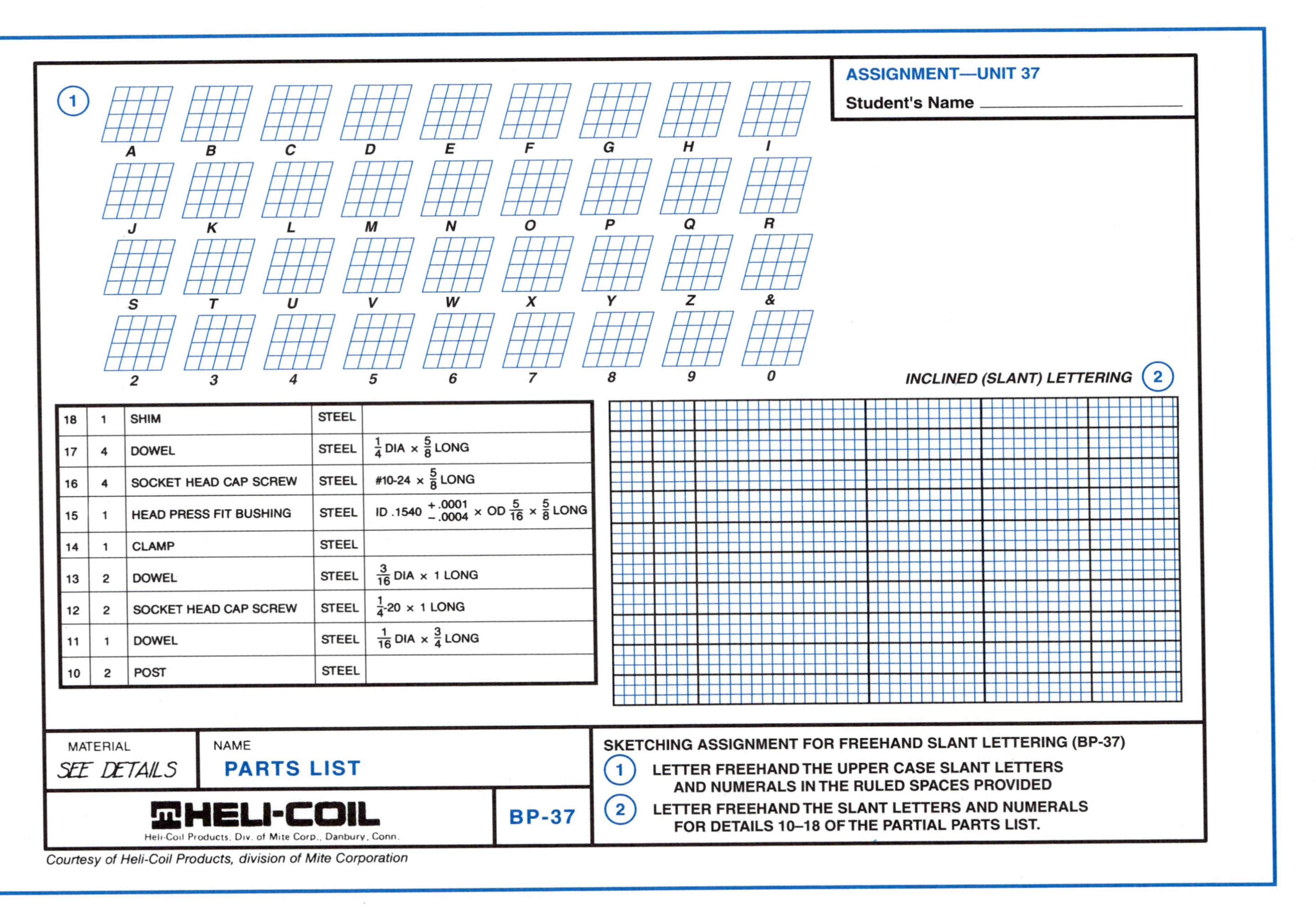

ASSIGNMENT—UNIT 37

Student's Name ____________

① A B C D E F G H I J K L M N O P Q R S T U V W X Y Z & 2 3 4 5 6 7 8 9 0

INCLINED (SLANT) LETTERING ②

18	1	SHIM	STEEL	
17	4	DOWEL	STEEL	$\frac{1}{4}$ DIA × $\frac{5}{8}$ LONG
16	4	SOCKET HEAD CAP SCREW	STEEL	#10-24 × $\frac{5}{8}$ LONG
15	1	HEAD PRESS FIT BUSHING	STEEL	ID .1540 $^{+.0001}_{-.0004}$ × OD $\frac{5}{16}$ × $\frac{5}{8}$ LONG
14	1	CLAMP	STEEL	
13	2	DOWEL	STEEL	$\frac{3}{16}$ DIA × 1 LONG
12	2	SOCKET HEAD CAP SCREW	STEEL	$\frac{1}{4}$-20 × 1 LONG
11	1	DOWEL	STEEL	$\frac{1}{16}$ DIA × $\frac{3}{4}$ LONG
10	2	POST	STEEL	

MATERIAL SEE DETAILS

NAME PARTS LIST

HELI-COIL

Heli-Coil Products, Div. of Mite Corp., Danbury, Conn.

BP-37

SKETCHING ASSIGNMENT FOR FREEHAND SLANT LETTERING (BP-37)

① LETTER FREEHAND THE UPPER CASE SLANT LETTERS AND NUMERALS IN THE RULED SPACES PROVIDED

② LETTER FREEHAND THE SLANT LETTERS AND NUMERALS FOR DETAILS 10–18 OF THE PARTIAL PARTS LIST.

Courtesy of Heli-Coil Products, division of Mite Corporation

SECTION 13 Shop Sketching: Pictorial Drawings

Unit 38

ORTHOGRAPHIC SKETCHING

PICTORIAL DRAWINGS

Pictorial drawings are very easy to understand because they show an object as it appears to the person viewing it. Pictorial drawings show the length, width, and height of an object in a single view. The use of a pictorial drawing enables an individual, inexperienced in interpreting drawings, to visualize quickly the shape of single parts or various components in a complicated mechanism.

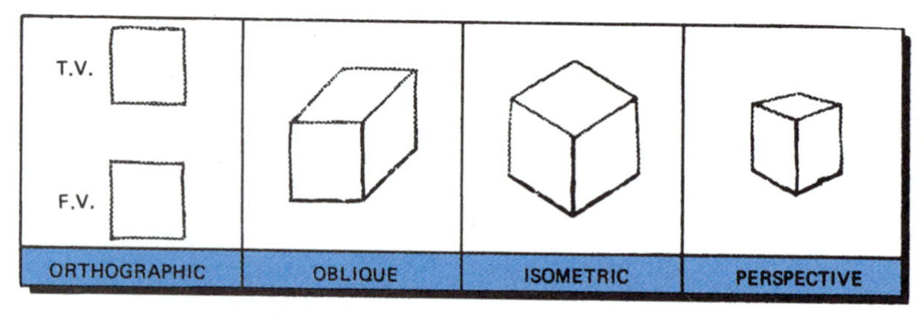

FIGURE 38-1 Four common types of sketches.

There are three general types of pictorial drawings in common use: (1) oblique, (2) isometric and (3) perspective. A fourth type of freehand drawing is the orthographic sketch, figure 38-1. The advantages and general principles of making orthographic sketches are described in this unit.

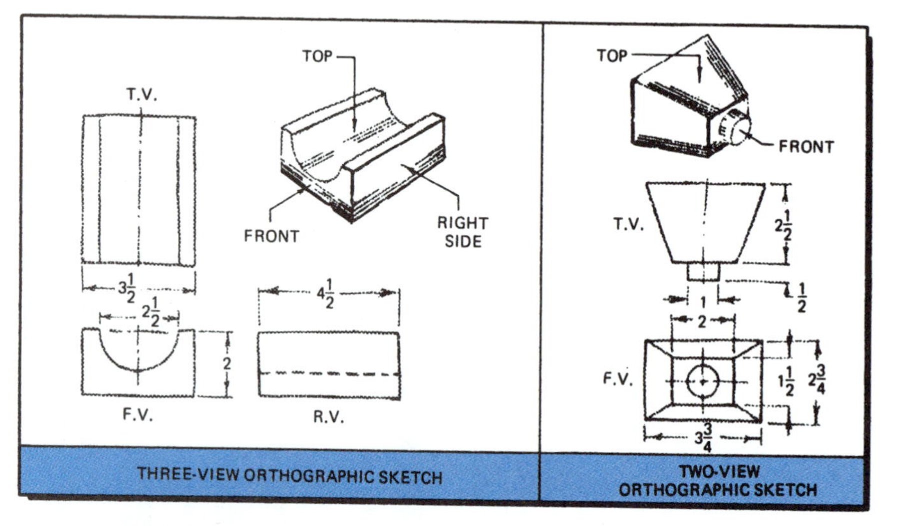

FIGURE 38-2 Examples of orthographic sketches.

MAKING ORTHOGRAPHIC SKETCHES

The orthographic sketch is the simplest type to make of the four types of sketches. The views (figure 38-2) are developed as in any regular mechanical drawing and the same types of lines are used. The only difference is that orthographic sketches are drawn freehand. In actual practice, on-the-spot sketches of small parts are made as near the actual size as possible.

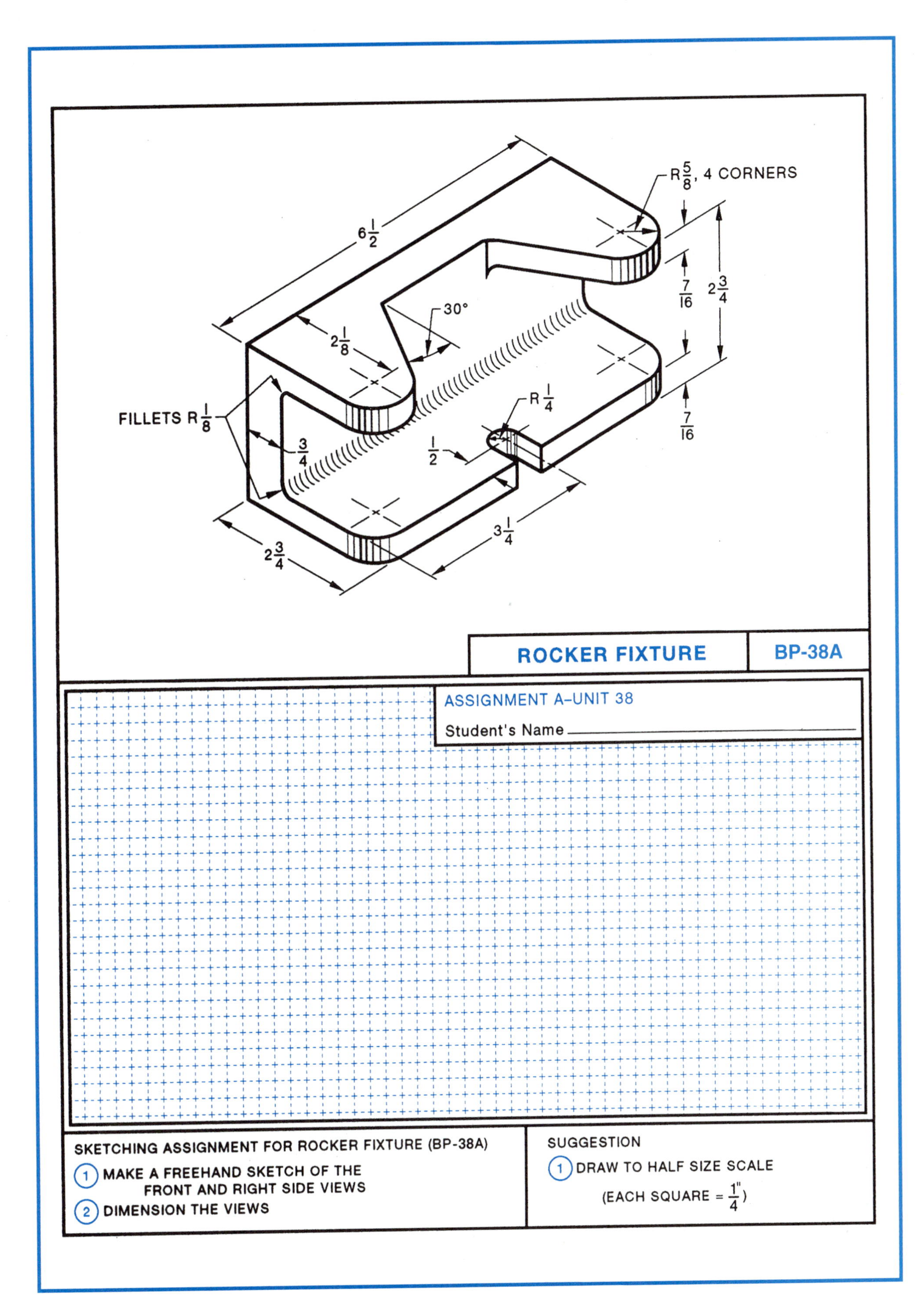

R 5/8, 4 CORNERS
6 1/2
30°
2 1/8
7/16
2 3/4
FILLETS R 1/8
3/4
R 1/4
1/2
7/16
2 3/4
3 1/4
ROCKER FIXTURE
BP-38A
ASSIGNMENT A–UNIT 38
Student's Name
SKETCHING ASSIGNMENT FOR ROCKER FIXTURE (BP-38A)
1 MAKE A FREEHAND SKETCH OF THE FRONT AND RIGHT SIDE VIEWS
2 DIMENSION THE VIEWS
SUGGESTION
1 DRAW TO HALF SIZE SCALE
(EACH SQUARE = 1/4")

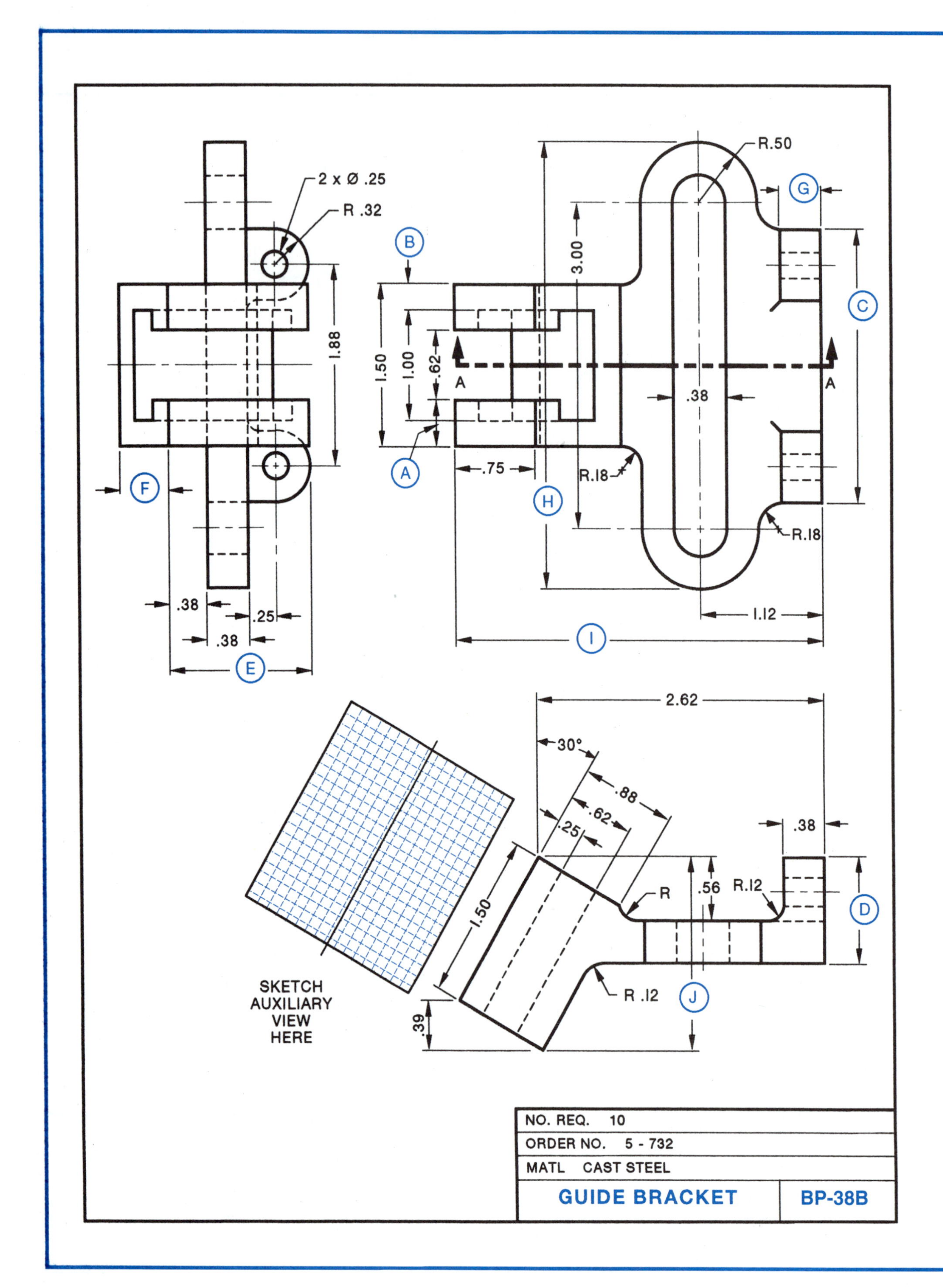

2 x Ø .25
R .32
1.88
B
A
1.50
1.00
.62
A
A
F
.38
.25
.38
E
R.50
G
C
3.00
.38
.75
R.18
H
R.18
1.12
I
2.62
30°
.88
.62
.25
.38
1.50
R
.56
R.12
D
R .12
J
.39
SKETCH
AUXILIARY
VIEW
HERE
NO. REQ. 10
ORDER NO. 5 - 732
MATL CAST STEEL
GUIDE BRACKET
BP-38B

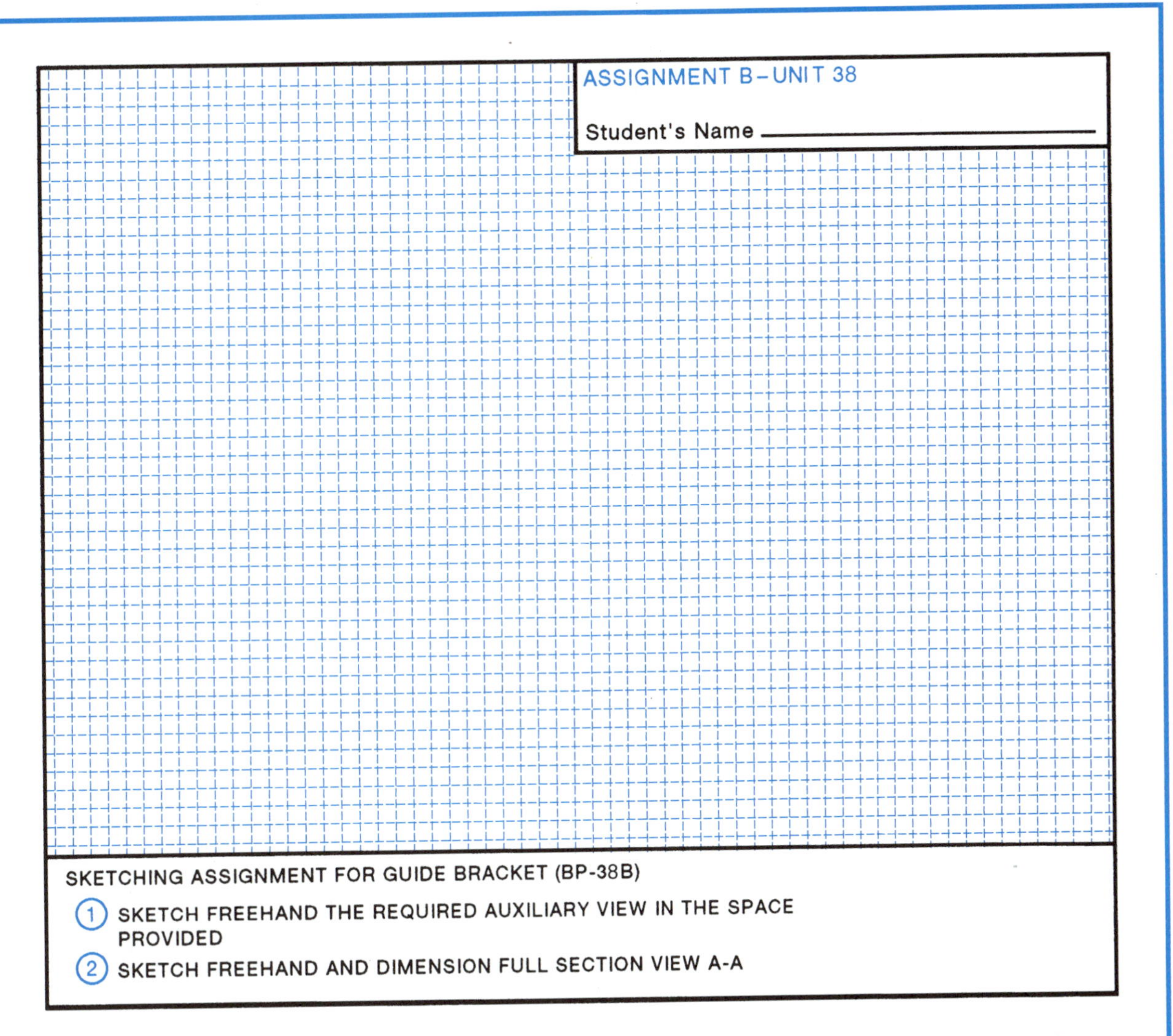

GUIDE BRACKET (BP-38B)

1. Name each of the three views.
2. Give the number and diameter of the drilled holes.
3. Determine dimensions (A) and (B).
4. Compute dimension (C).
5. What is the dimension of the leg (D)?
6. Determine dimension (E)?
7. Give dimension (F) and (G).
8. What is the overall length of (H)?
9. Determine overall height of (I).
10. Give overall width of (J).

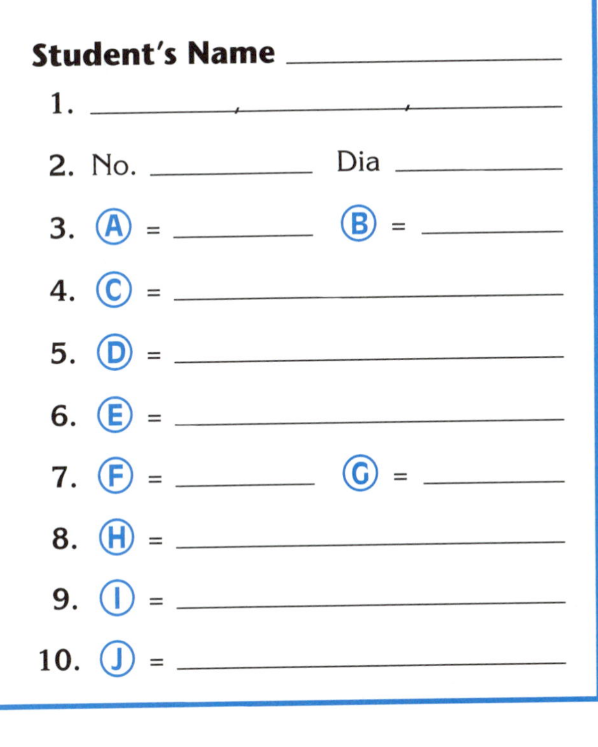

Unit 39

OBLIQUE SKETCHING

An oblique sketch is a type of pictorial drawing on which two or more surfaces are shown at one time on one drawing. The front face of the object is sketched in the same manner as the front view of either an orthographic sketch or a mechanical drawing. All of the straight, inclined, and curved lines on the front plane of the object will appear in their true size and shape on this front face. Since the other sides of the object are sketched at an angle, the surfaces and lines are not shown in their true size and shape.

MAKING OBLIQUE SKETCHES WITH STRAIGHT LINES

The steps in making an oblique sketch are simple. For example, if an oblique sketch of a rectangular die block is needed, seven basic steps are followed as shown in figure 39-1.

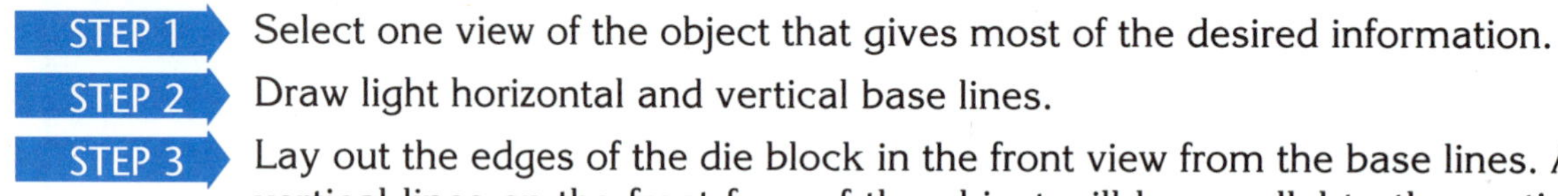

STEP 1 Select one view of the object that gives most of the desired information.

STEP 2 Draw light horizontal and vertical base lines.

STEP 3 Lay out the edges of the die block in the front view from the base lines. All vertical lines on the front face of the object will be parallel to the vertical base line; all horizontal lines will be parallel to the horizontal base line. All lines will be in their true size and shape in this view.

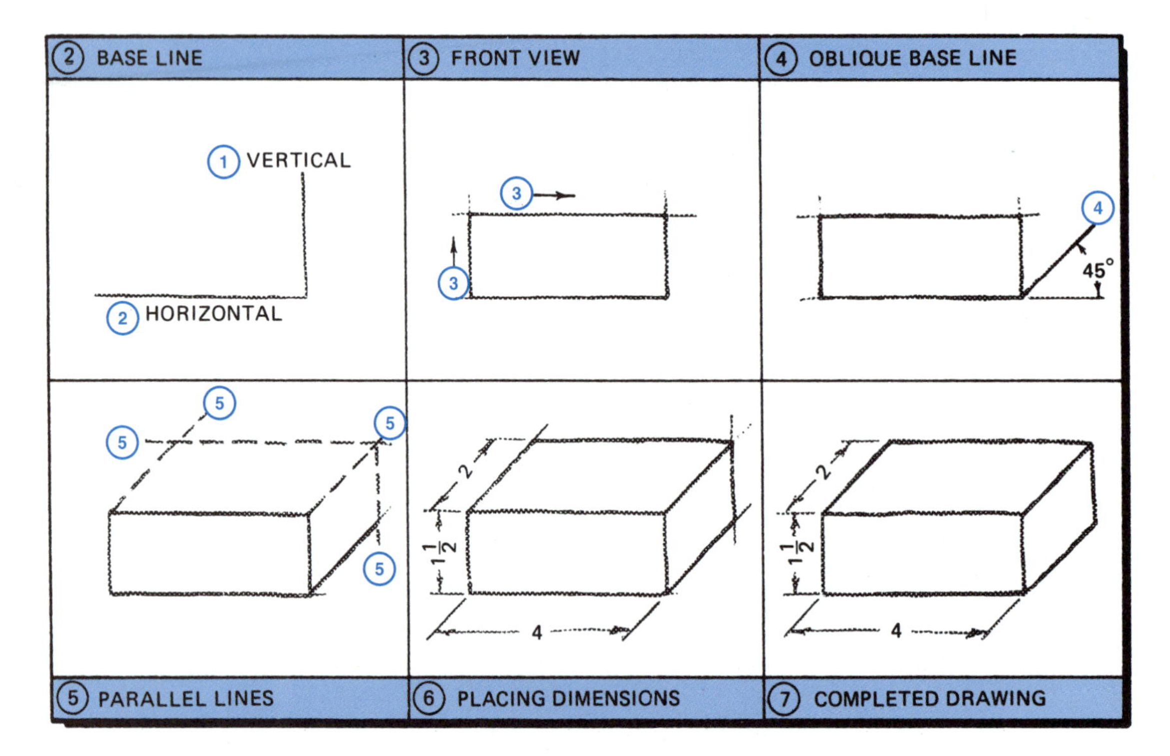

FIGURE 39-1 Steps in making an oblique sketch.

STEP 4 Start at the intersection of the vertical and horizontal base lines and draw a line at an angle of 45° to the base line. This line is called the *oblique base line.*

STEP 5 Draw the remaining lines for the right side and top view parallel to either the oblique base line, or to the horizontal or vertical base lines, as the case may be. Since these lines are not in their true size or shape, they should be drawn so they appear in proportion to the front view.

STEP 6 Place dimensions so they are parallel to the axis lines.

STEP 7 Erase unnecessary lines. Darken object lines to make the sketch clearer and easier to interpret.

SKETCHING CIRCLES IN OBLIQUE

A circle or arc located on the front face of an oblique sketch is drawn in its true size and shape. However, since the top and side views are distorted, a circle or arc will be elliptical in these views. Three circles drawn in oblique on the top, right side, and front of a cube are shown in figure 39-2.

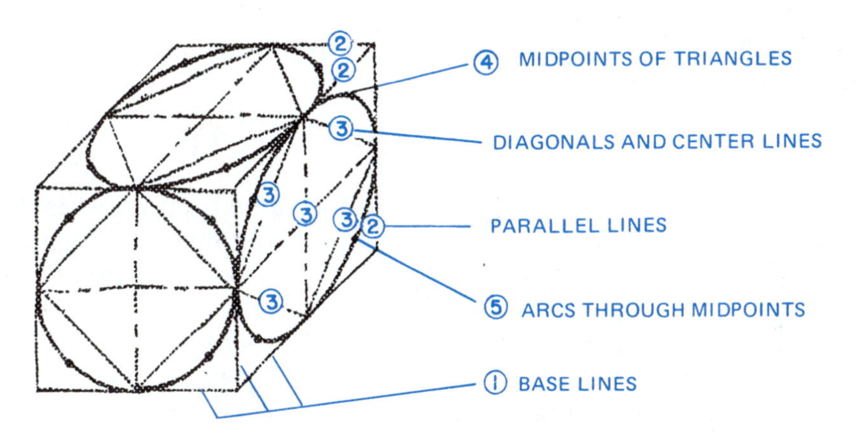

FIGURE 39-2 Sketching circles in oblique.

STEP 1 Draw horizontal, vertical, and oblique base lines.

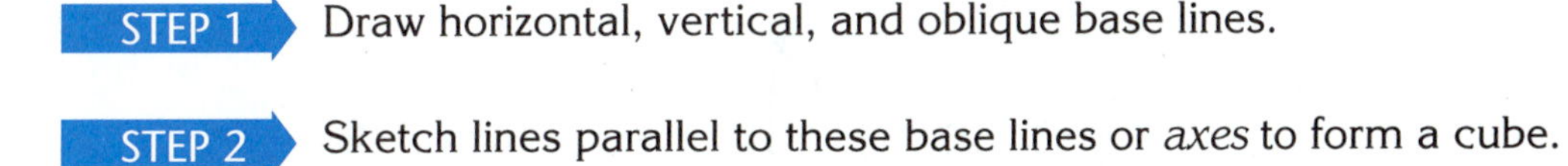

STEP 2 Sketch lines parallel to these base lines or *axes* to form a cube.

STEP 3 Draw center lines and diagonals in the front, right side, and top faces.

STEP 4 Locate the midpoints of triangles formed in the three faces.

STEP 5 Draw curved lines through these points. NOTE: The circle appears in its true size and shape in the front face, and as an ellipse in the right side and top faces.

STEP 6 Touch up and darken the curved lines. Erase guidelines where they are of no value in reading the sketch.

RIGHT AND LEFT OBLIQUE SKETCHES

Up to this point, the object has been viewed from the right side. In many cases, the left side of the object must be sketched because it contains better details. In these instances, the oblique base line or axis is 45° from the horizontal base line, starting from the left edge of the object. A rectangular die block sketched from the right side is shown in figure 39-3A. The same rectangular die block is shown in figure 39-3B as it would appear when sketched from the left side.

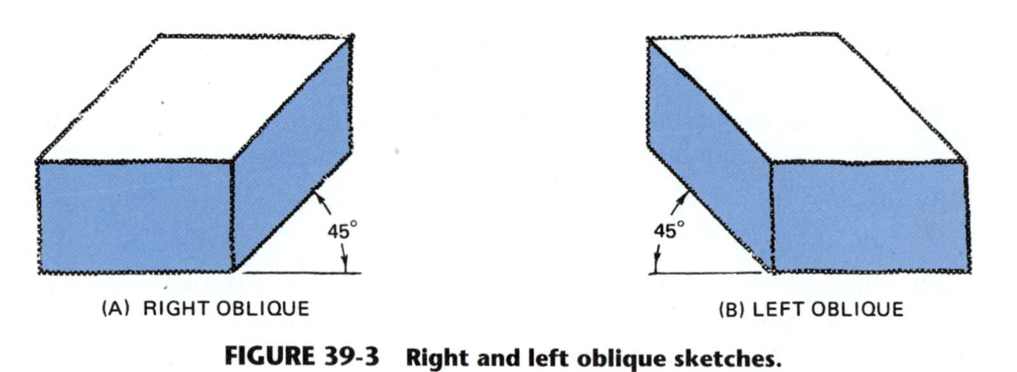

FIGURE 39-3 Right and left oblique sketches.

The same principles and techniques of making oblique sketches apply, regardless of whether the object is drawn in the right or left position.

FORESHORTENING

When a line in the side and top views is drawn in the same proportion as lines are in the front view, the object may appear to be distorted and longer than it actually is. To correct the distortion, the lines are drawn shorter than actual size so the sketch of the part looks balanced. This drafting technique is called *foreshortening.*

In figure 39-4, the oblique sketch shows a part as it would look before and after foreshortening. The foreshortened version is preferred. The amount the sketch is foreshortened depends on the ability of the individual to make the sketch resemble the part as closely as possible.

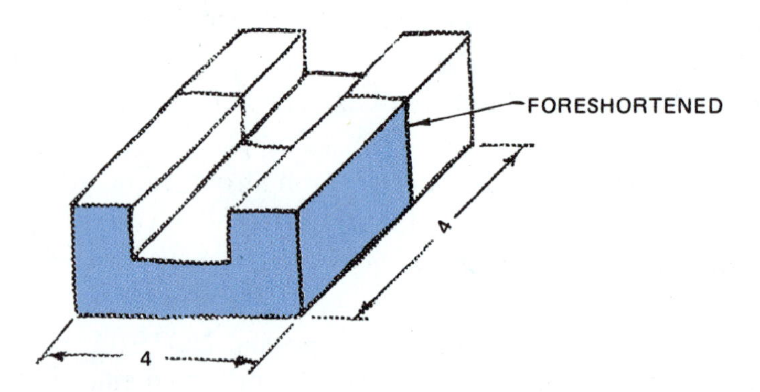

FIGURE 39-4 Foreshortening corrects distortion.

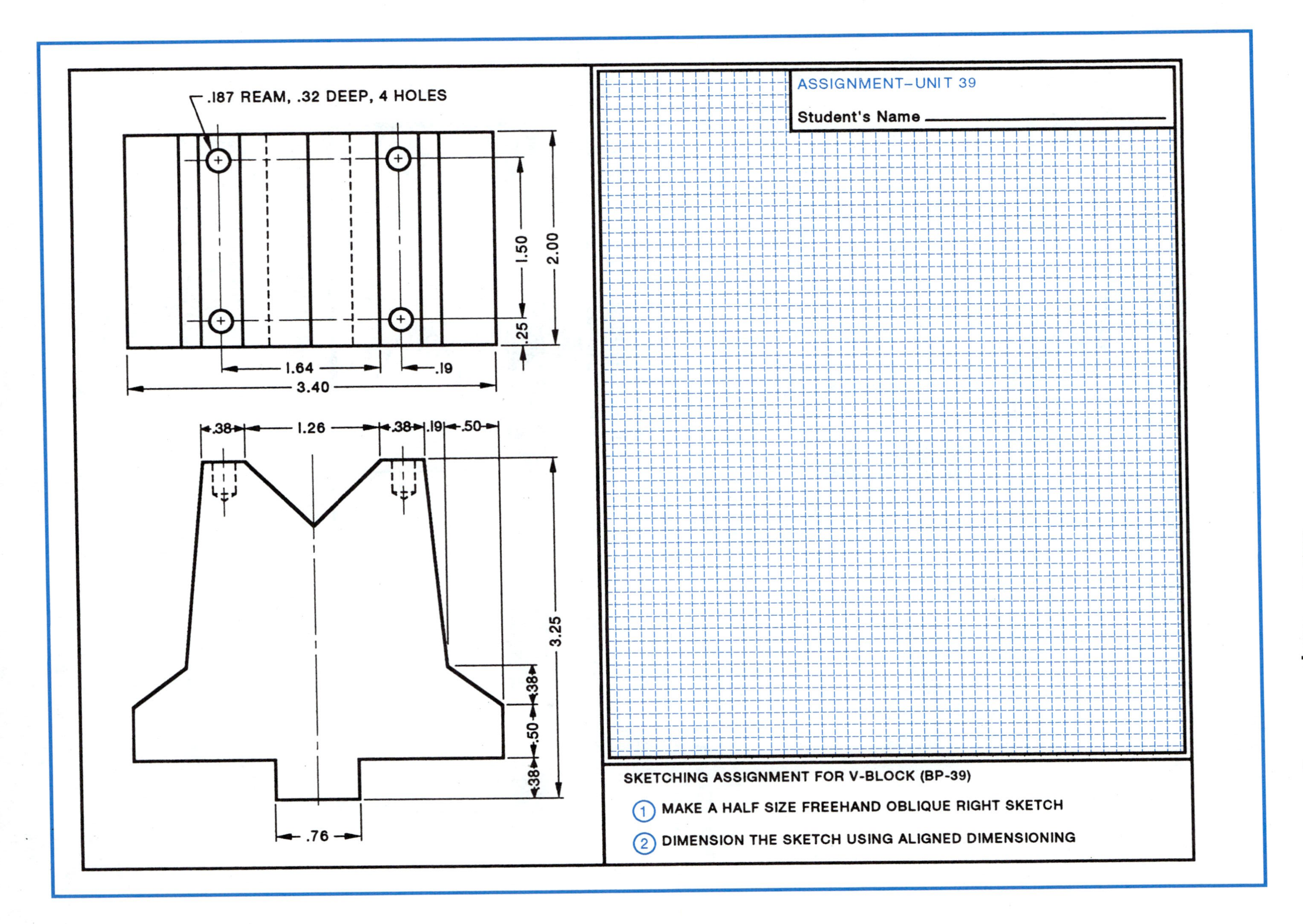
.187 REAM, .32 DEEP, 4 HOLES
1.50
2.00
.25
1.64
.19
3.40
.38
1.26
.38
.19
.50
3.25
.38
.50
.38
.76
ASSIGNMENT–UNIT 39
Student's Name
SKETCHING ASSIGNMENT FOR V-BLOCK (BP-39)
1 MAKE A HALF SIZE FREEHAND OBLIQUE RIGHT SKETCH
2 DIMENSION THE SKETCH USING ALIGNED DIMENSIONING

Unit 40

ISOMETRIC SKETCHING

Isometric and oblique sketches are similar in that they are another form of pictorial drawing in which two or more surfaces may be illustrated in one view. The isometric sketch is built around three major lines called isometric base lines or axes, figure 40-1. The right side and left side isometric base lines each form an angle of 30° with the horizontal, and the vertical axis line forms an angle of 90° with the horizontal base line.

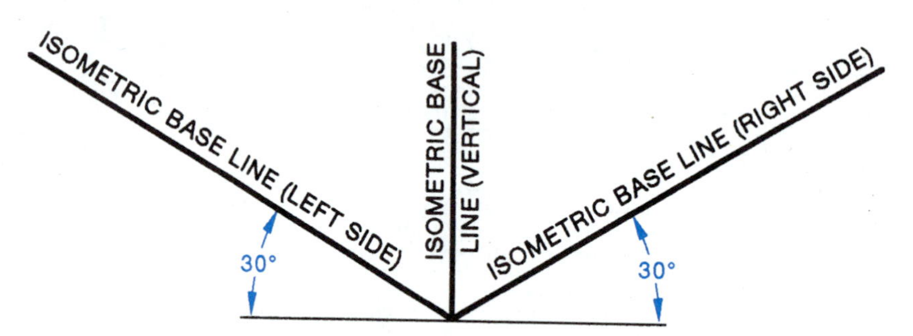

FIGURE 40-1 Axes for isometric drawings.

MAKING A SIMPLE ISOMETRIC SKETCH

An object is sketched in isometric by positioning it so that the part seems to rest on one corner. When making an isometric sketch of a part where all the surfaces or corners are parallel or at right angles to each other, the basic steps used are shown in figure 40-2. A rectangular steel block 1" thick, 2" wide, and 3" long is used as an example.

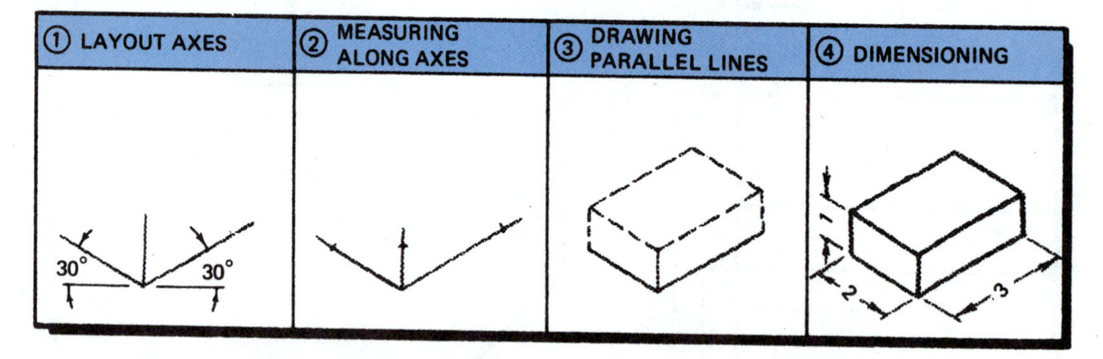

FIGURE 40-2 Steps in making an isometric sketch.

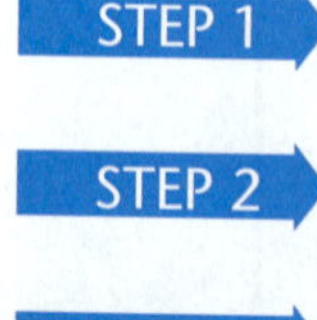

STEP 1 Sketch the three isometric axes. If a ruled isometric sheet is available, select three lines for the major axes.

STEP 2 Lay off the 3" length along the right axis, the 2" width on the left axis line, and the 1" height on the vertical axis line.

STEP 3 Draw lines from these layout points parallel to the three axes. Note that all parallel lines on the object are parallel on the sketch.

STEP 4 Dimension the sketch. On isometric sketches, the dimensions are placed parallel to the edges.

SKETCHING SLANT LINES IN ISOMETRIC

Only those lines that are parallel to the axes may be measured in their true lengths. Slant lines representing inclined surfaces are not shown in their true lengths in isometric sketches. In most cases, the slant lines for an object (such as the casting shown in figure 40-3) may be drawn by following the steps shown in figure 40-4.

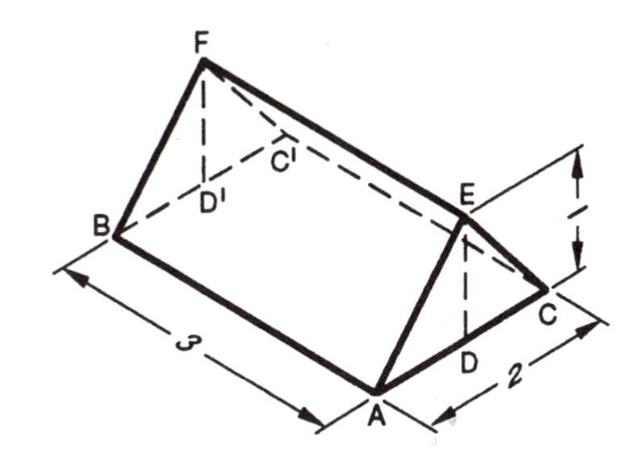

FIGURE 40-3 Part to be sketched in isometric.

STEP 1 Draw the three major isometric axes.

STEP 2 Measure distance A-B on the left axis and A-C on the right axis.

STEP 3 Measure distance A-D on the right axis and draw a vertical line from this point. Lay out distance D-E on this line.

STEP 4 Draw parallel lines B-C', D-D', E-F, and D'-F.

STEP 5 Connect points A-E, B-F, and C-E.

STEP 6 Darken lines and dimension the isometric sketch.

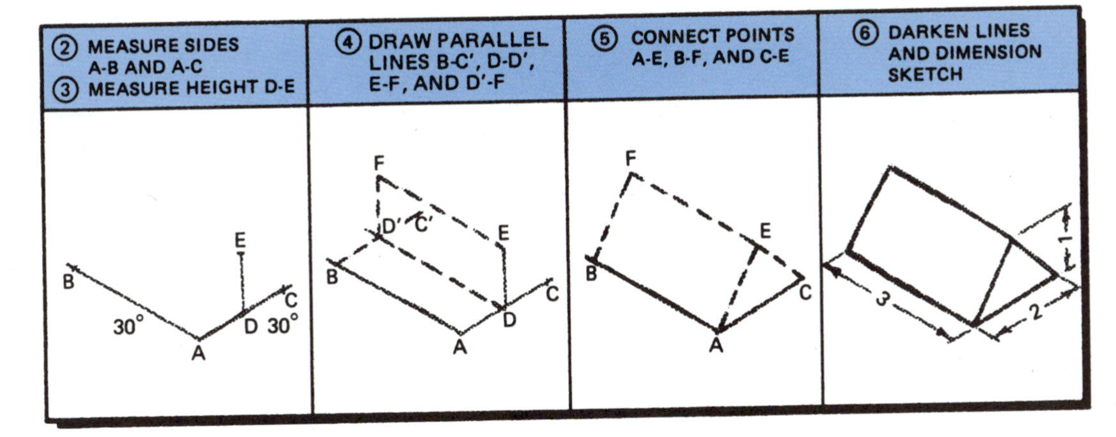

FIGURE 40-4 Isometric sketch requiring use of slant lines.

SKETCHING CIRCLES AND ARCS IN ISOMETRIC

The techniques used to sketch arcs and circles in oblique also may be used for sketching arcs and circles in isometric. Each step is illustrated in figure 40-5. Note that in each face the circle appears as an ellipse. This is also shown in figure 40-6.

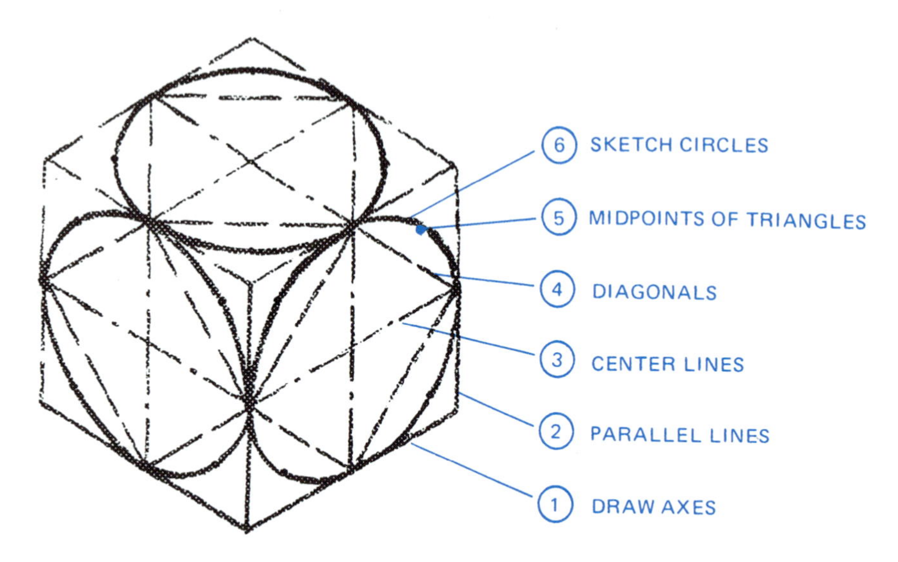

FIGURE 40-5 Sketching circles in isometric.

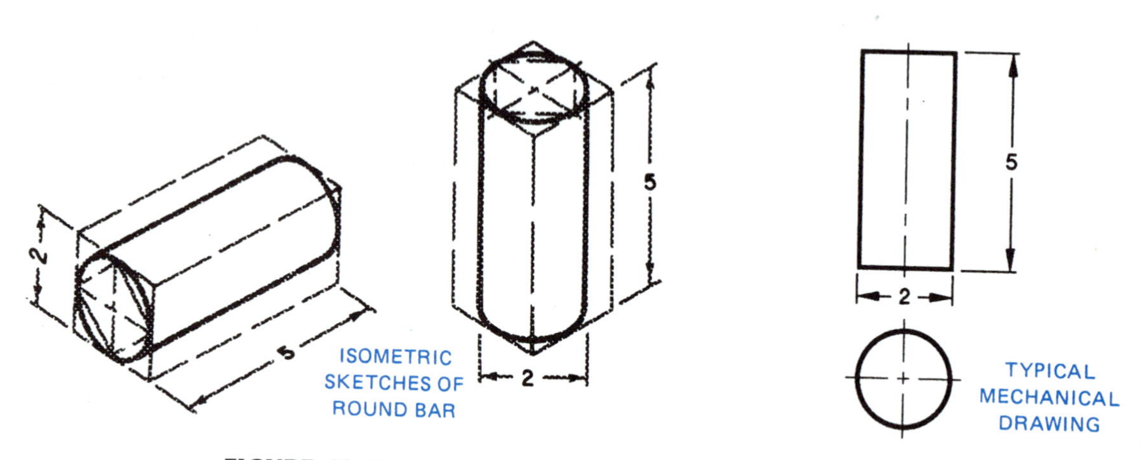

FIGURE 40-6 Application of circles and arcs in isometric.

LAYOUT SHEETS FOR ISOMETRIC SKETCHES

The making of isometric drawings can be simplified if specially ruled isometric layout sheets are used. Considerable time may be saved as these graph sheets provide guide lines which run in three directions. The guide lines are parallel to the three isometric axes and to each other. The diagonal lines of the graph paper usually are at a given distance apart to simplify the making of the sketch and to ensure that all lines are in proportion to the actual size. The ruled lines are printed on a heavy paper which can be used over and over again as an underlay sheet for tracing paper.

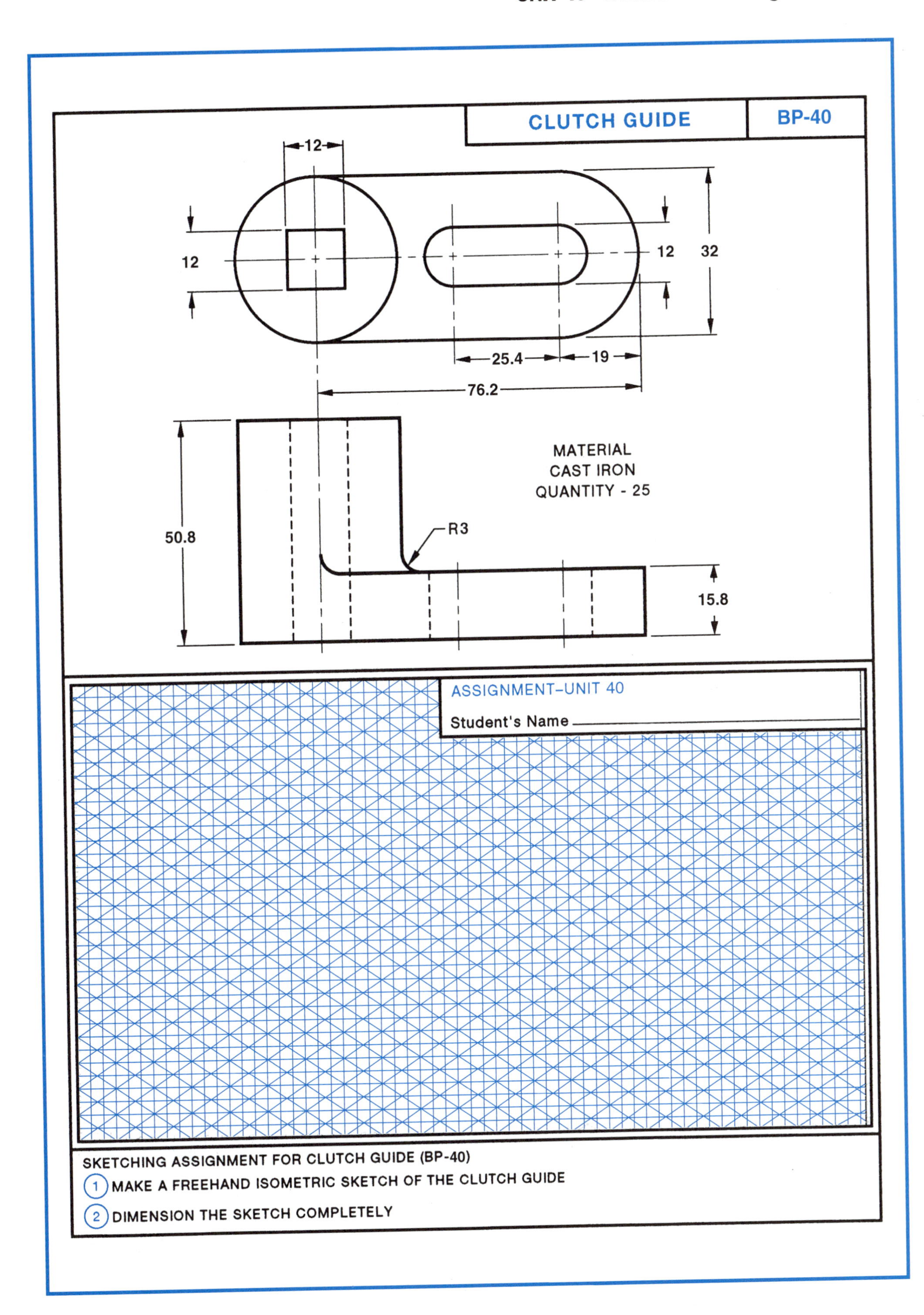
CLUTCH GUIDE
BP-40
12
12
12
32
25.4
19
76.2
MATERIAL
CAST IRON
QUANTITY - 25
R3
50.8
15.8
ASSIGNMENT–UNIT 40
Student's Name
SKETCHING ASSIGNMENT FOR CLUTCH GUIDE (BP-40)
1 MAKE A FREEHAND ISOMETRIC SKETCH OF THE CLUTCH GUIDE
2 DIMENSION THE SKETCH COMPLETELY

Unit 41

PERSPECTIVE SKETCHING

Perspective sketches show two or three sides of an object in one view. As a result, they resemble a photographic picture. In both the perspective sketch and the photograph, the portion of the object which is closest to the observer is the largest. The parts that are farthest away are smaller. Lines and surfaces on perspective sketches become smaller and come closer together as the distance from the eye increases. Eventually they seem to disappear at an imaginary horizon.

SINGLE POINT PERSPECTIVE

There are two types of perspective sketches commonly used. The first type is known as *single point* or *parallel perspective*. In this case, one face of the object in parallel perspective is sketched in its true size and shape, the same as in an orthographic sketch. For example, the edges of the front face of a cube as shown in parallel perspective in figure 41-1 at Ⓐ, Ⓑ, Ⓒ, and Ⓓ, are parallel and square.

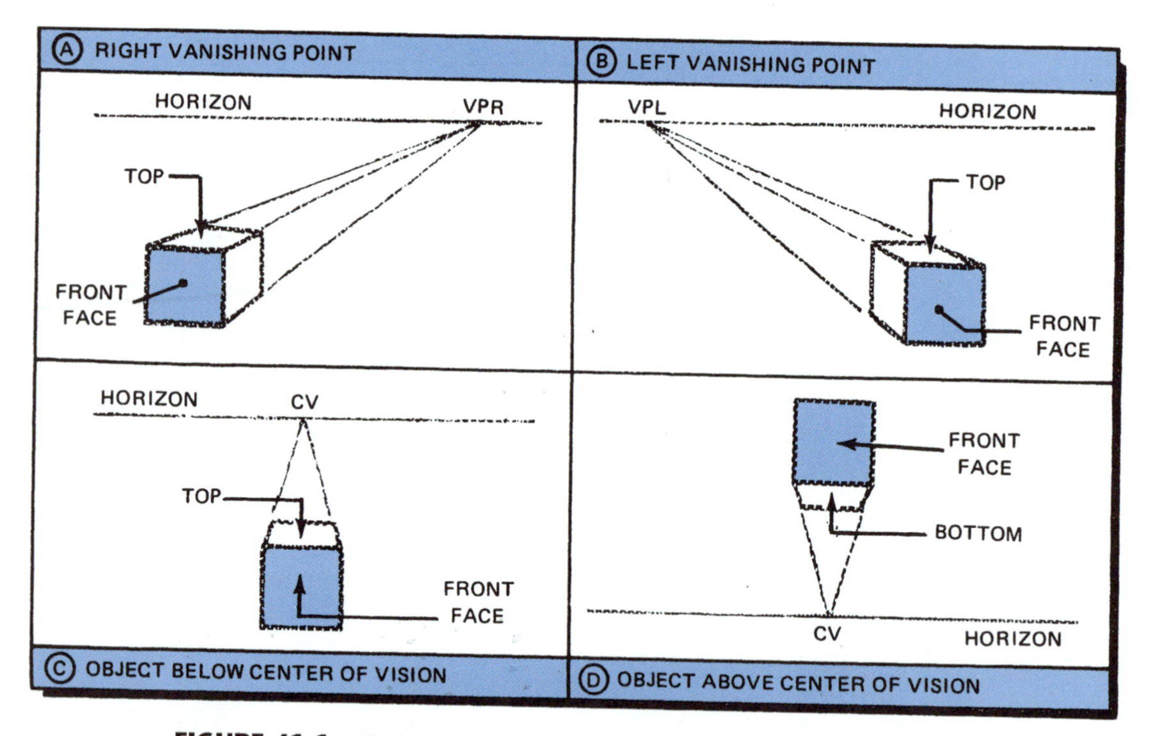

FIGURE 41-1 Single point, parallel perspective sketches of a cube.

To draw the remainder of the cube, the two sides decrease in size as they approach the horizon. On this imaginary horizon, which is supposed to be at eye level, the point where the lines come together is called the *vanishing point.* This vanishing point may be above or below the object, or to the right or left of it, figure 41-1. Vertical lines on the object are vertical on parallel perspective sketches and do not converge. Single point parallel perspective is the simplest type of perspective to understand and the easiest to sketch.

Parallel prospective CAD drawings are used to show3-D objects in a dimensionally correct manner. Relative and scaled dimensions may be taken from the drawings. The down side is that parallel perspective drawings do not appear as realistic as isometric or angular perspective drawings.

ANGULAR OR TWO-POINT PERSPECTIVE

The second type of perspective sketch is the two-point perspective, also known as *angular perspective.* As the name implies, two vanishing points on the horizon are used and all lines converge toward these points. Usually, in a freehand perspective sketch, the horizon is in a horizontal position. Seven basic steps are required to make a two-point perspective of a cube, using right and left vanishing points. The application of each of these steps is shown in figure 41-2.

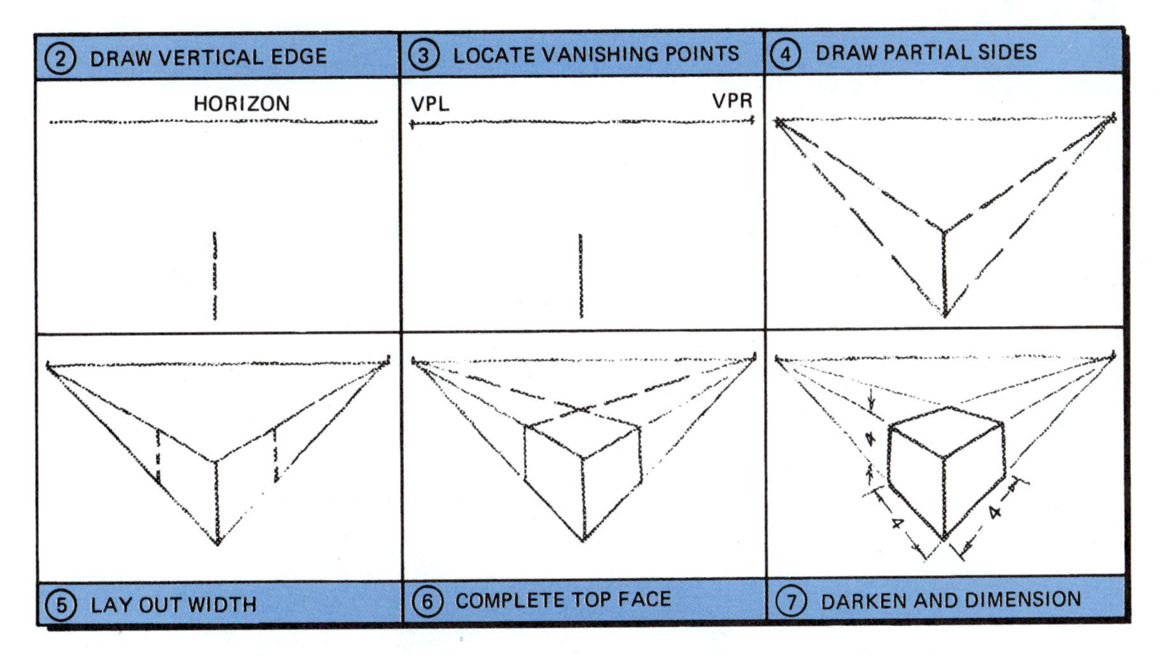

FIGURE 41-2 Making an angular perspective sketch of a cube.

STEP 1 Sketch a light horizontal line for the horizon. Then position the object above or below this line so the right vertical edge of the cube becomes the center of the sketch.

STEP 2 Draw a vertical line for this edge of the cube.

STEP 3 Place two vanishing points on the horizon: one to the right and the other to the left of the object.

STEP 4 Draw light lines from the corners of the vertical line to the vanishing points.

STEP 5 Lay out the width of the cube and draw parallel vertical lines.

STEP 6 Sketch the two remaining lines for the top, starting at the points where the two vertical lines intersect the top edge of the cube.

STEP 7 Darken all object lines and dimension. Once again, each dimension is placed parallel to the edge which it measures.

SKETCHING CIRCLES AND ARCS IN PERSPECTIVE

Circles and arcs are distorted in all views of perspective drawings except in one face of a parallel perspective drawing. Circles are drawn in perspective, using the same techniques of blocking-in which apply to orthographic, oblique, and isometric sketches. The steps in drawing circles in perspective are summarized in figure 41-3. The same practices may be applied to sketching arcs.

STEP 1 Lay out the four sides of the square which correspond to the diameter of the required circle. Note that two of the sides converge toward the vanishing point.

STEP 2 Draw the center lines.

STEP 3 Draw the diagonals.

STEP 4 Locate the midpoints of the triangles which are formed.

STEP 5 Sketch the circle through the points where the center lines touch the sides of the square and the midpoints.

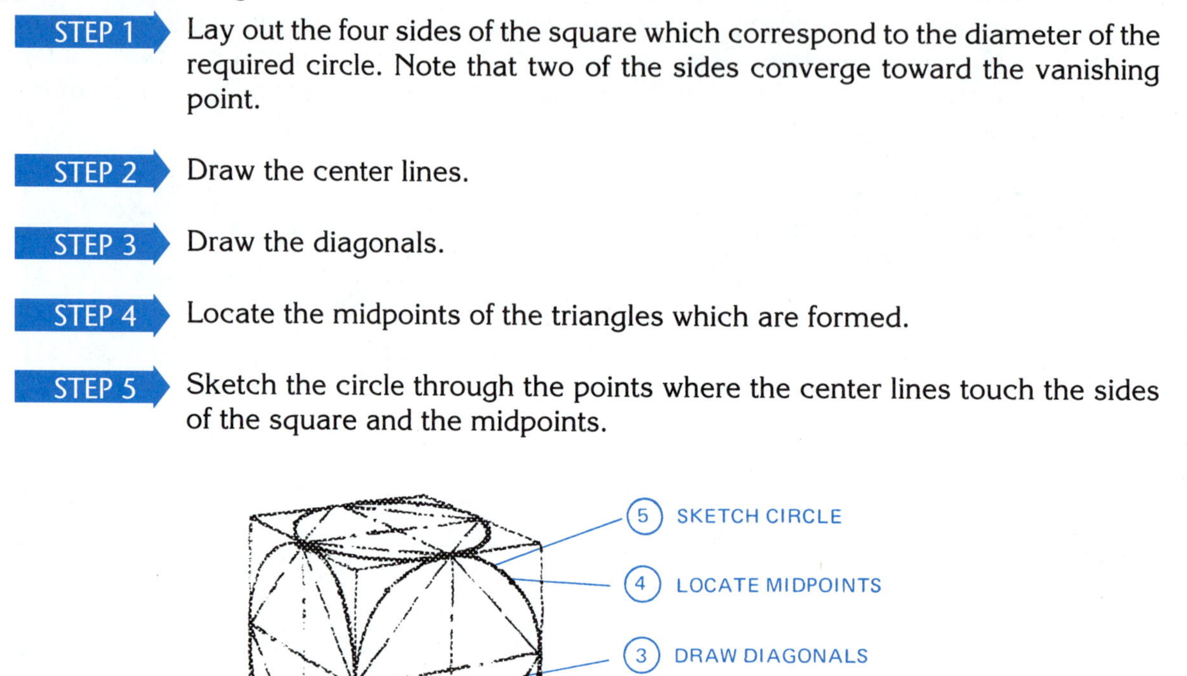

FIGURE 41-3 Sketching circles in perspective.

SHADING PERSPECTIVE SKETCHES

Many perspective sketches are made of parts which include fillets, rounds, chamfers, and similar construction. By bringing out some of these details, the drawing becomes easier to read and is more attractive. The same techniques that are used for shading regular mechanical drawings or other pictorial sketches may be applied to perspective drawings. This shading brings out some of the construction details which otherwise might not be included, figure 41-4.

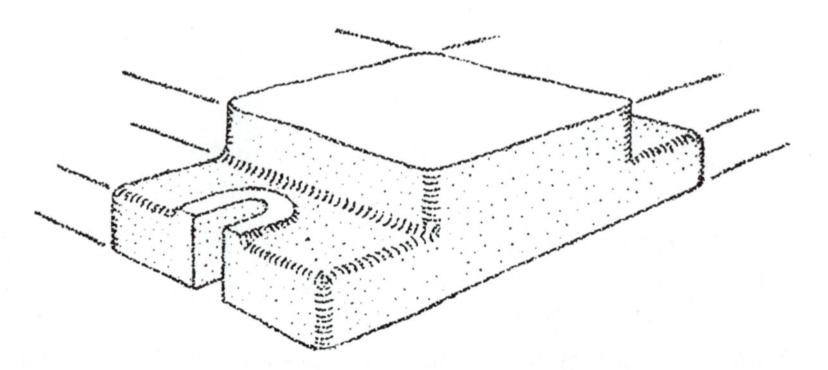

FIGURE 41-4 Shading applied to perspective sketch of die block.

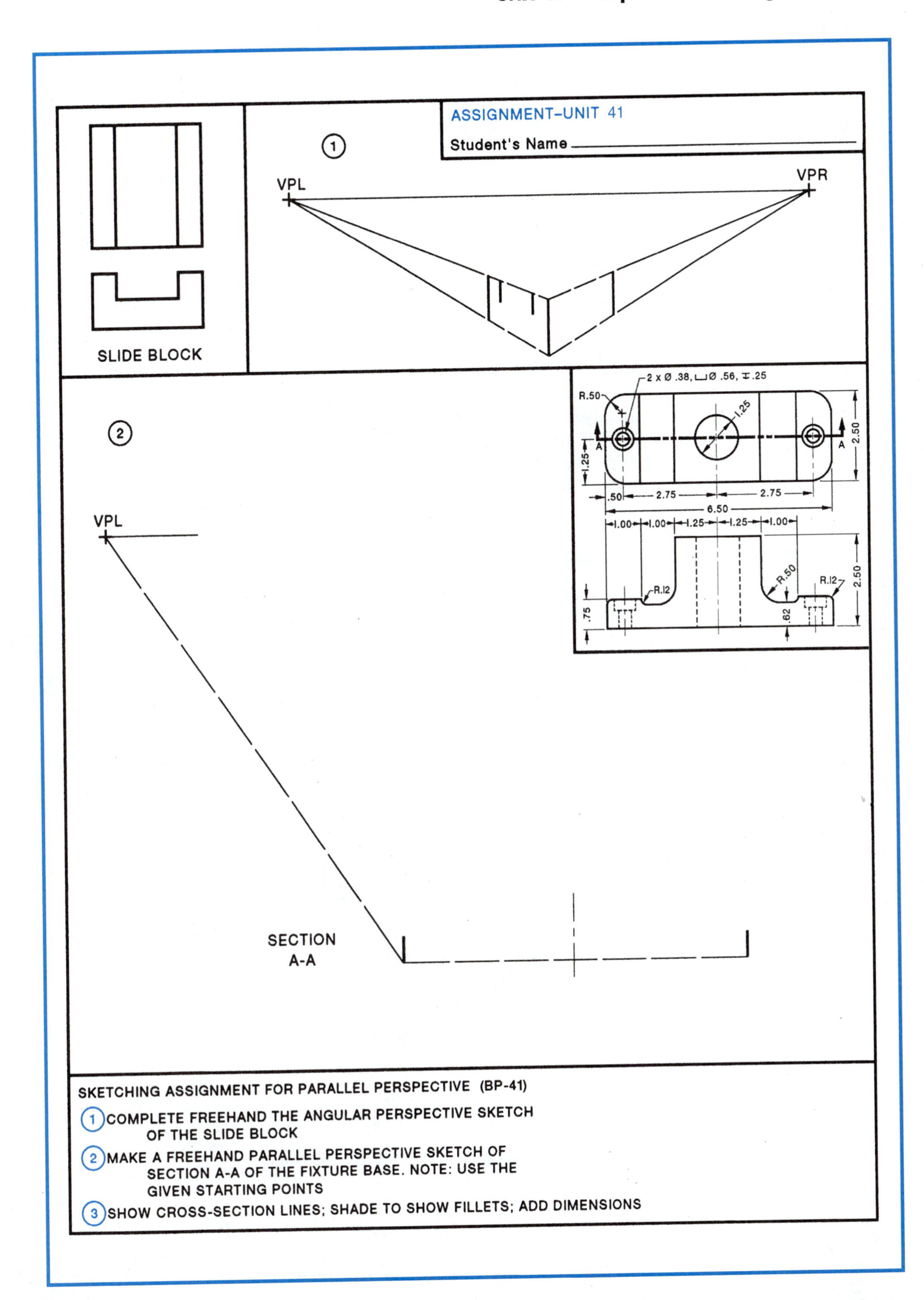
ASSIGNMENT–UNIT 41
Student's Name
SLIDE BLOCK
1
VPL
VPR
2
VPL
SECTION
A-A
2 X Ø .38, ⌴Ø .56, ↧.25
R.50
1.25
A
A
2.50
1.25
.50
2.75
2.75
6.50
1.00
1.00
1.25
1.25
1.00
R.12
R.50
R.12
.75
.62
2.50
SKETCHING ASSIGNMENT FOR PARALLEL PERSPECTIVE (BP-41)
1 COMPLETE FREEHAND THE ANGULAR PERSPECTIVE SKETCH OF THE SLIDE BLOCK
2 MAKE A FREEHAND PARALLEL PERSPECTIVE SKETCH OF SECTION A-A OF THE FIXTURE BASE. NOTE: USE THE GIVEN STARTING POINTS
3 SHOW CROSS-SECTION LINES; SHADE TO SHOW FILLETS; ADD DIMENSIONS

Unit 42

PICTORIAL DRAWINGS AND DIMENSIONS

PICTORIAL WORKING DRAWINGS

Many industries, engineering, design, sales, and other organizations use pictorial working drawings that are made freehand. In a great many cases, this method is preferred to instrument drawings. Technicians are engineers, and tradespersons are making sketches on the job. These sketches are used from exploring project ideas to the completion and revision stages of product development. The fact that no drawing instruments are required and sketches may be made quickly and on-the-spot makes freehand drawings practical.

A pictorial drawing is considered a working drawing when dimensions and other specifications that are needed to produce the part or assemble a mechanism are placed on the sketch. There are two general systems of dimensioning pictorial sketches: (1) *pictorial plane* (aligned) and (2) *unidirectional.*

PICTORIAL PLANE (ALIGNED) AND UNIDIRECTIONAL DIMENSIONS

In the aligned or pictorial plane dimensioning system, the dimension lines, extension lines, and certain arrowheads are positioned parallel to the pictorial planes. These features, as they are positioned on a sketch, are illustrated in figure 42-1.

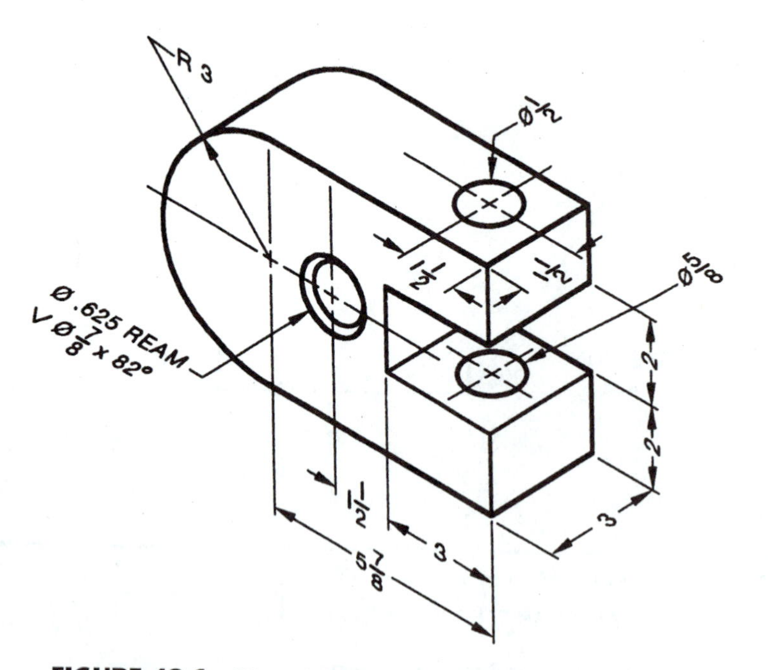

FIGURE 42-1 Pictorial plane (aligned) dimensioning.

By contrast, dimensions, notes, and technical details are lettered vertically in unidirectional dimensioning, figure 42-2. In this vertical position the dimensions are easier to read.

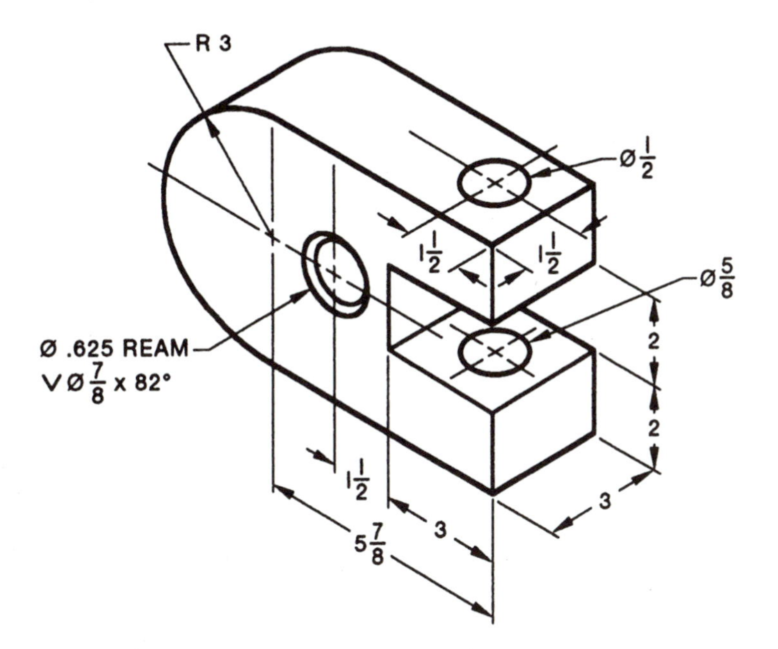

FIGURE 42-2 Unidirectional dimensioning.

Pictorial drawings are prepared from both first and third projection systems (see Unit 23) of representation. Thus, the dimensions, notes, and important data as included on the sketch, conform to the standards of the system being used.

PICTORIAL DIMENSIONING RULES

Essentially, the same basic rules for dimensioning a pictorial sketch are followed as for a multiview drawing. The basic pictorial dimensioning rules apply to both the aligned and unidirectional dimensioning systems.

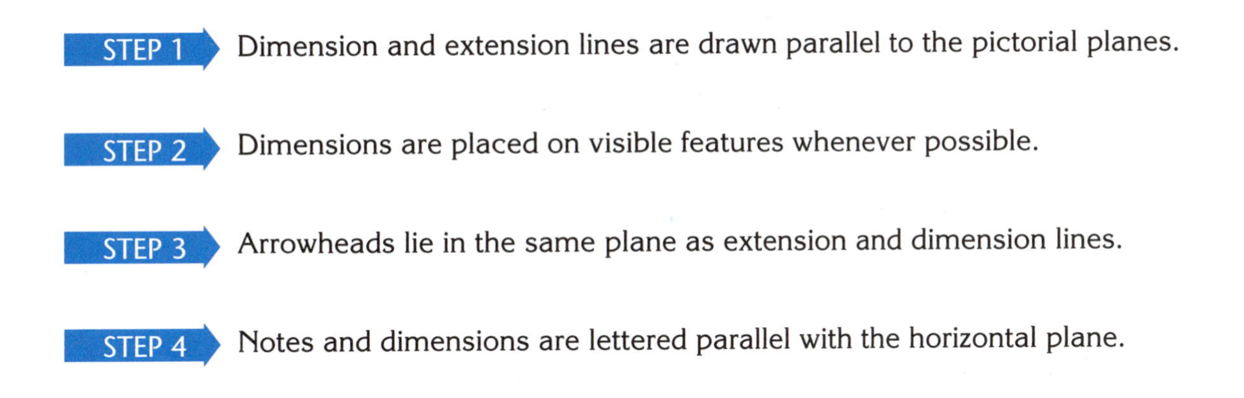

STEP 1 Dimension and extension lines are drawn parallel to the pictorial planes.

STEP 2 Dimensions are placed on visible features whenever possible.

STEP 3 Arrowheads lie in the same plane as extension and dimension lines.

STEP 4 Notes and dimensions are lettered parallel with the horizontal plane.

Since a pictorial drawing is a one-view drawing, it is not always possible to avoid dimensioning on the object, across dimension lines, or on hidden surfaces. These practices should be avoided in order to prevent errors in reading dimensions or interpreting particular features of the part.

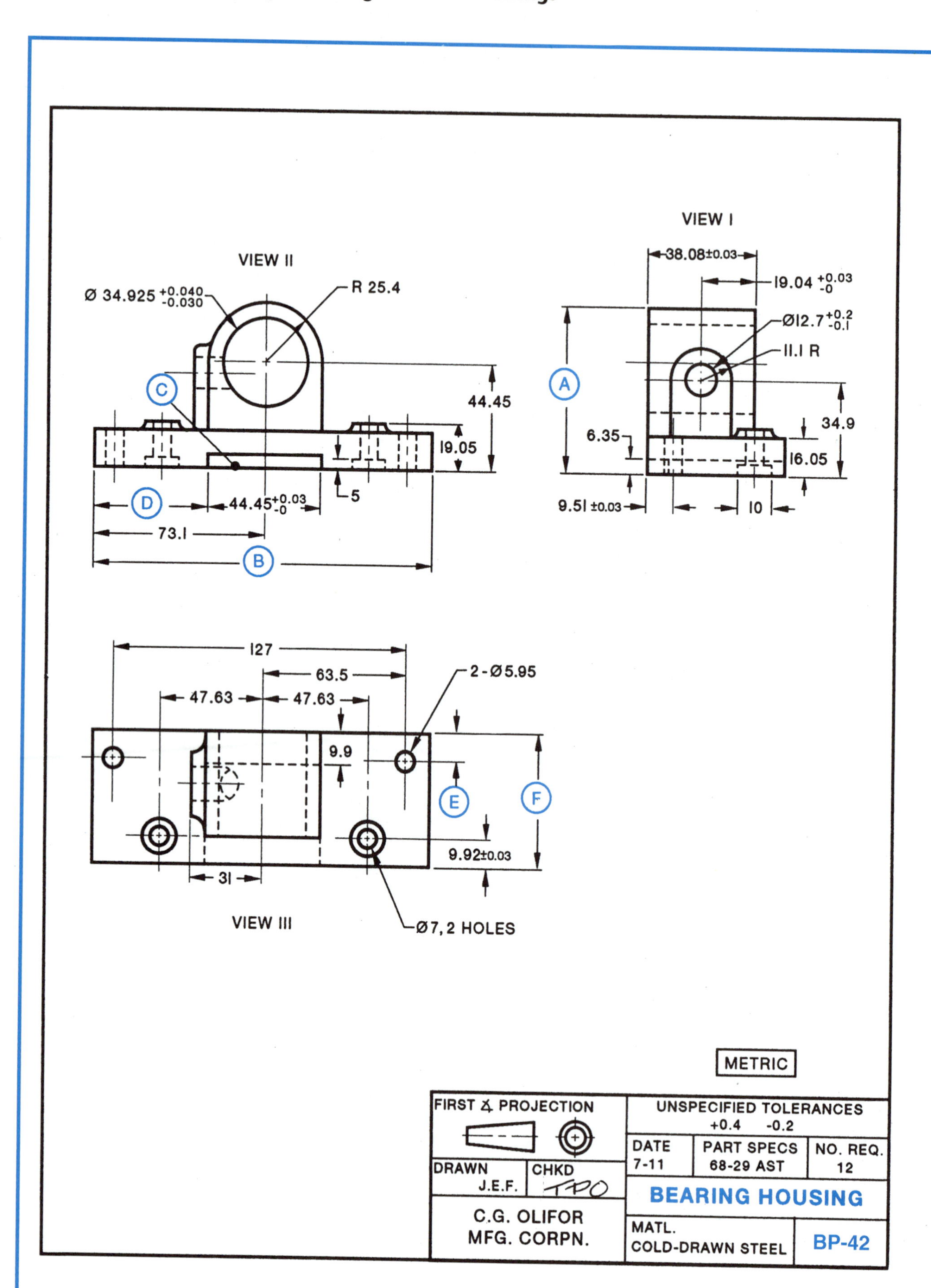

VIEW II
Ø 34.925 +0.040 -0.030
R 25.4
44.45
19.05
5
44.45 +0.03 -0
73.1
VIEW I
38.08±0.03
19.04 +0.03 -0
Ø12.7 +0.2 -0.1
11.1 R
6.35
34.9
16.05
9.51 ±0.03
10
127
63.5
47.63
47.63
2-Ø5.95
9.9
9.92±0.03
31
VIEW III
Ø7, 2 HOLES
A
B
C
D
E
F
METRIC
FIRST ∡ PROJECTION
UNSPECIFIED TOLERANCES
+0.4 -0.2
DRAWN
J.E.F.
CHKD
DATE
7-11
PART SPECS
68-29 AST
NO. REQ.
12
BEARING HOUSING
C.G. OLIFOR
MFG. CORPN.
MATL.
COLD-DRAWN STEEL
BP-42

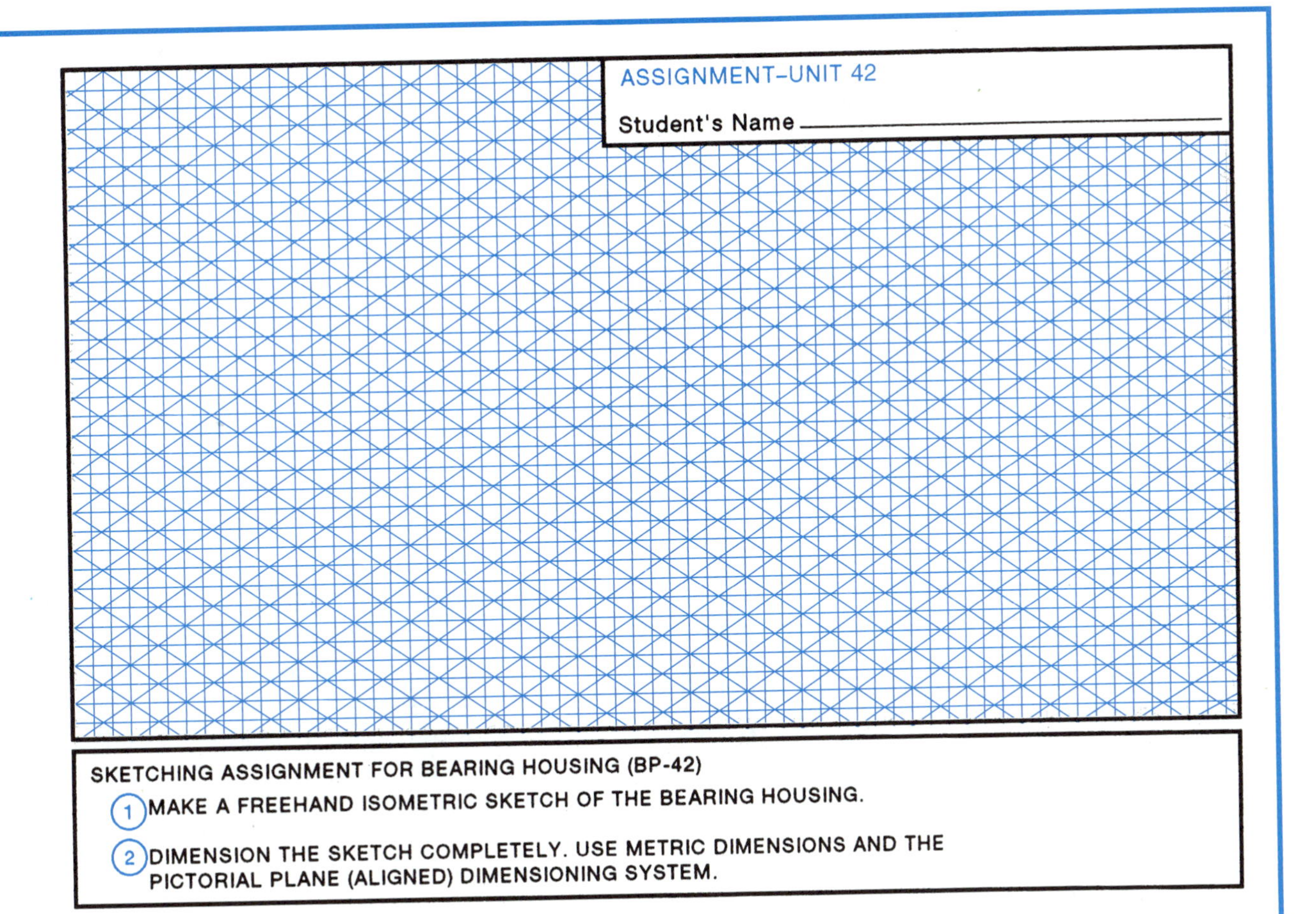

BEARING HOUSING (BP-42)

1. State what angle of projection is used.
2. Name Views I, II, and III.
3. Compute maximum overall height Ⓐ.
4. Determine the lower limit of Ⓑ.
5. Determine the maximum height and depth of slot Ⓒ.
6. Determine the maximum distance Ⓓ.
7. Compute the minimum and maximum center line distance Ⓔ.
8. Compute the maximum overall width Ⓕ.
9. What does the oval hole in View III represent?
10. Give the specifications for the counterbored holes.

ASSIGNMENT—UNIT 42

Student's Name ____________

1. ____________
2. I = ________ II = ________

   III = ____________
3. Ⓐ = ________ 4. Ⓑ = ________
5. Ⓒ = ____________

   ____
6. Ⓓ = ____________
7. Ⓔ = ____________
8. Ⓕ = ____________
9. ____________
10. ____________

SECTION 14 Sketching for CAD/CNC

Unit 43

TWO-DIMENSIONAL AND THREE-DIMENSIONAL CAD SKETCHING

Both CNC machines and CADD (Computer-Aided Drafting & Design) use two- and three-dimensional Cartesian coordinates in design and manufacturing of detailed and assembled structures.

In addition to understanding the Cartesian X, Y, & Z axes (Unit 28) and Datums (Unit 29), CAD/CADD/CNC drawing must take into account the zero origin point. This point is called the *machine zero*. From this point all machine movement starts to take place. CNC machines use a "G" code programming to define a tool path to draw or make a part.

Two methods of indicating movement to points or positions are used: absolute and incremental. Absolute coordinates use the same concept for defining 2D and 3D points.

The absolute method defines a point from machine zero. Example: Point A (figure 43-1) would be located at 2,2,3; point B would be located at 5,2,3. The incremental movement defines a point from the machine's current position. In this example, to move from point A to point B, the command would be 3,0,0.

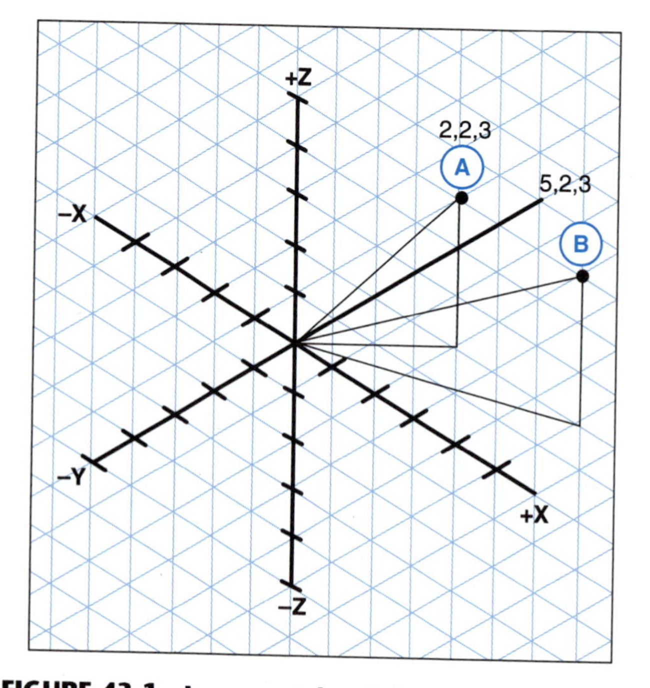

FIGURE 43-1 Incremental and absolute coordinates.

Both wire-frame and solid 3D objects can be sketched using these dimensioning concepts.

A wire-frame model appears to be transparent, using lines to represent all edges. Hidden edges are indicated with broken lines. (See figure 43-2 for a partial wire-frame model.)

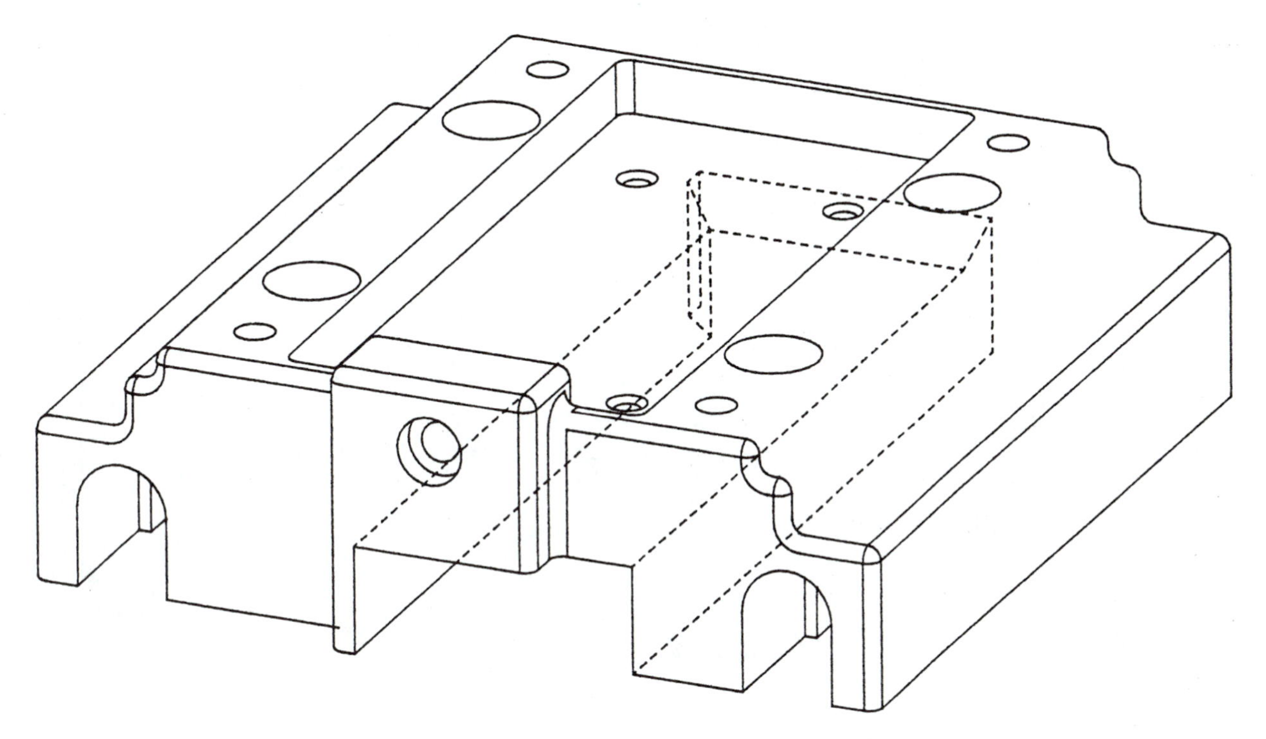

FIGURE 43-2 Partial wire-frame model.

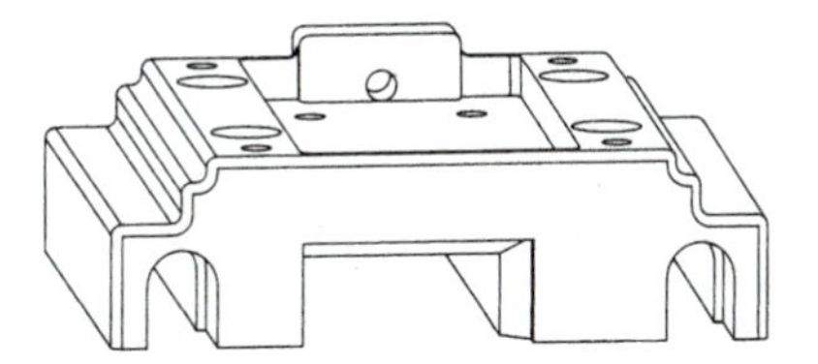

FIGURE 43-3 Surface model.

A surface model can be described as a wire-frame model with a sheet of material stretched over it (figure 43-3). In a solid model, no hidden surfaces or edges are shown.

Sketches of 2D and 3D models are frequently used in the "idea" stages of product development and revision. By sketching a mental idea on paper, it can be communicated to others and adjustments incorporated.

CAD SKETCHING

An option exists within CAD programs which allows the equivalent of freehand sketching by the movement of the mouse or puck. The created line length is specified by the user who can generate irregular shapes which are not easily created by more conventional methods of placing entities. This is useful to portray break lines in a broken-out sectional view or any other features which the draftsperson may wish to create so that they don't look as if they came from a plotter.

Two-dimensional (2-D) CAD drawings appear as a normal orthographic view. Every point is defined by a position along the *x* and *y* axes. Three-dimensional (3-D) CAD drawings can be drawn in either of two coordinate systems.

In the first system, View Coordinates, values are drawn so the *x* direction is the horizontal axis, *y* is drawn in the vertical axis direction, and *z* perpendicular to the *x* and *y* axis. No matter what view is used as the front view, the axes stay the same (figure 43-4).

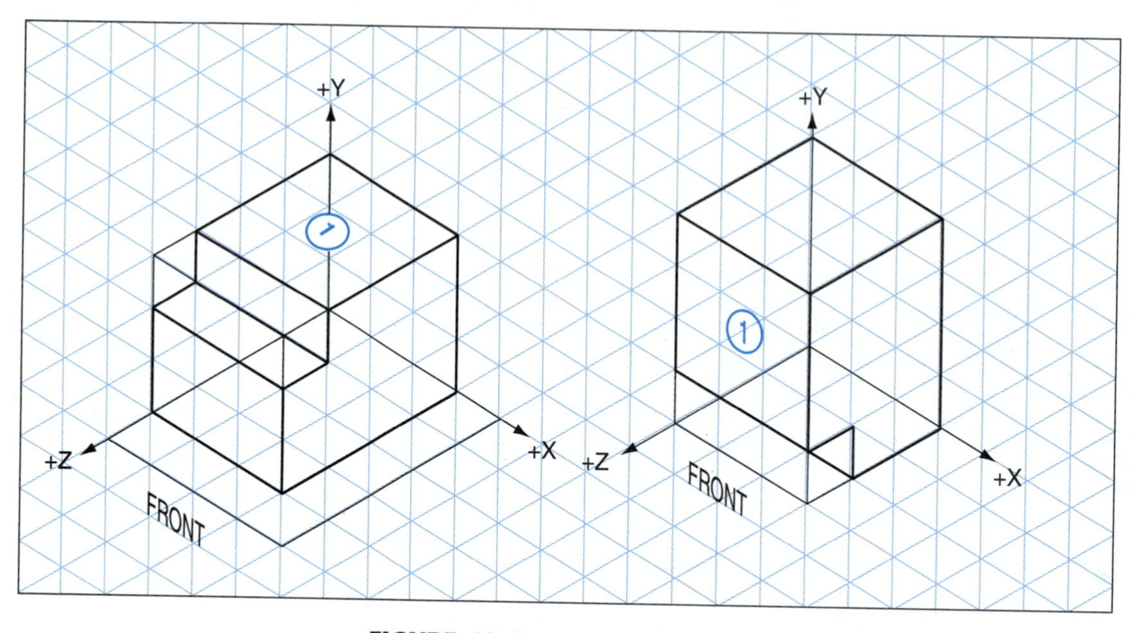

FIGURE 43-4 **View coordinates.**

With World Coordinates, the second system, the axes of the object remain attached to the same view even if the object is rotated (figure 43-5).

World Views are very useful when programming CNC machines. Axis designations are necessary for electronic systems to assign movement to machine elements. For example, the depth axis (*z*) will be horizontal on a profile milling machine while it will be vertical on a drilling machine.

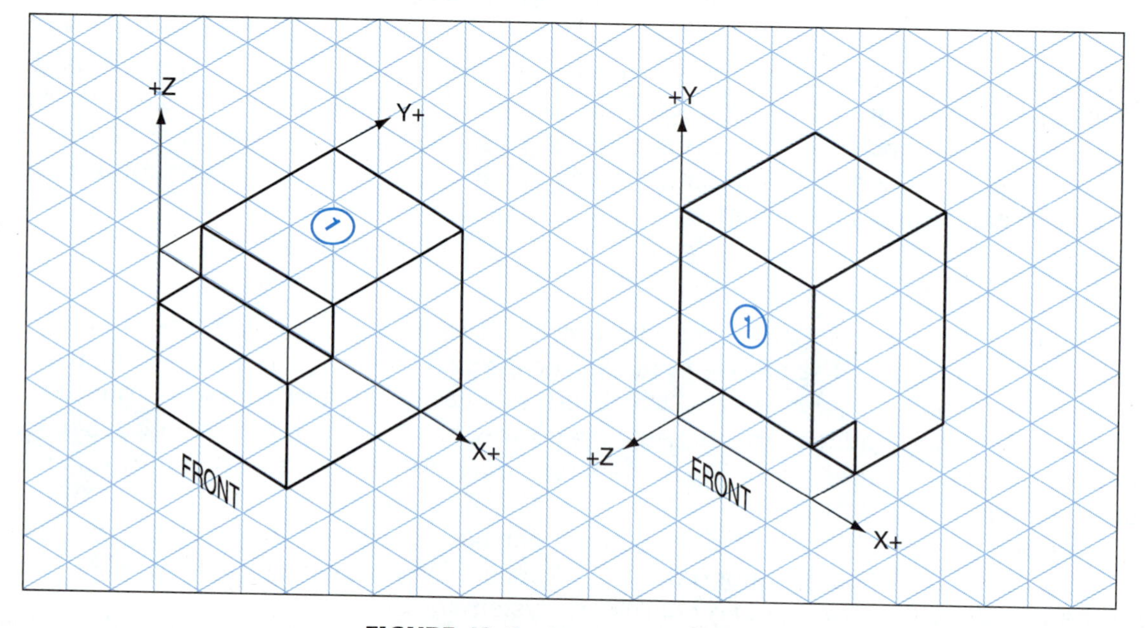

FIGURE 43-5 **World coordinates.**

ASSIGNMENT—UNIT 43A

Student's Name ______________________

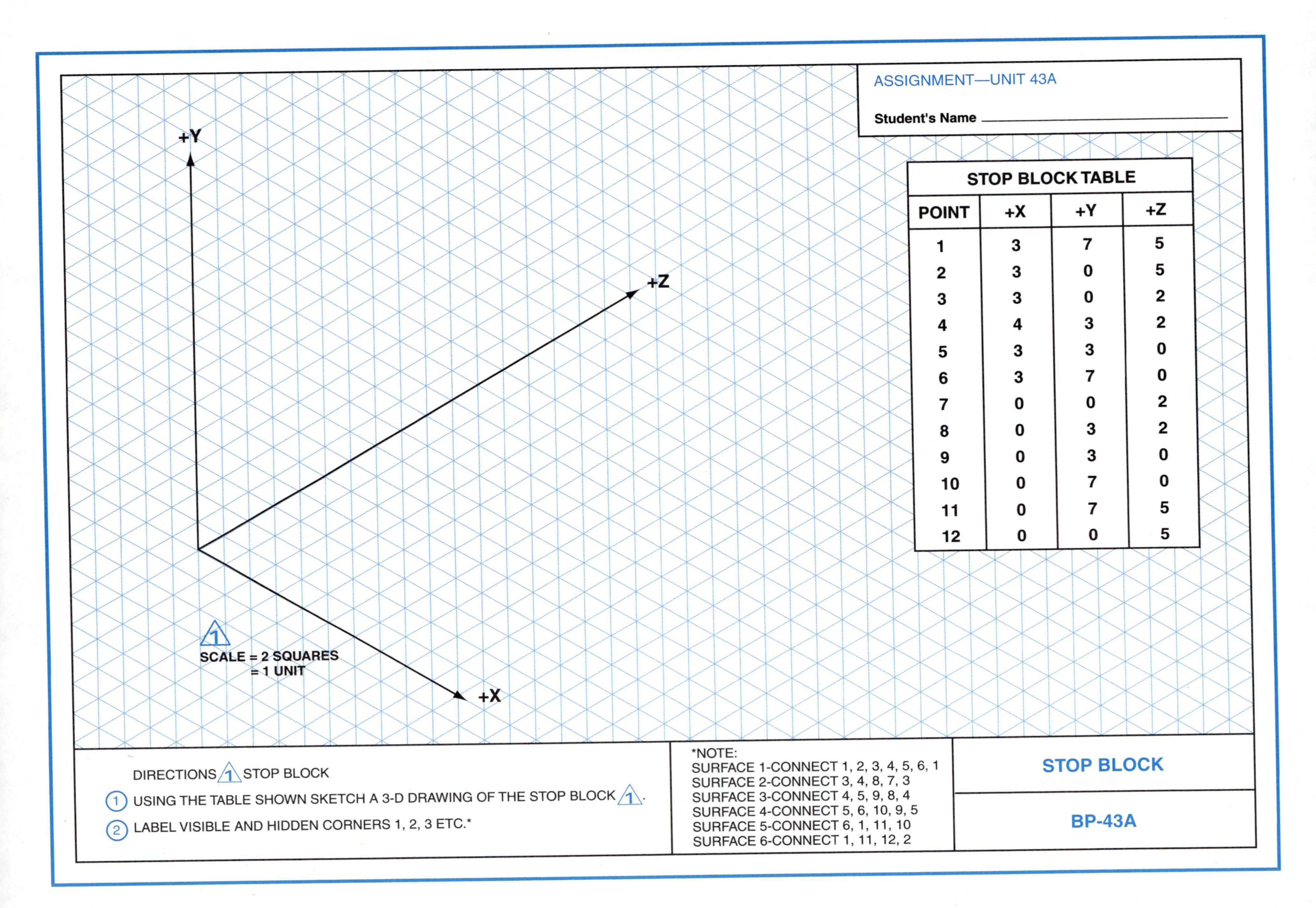

STOP BLOCK TABLE

POINT	+X	+Y	+Z
1	3	7	5
2	3	0	5
3	3	0	2
4	4	3	2
5	3	3	0
6	3	7	0
7	0	0	2
8	0	3	2
9	0	3	0
10	0	7	0
11	0	7	5
12	0	0	5

DIRECTIONS 1 STOP BLOCK

1. USING THE TABLE SHOWN SKETCH A 3-D DRAWING OF THE STOP BLOCK 1.
2. LABEL VISIBLE AND HIDDEN CORNERS 1, 2, 3 ETC.*

*NOTE:
SURFACE 1-CONNECT 1, 2, 3, 4, 5, 6, 1
SURFACE 2-CONNECT 3, 4, 8, 7, 3
SURFACE 3-CONNECT 4, 5, 9, 8, 4
SURFACE 4-CONNECT 5, 6, 10, 9, 5
SURFACE 5-CONNECT 6, 1, 11, 10
SURFACE 6-CONNECT 1, 11, 12, 2

STOP BLOCK

BP-43A

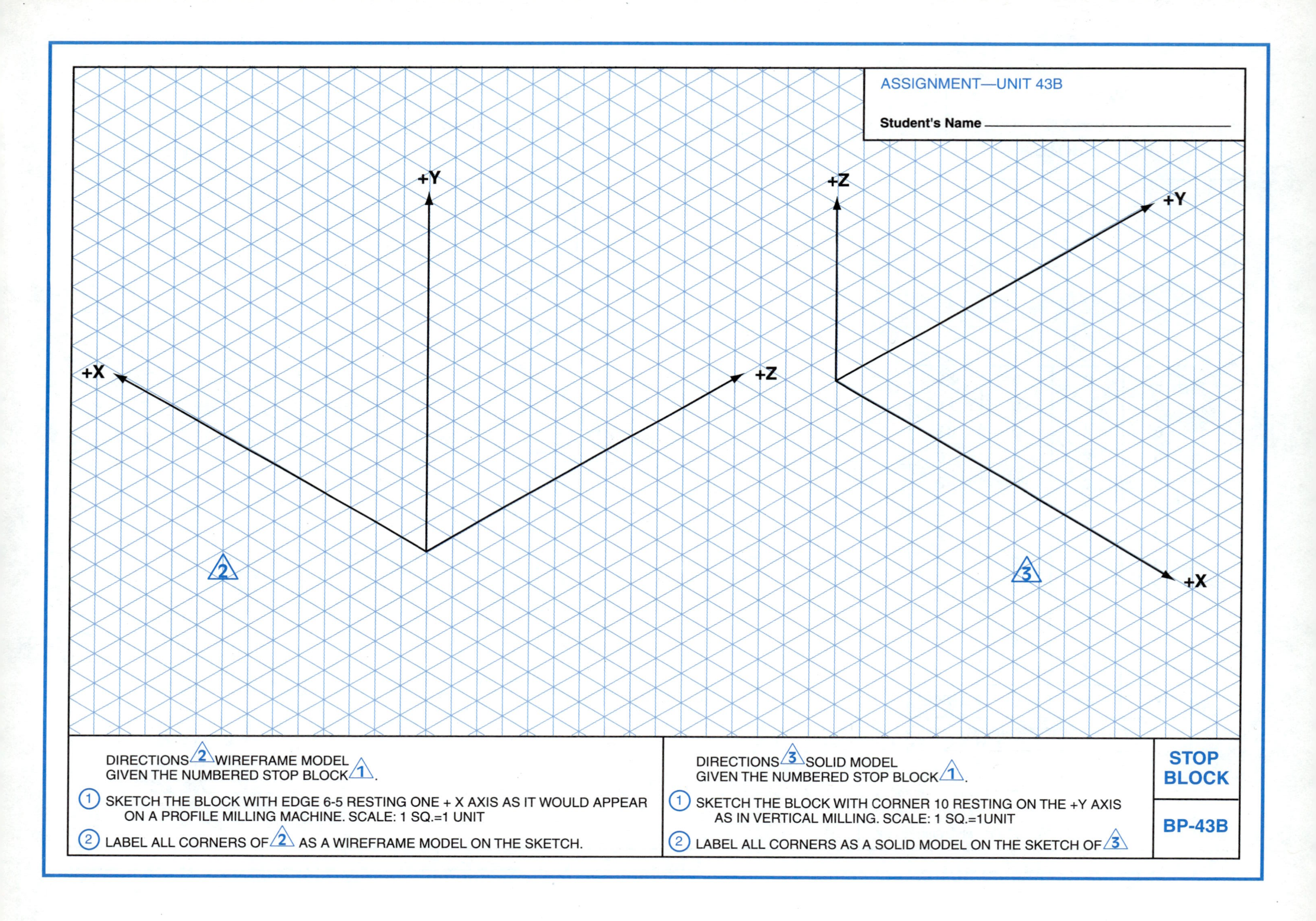
ASSIGNMENT—UNIT 43B
Student's Name
+Y
+X
+Z
2
+Z
+Y
+X
3
DIRECTIONS 2 WIREFRAME MODEL
GIVEN THE NUMBERED STOP BLOCK 1.
1 SKETCH THE BLOCK WITH EDGE 6-5 RESTING ONE + X AXIS AS IT WOULD APPEAR ON A PROFILE MILLING MACHINE. SCALE: 1 SQ.=1 UNIT
2 LABEL ALL CORNERS OF 2 AS A WIREFRAME MODEL ON THE SKETCH.
DIRECTIONS 3 SOLID MODEL
GIVEN THE NUMBERED STOP BLOCK 1.
1 SKETCH THE BLOCK WITH CORNER 10 RESTING ON THE +Y AXIS AS IN VERTICAL MILLING. SCALE: 1 SQ.=1UNIT
2 LABEL ALL CORNERS AS A SOLID MODEL ON THE SKETCH OF 3
STOP BLOCK
BP-43B

Unit 44

PROPORTIONS AND ASSEMBLY DRAWINGS

SKETCHING IN CORRECT PROPORTIONS

Freehand sketching requires judgment about how the sizes of different features are to be correctly proportioned. Usually, an approximation by eye is adequate. If a grid paper is used, each block generally represents a definite measurement. Still more precise sketches may be made by the use of measuring instruments.

One simple technique of sketching a part in correct proportion is illustrated in figure 44-1. In this case, the handle diameter is equivalent to three units for height and depth and four units for length (width). The metal ferrule is proportionally 1 1/4 units wide, and the blade is 1 3/4 units long.

The three major steps in drawing this short blade screwdriver in proportion include:

- Blocking out the major sizes
- Laying out the units for each detail
- Roughing out and then completing the sketch.

PICTORIAL ASSEMBLY DRAWINGS

Most mechanisms and structures are made up of several individual parts that are connected together into subassembly and assembly drawings (see Unit 25). Examples of three types of assembly drawings follow.

The piston pin system shown in figure 44-2 is an example of a sketched pictorial assembly drawing.

The housing clamp assembly drawing shows a sketched top view (figure 44-3).

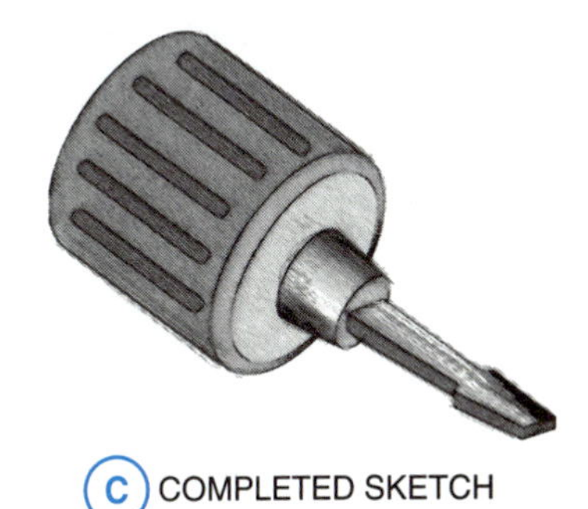

(C) COMPLETED SKETCH

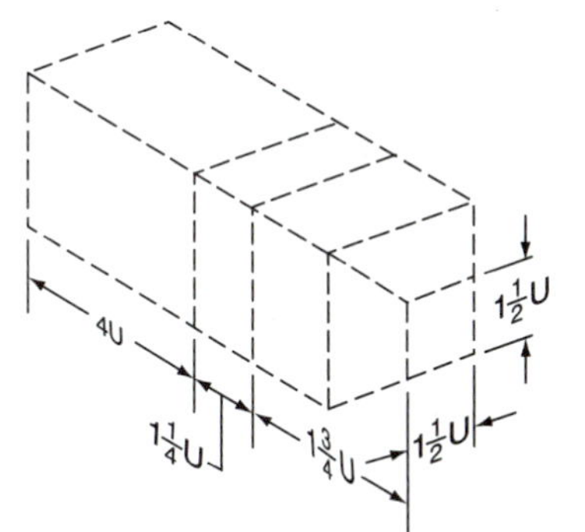

(B) BLOCK OUT DETAILS

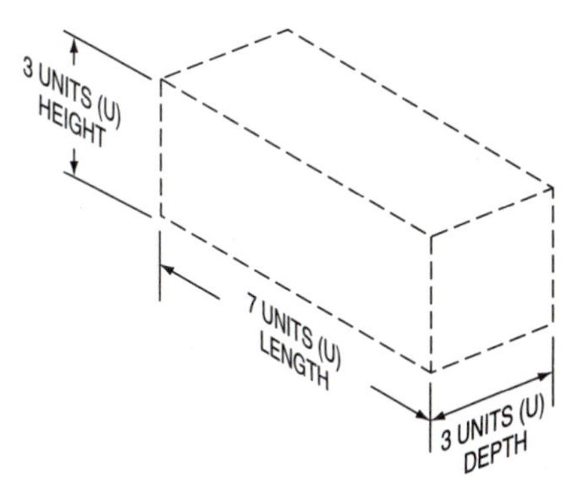

FIGURE 44-1 Steps in sketching a stub nose screwdriver in correct proportion.

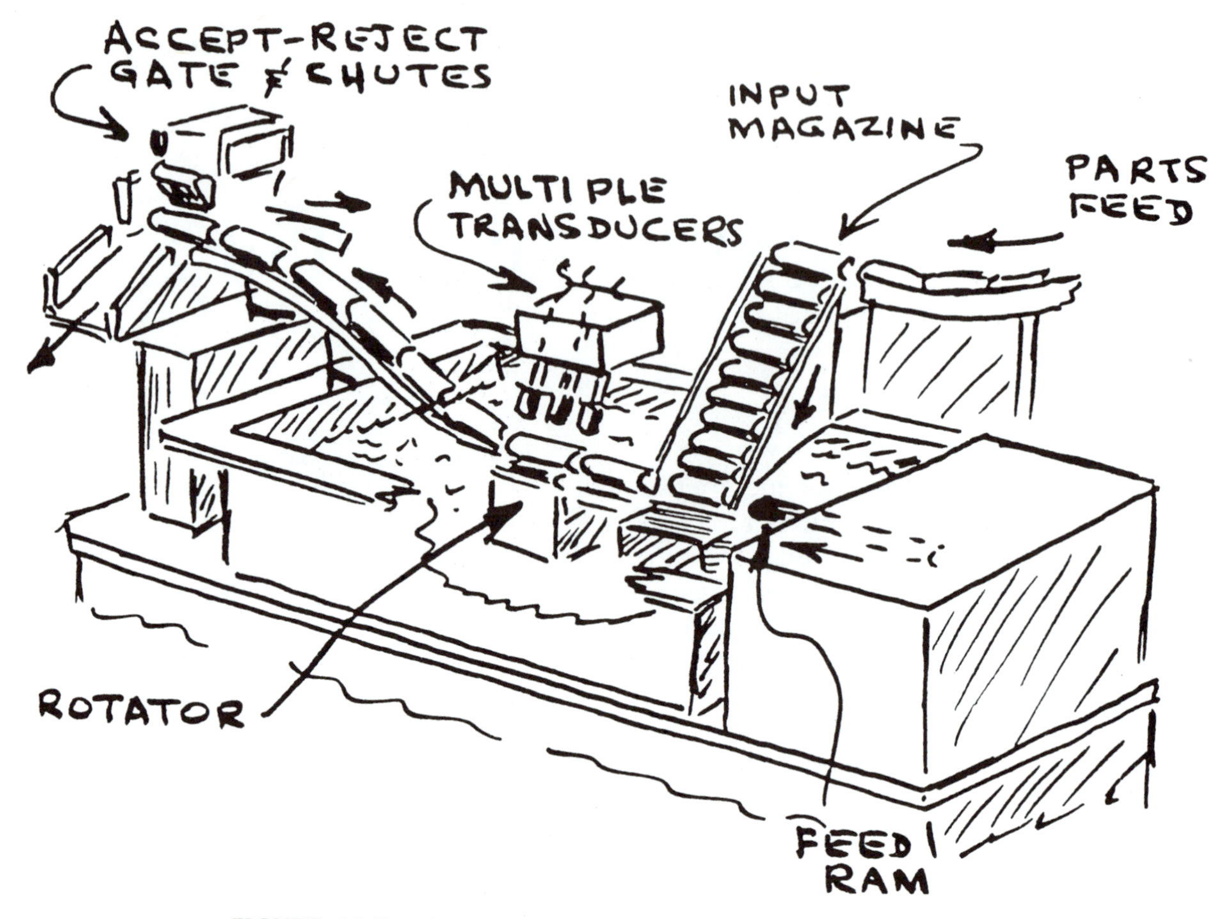

FIGURE 44-2 Piston pin assembly (copyright Thomas P. Olivo).

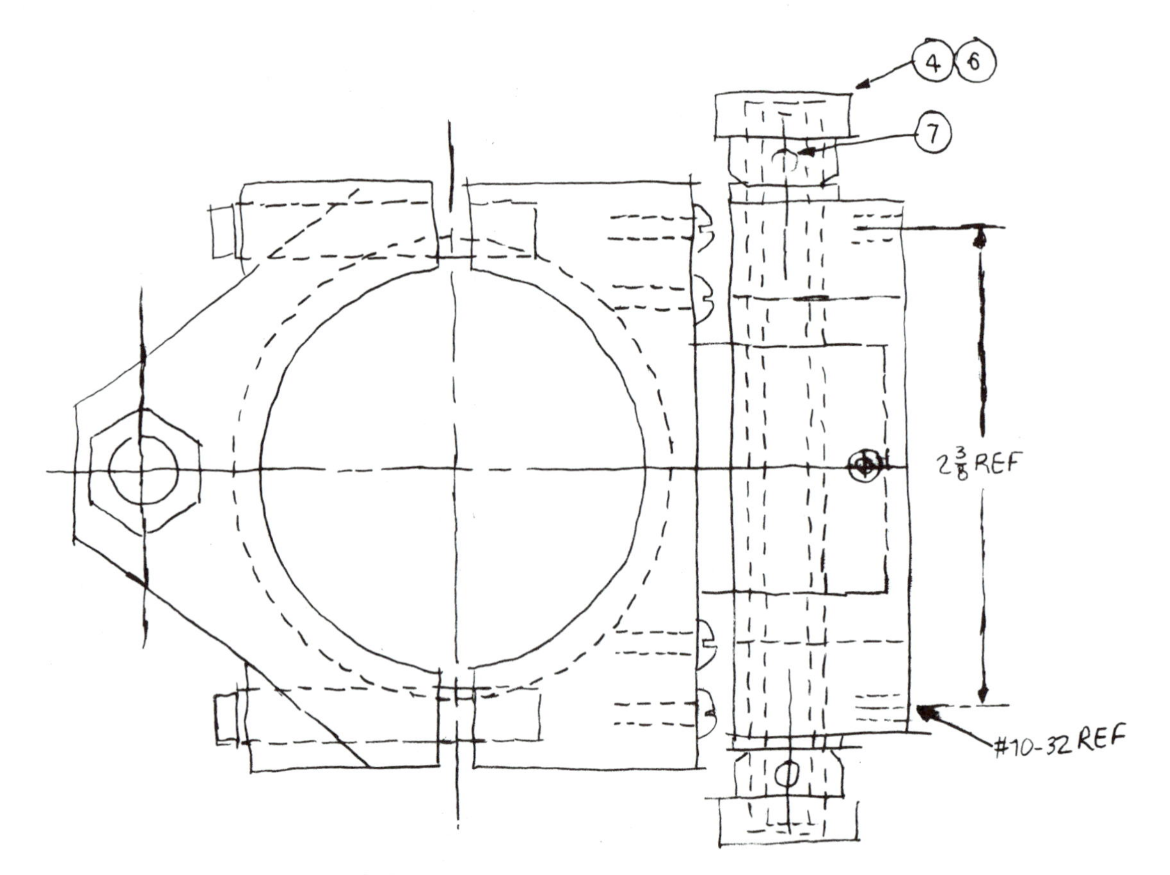

FIGURE 44-3 House clamp assembly.

At other times a sectioned assembly drawing is the best way to depict a mechanism, as shown in figure 44-4.

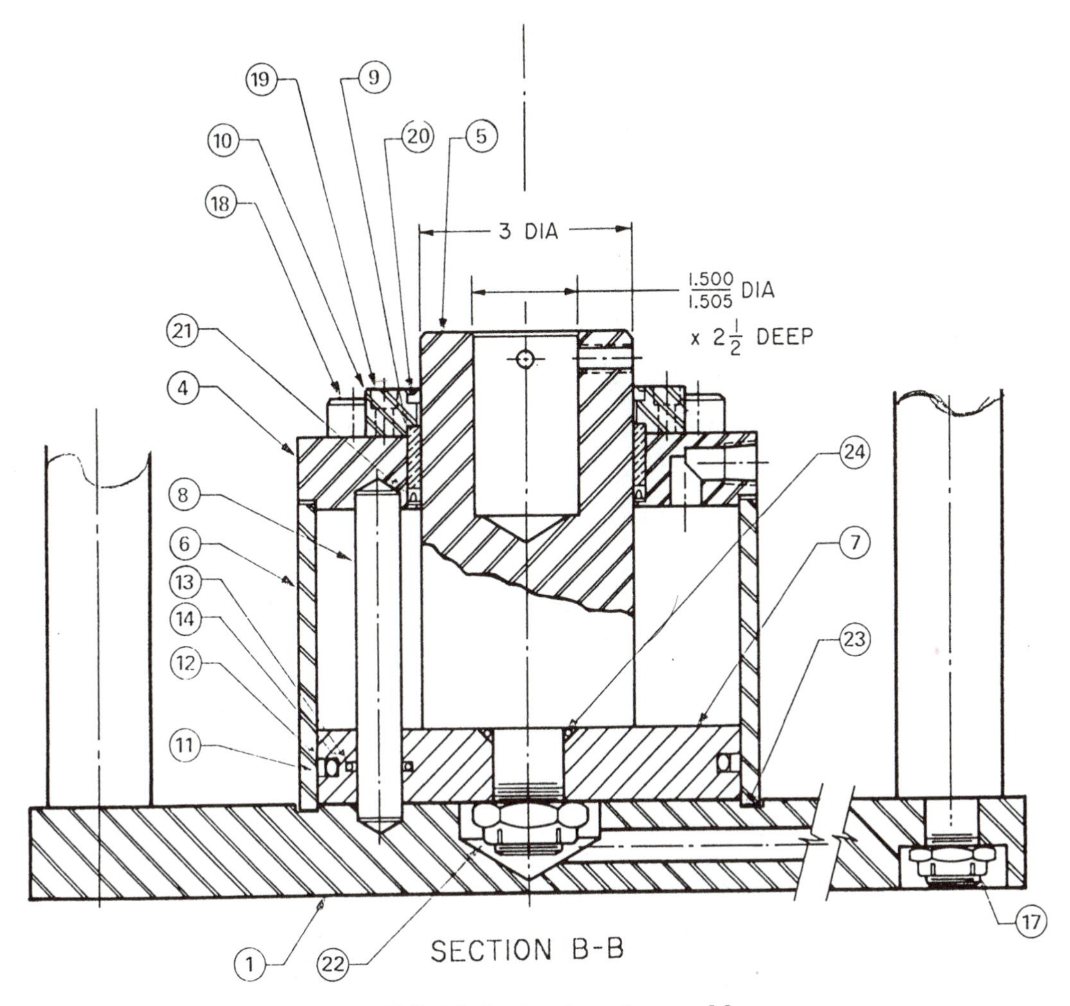

FIGURE 44-4 Sectioned assembly.

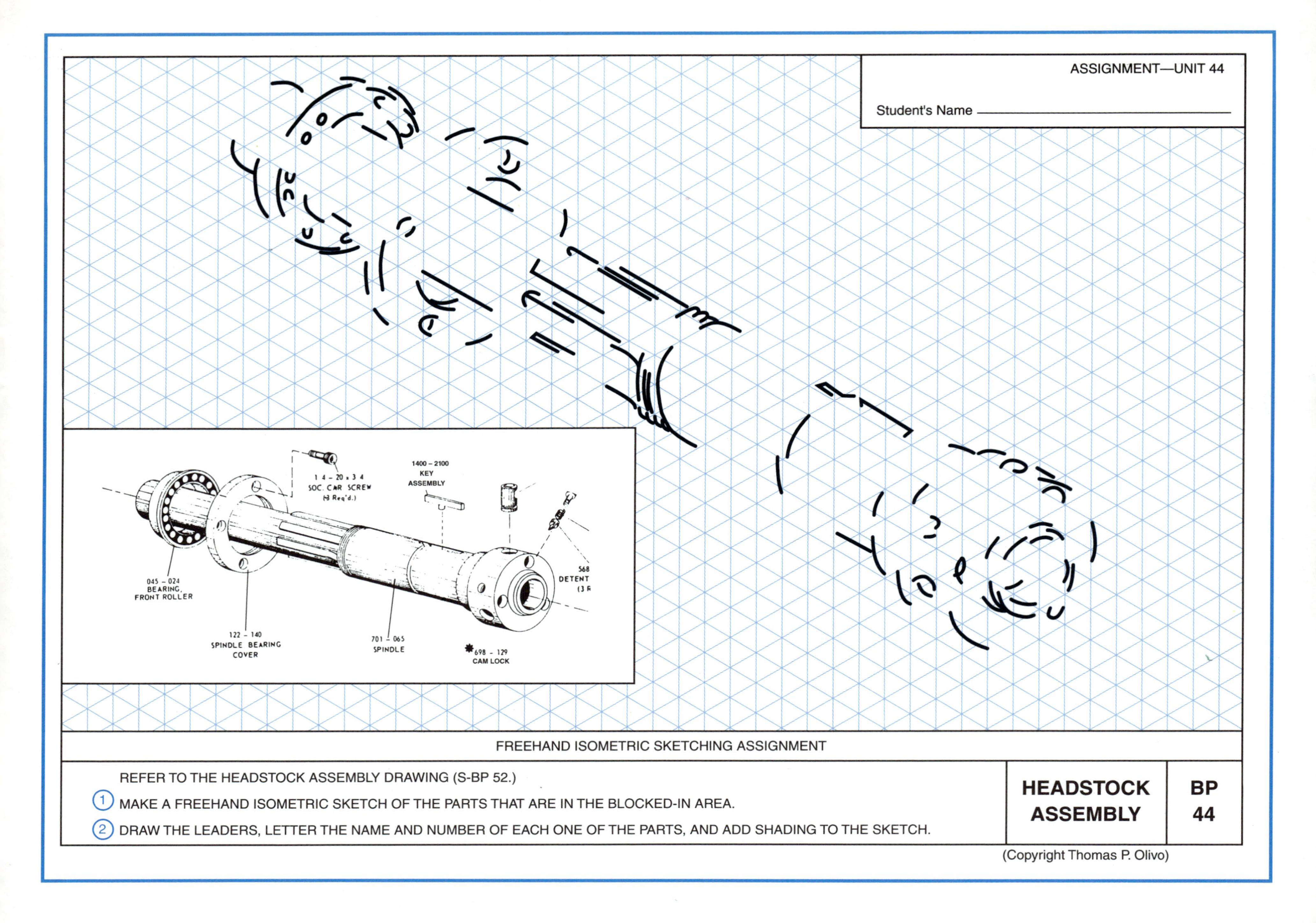
ASSIGNMENT—UNIT 44
Student's Name
1 4 - 20 x 3 4
SOC. CAR SCREW
(3 Req'd.)
1400 - 2100
KEY
ASSEMBLY
568
DETENT
045 - 024
BEARING,
FRONT ROLLER
122 - 140
SPINDLE BEARING
COVER
701 - 065
SPINDLE
698 - 129
CAM LOCK
FREEHAND ISOMETRIC SKETCHING ASSIGNMENT
REFER TO THE HEADSTOCK ASSEMBLY DRAWING (S-BP 52.)
1 MAKE A FREEHAND ISOMETRIC SKETCH OF THE PARTS THAT ARE IN THE BLOCKED-IN AREA.
2 DRAW THE LEADERS, LETTER THE NAME AND NUMBER OF EACH ONE OF THE PARTS, AND ADD SHADING TO THE SKETCH.
HEADSTOCK
ASSEMBLY
BP
44

INDEX